SPRING DESIGNER'S HANDBOOK

MECHANICAL ENGINEERING

A Series of Textbooks and Reference Books

1. Spring Designer's Handbook, *by Harold Carlson*

OTHER VOLUMES IN PREPARATION

SPRING DESIGNER'S HANDBOOK

HAROLD CARLSON
Consulting Engineer
Lakewood, New Jersey

MARCEL DEKKER, INC. New York and Basel

Library of Congress Cataloging in Publication Data

Carlson, Harold.
Spring designer's handbook.

(Mechanical engineering ; 1)
Includes index.
1. Springs (Mechanism)–Design and construction–
Handbooks, manuals, etc. I. Title. II. Series.
TJ210.C37 621.8'24 77-27436
ISBN 0-8247-6623-7

MARCEL DEKKER, INC.

270 Madison Avenue, New York, New York 10016

Current Printing (last digit):

10 9 8 7 6 5 4 3 2 1

PRINTED IN THE UNITED STATES OF AMERICA

Contents

Foreword

The literature of spring technology consists of a large number of technical papers devoted mostly to theoretical derivations and the results of tests, manufacturer's catalogs emphasizing products manufactured along with design data, and a few textbooks of design principles. In this volume the author has endeavored to fill the need for a comprehensive treatise on the practical application of these design principles simplified, based on the metallurgy of spring materials and economical manufacturing methods.

Just as a machinist needs certain tools in his tool box, an engineering designer needs certain tools of design or reference data conveniently at hand. This book fills that need. The author has furnished an extensive and convenient compilation of the metallurgy of the most popular spring materials, engineering design data, and manufacturing methods for spring production. Much of this information has never before been available in any one reference work, although a large portion has been the subject of scattered but limited treatment in hundreds of technical papers, including more than three dozen written by this author. A large portion also has been gleaned from the author's 45 years of experience in the field of engineering design and spring technology.

The book comprises three main parts covering spring materials, spring design data, and methods of spring manufacture. Each part is complete in itself and contains many useful tables, charts, and curves frequently used by spring designers.

Since the information presented is a complete working guide for the engineer concerned with the practical problems of spring design, I recommend this book to the whole family of the engineering profession including design engineers, metallurgists, spring manufacturers, and engineering instructors as well as to students of machine design.

L. D. "Jack" Seymour
Formerly Manufacturing Manager
John A. Roebling's Sons Co.

Preface

Spring manufacturers are in an unusual industry beset with many problems, difficulties, peculiarities, and misconceptions. The spring industry was created in the middle 1850s as a result of the need for specialized knowledge, particularly in the field of heat treatment, which was not readily available to the engineers and machinists of that time. In those early days, product manufacturers and machine builders made their own springs on lathes, and the hardening and tempering was done by blacksmiths. Frequent spring breakage, lack of uniformity in manufacture, inability to produce large quantities quickly, and limited knowledge of materials or design hampered industry tremendously. Enterprising businessmen and ingenious mechanics were quick to seize upon this advantage and soon established companies specializing in spring manufacture, but the first patent for a spring coiling machine that would make both close-coiled extension and open-coiled compression springs with closed ends was not granted until 1908.

The spring companies of these early days encouraged product manufacturers to give them all their spring problems. As a result, even today, many spring manufacturers have excellent reputations for solving some of the complex problems encountered in machine design. Whether it is fair or not for product manufacturers to expect spring companies to be responsible for the selection of the right material and the design of a desired spring to meet required loads and deflections is a debatable question. In this day of broadened scope in engineering training we should perhaps expect design engineers to have sufficient ability to design springs or to have access to design information.

Some spring companies, under the guidance of businessmen trained in other industries, feel that the responsibility for spring design should be returned to the customer. By doing this, these companies expect to reduce the prices of their springs. However, the more progressive spring companies still find it necessary to employ competent spring design engineers, particularly to satisfy the requirements of customers of long standing. Moreover, because many of the developments in new materials, designs, and production methods were invented or sponsored by spring engineers, their continued accomplishments are a decided advantage to both the spring manufacturer and his customers.

Many progressive product manufacturers employ engineers qualified to design springs, and can often obtain lower quotations on their spring requirements by sending complete specifications to small spring companies who do not employ engineers. If the small spring company is reliable and uses the proper materials and production methods, this policy may save money. However, there are occasions where unreliable spring companies have reduced prices by skimping on material, or used substitute materials or eliminated heat treating and other operations vitally needed to produce a good product.

To avoid such compromise with standards of proper design and production, there can be no substitute for a qualified spring engineer. Such a man must have his ability at stress analysis enhanced by a well-rounded knowledge of the metallurgy of spring materials and be supported by a good knowledge of spring manufacturing methods. All this, plus actual shop experience, is essential to being competent in this little-known but important phase of machine design.

The purpose of this book is to place at the designer's hand a volume of data as complete as possible in all of these respects: to provide data for proper selection of materials, to contain design formulas and tables in readily accessible form, and to describe manufacturing operations with production equipment required for spring manufacture. Other data incident to a well-equipped designer's file is also included. Derivations of formulas, complicated equations, and complex deductions are either eliminated or reduced to simple forms wherever possible. Little-used data is omitted, and the charts and tables contain only that data which is referred to frequently.

All recommended design stresses are in both pounds per square inch (psi) and in metric (SI) megapascals (MPa), with wire diameters in decimal inches and in millimetres. Also, all design problems are shown in both systems.

Harold Carlson, P.E.
Consulting Engineer
Lakewood, N. J.

Acknowledgments

The author acknowledges that this book could not have been written without the help and cooperation of many individuals, spring manufacturing companies, technical and trade organizations. In particular, acknowledgment is given to the American Iron and Steel Institute for providing the whole series of photographs covering mining and steel mills, to the Spring Manufacturers Institute for permission to reproduce their spring tolerances, to the American Society for Testing and Materials for permission to reproduce excerpts of their standards for spring materials, and to the American Society of Mechanical Engineers, who sponsored many technical papers prepared by the author.

In addition, many of the technical members of the Spring Manufacturers Institute provided photographs of their equipment for the section on spring manufacturing. Thanks are given to these and other companies that contributed data, pictures, and materials for testing, including:

Anaconda American Brass
Armco Steel Corp.
Associated Spring Corp.
U.S. Baird Corp.
Wallace Barnes Co.
Bendix-Besly Grinder Div.
Bennett Tools Ltd.
Bethlehem Steel Corp.
The Carlson Co.
C. Faust & Sohn
Link Engineering Co.
Little Falls Alloys
Morgan Machine Co.
National Standard Corp.
A. H. Nilson Machine Co.
E. A. Samuel Machine Co.
Sleeper & Hartley Corp.
Swedish Wire Co.
Techalloy Co.
Torin Corp.

Gardner Machine Co.
Hunter Spring Co.
International Nickel Co.
Johnson Steel & Wire Co.
Tormet Industries
United States Steel Corp.
Wafios Machinery Corp.
Fansteel–V. R. Wesson, Div.

The author is grateful to his wife for her painstaking work in typing the manuscript, and to Graham Garratt for his encouragement and critical analysis of the various sections.

List of Symbols

Symbol	Definition	Units
A	constant or a designated point	—
B	constant	—
B_m	bending moment	in. lb or N mm
b	breadth or width	in. or mm
°C	degrees Celsius (formerly centigrade)	5/9 (°F – 32) degrees
CL	compressed length	in. or mm
cm	centimeter	10 mm or 0.3937 in.
D	mean diameter = OD – d	in. or mm
D/d	spring index	ratio
d	wire diameter	in. or mm
E	modulus of elasticity in tension	psi or MPa
F	deflection	in. or mm or degrees
F	force	lb or N
°F	degrees Fahrenheit	9/5°C + 32 degrees

Symbol	Definition	Units
FL	free length	in. or mm
f	deflection of one coil	in. or mm or degree
G	modulus of elasticity in shear	psi or MPa
g	gram, a metric weight	0.035 oz or 0.0098 N
h	height	in. or mm
ID	inside diameter = OD – 2d	in. or mm
IT	initial tension	lb or N
in.	inch	25.4 mm
in. lb	torque	0.112 985 Nm
K	stress multiplication factor	–
kg	kilogram = 1000 g = 2.2046 lb = 35.273 97 oz	–
L or l	length	in. or mm
lb in.	same as in. lb	–
lb	pound = 453.59 g or 4.448 N	–
M	mega	million
M	bending moment	in. lb or N mm
MPa	megapascal = million pascals = 1 MN/m^2 = 1 N/mm^2	–
mm	millimetre	0.039 37 in.
N	newton = 102 g = 3.5969 oz = 0.2248 lb .	–
N	number of active coils	–
n′	natural period of vibration	per second
OD	outside diameter	in. or mm
oz	ounce = 0.0625 lb = 28.35 g = 0.278 N. .	–

Symbol	Definition	Units
P	force	lb or N
Pa	pascal = a newton per square metre . . .	–
PH	precipitation hardness	–
psi	pounds per square inch	0.006 895 MPa
p	pitch	in. or mm
R	radius or distance to center line	in. or mm
rpm	revolutions per minute	–
r	rate = force per unit of deflection . . .	lb/in. or N/mm
S	stress, torsional	psi or MPa
S_b	stress, bending	psi or MPa
S_{IT}	stress, initial tension	psi or MPa
SH	solid height	in. or mm
S_m	section modulus	$in.^3$
TC	total coils	–
T	torque	in. lb or N mm
t	thickness or side of square	in. or mm
U	number of revolutions	–
W	weight	oz or lb or g or N
π	pi = 3.1 416 or 3.141 592 653 5898 . . .	–

SPRING MATERIALS

1 Metallurgy of Spring Steels

How Steel is Made

Many millions of years ago, the earth contained multitudes of small organisms that depended for survival upon water containing iron in suspension. When this iron water receded, the skeletons of these bacteria, now high in iron content, remained to become the basic element of certain iron ores. Thus even iron ore is generally considered to be of organic origin; particularly the Alabama, Tennessee, and New York deposits. Nature required a span of over 2 billion years to accomplish this feat so necessary for our steel making requirements of today. From these ores, iron is extracted by melting the ore along with limestone and coke. Steel is made by further refining iron after it comes from the blast furnace.

MINING

Iron ore is found in nearly every state of the Union. However, much of it has come from open pit mines on great ranges such as the Mesabi Range of Minnesota and from mines in Michigan and Wisconsin in the Lake Superior region. At these mines the ore is scooped up at the earth's surface by huge power shovels and deposited in trucks and railroad cars for transfer to large ore boats which take it to the steel mills. In shaft mining, the ore is dug from underground tunnels branching off from a vertical shaft which may be as much as a mile deep. All mines produce ores containing chemical elements in widely varying proportions which are blended to meet requirements. At the mills,

FIG. 1. Open pit iron ore mine in Minnesota, showing power shovels, trucks, and other equipment. (Photo courtesy American Iron and Steel Institute.)

large stockpiles of this iron ore, along with stocks of quarried limestone and bituminous coal, are maintained. See Fig. 1.

The steel industry in the United States has relied heavily on the high grade ores containing 50 to 55 percent iron, mined in the Lake Superior district, but this supply of high grade ore is rapidly diminishing.

Low-grade ores such as taconite, containing 25 to 35 percent iron, and jasper, averaging slightly higher iron content, can be beneficiated and formed into pellets ¼ to ½ in. (6 to 13 mm) diam. containing 62 to 65 percent iron. Such ores are now extensively used. In an effort to conserve the high-quality ore reserves in the United States, low-grade ores can be processed in sintering machines. These machines help to get more tonnage from a steel mill by lumping fine ore dust into clinker-like chunks so that they can be fed to blast furnaces. The low-grade ores are first powdered so that much of the foreign substances can be removed. The sintered powder then has a higher percentage of iron.

New deposits of iron ore are also being mined in Venezuela, Quebec-Labrador, Liberia, Brazil, and other parts of South America. Sweden has large deposits of iron ore of such purity and high iron content that excellent steels can be produced with relatively little refining. This is one of the reasons for the high quality of Swedish steels. The purity of Swedish iron ores also result in Swedish steels having a much lower percentage of impurities, particularly phosphorus and sulfur.

IRON ORES

Iron is in almost limitless supply all over the world, as it is one of the most abundant elements in the earth's crust. Economics plays an important part in what is classified as "ore." Thus the proximity of the ore to coking coal and to an important market for steel give it a favorable price-quality relationship when compared with higher-grade but more distant ores.

Iron is derived from ores containing iron oxide. The principal ore, called hematite, is an oxide of iron, brick red in color, and is a heavy, stone-like mineral. Another ore, called magnetite, is black in color and is magnetic. Most of these ores contain 50 to 55 percent iron. Taconite and other low-grade ores are now used to a greater extent because of economically improved processing methods.

FLUXES

A metallurigcal process wherein a metal and its impurities are separated by fusion is called smelting. Many of the impurities found in iron ore are difficult to melt. To make these impurities fusible requires a flux. Certain types of limestone, so essential to the manufacture of steel, are composed of the shells and skeletons of prehistoric sea life. They contain calcium and other elements and are obtained from many quarries throughout the country. During the manufacture of steel, limestone floats on top of the molten metal and acts as a cleaner or flux to soak up and carry off the phosphorus, sulfur and non-metallic impurities called gangue.

COKE

To provide fuel for the blast furnace, bituminous coal is made into coke, which is principally carbon. Coal is made into coke by heating in large ovens for 14 to 17 hours, thus driving out the volatile liquids and gases containing the coal chemicals used for the manufacture of aspirin, vitamins, sulpha drugs, Nylon, and hundreds of other products. In the blast furnace, coke provides a free passage for air while supporting the iron ore and limestone. During heating, the coke combines with the oxygen in the iron ore, thus liberating the iron while also providing much of the carbon which is absorbed by the molten iron.

SCRAP

Old automobiles, steam engines, machine tools, rails, and other discarded products made of iron and steel, collected by scrap dealers, is remelted along with pig iron from the blast furnace in making steel. Scrap is of great practical value, as it helps save our deposits of iron ore, limestone, and coal and also speeds up the process of refining. Although steel can be made by melting scrap alone, the usual method is to use nearly 50 percent pig iron. See Fig. 2.

FIG. 2. Scrap steel in carefully assorted lots on the way to steel making furnaces. (Photo courtesy American Iron and Steel Institute.)

BLAST FURNACES

Huge structures called blast furnaces, 30 ft (9 m) in diameter and 100 ft (30 m) high, make pig iron; larger furnaces are also used. A blast furnace requiring 100 railroad cars of iron ore, coke, and limestone per day produced 1300 tons of pig iron daily in 1950. Output rose to 7000 tons in the 1960s and over 10 000 tons in the 1970s. Near a blast furnace are several tall, round cylindrical stoves that furnish the blast of hot air needed to produce the high temperatures required for melting the ore. Iron ore, coke, and limestone are deposited in the hopper at the top of a blast furnace where the temperature is 400°F (200°C). About half way down the temperature is nearly 1000°F (535°C),

FIG. 3. Four blast furnaces showing inclined skip hoists for carrying ore, coke, and limestone to the receiving hoppers. At left is a sintering plant where fine iron ore and flue dust are sintered for recharging into blast furnaces. Between the blast furnaces are several domed, cylindrical stoves for heating the air used to promote combustion inside the furnaces. (Photo courtesy American Iron and Steel Institute.)

and at the melting area it is 3500°F (1900°C). These furnaces in 1950 were tapped every 4 to 5 hours and 150 to 300 tons of iron were drawn off each time. Much larger production, by introducing oxygen into the furnace, was achieved in the 1970s. Spring steel represents about 2 percent of the total output. See Figs. 3 and 4.

PIG IRON

The molten metal coming out of a blast furnace is called pig iron. It is usually poured into a large horizontal container on wheels called a "mixer," and then transferred to a steel making furnace for refining. See Fig. 5.

Sometimes the molten iron is poured into molds capable of holding about 100 lb (45 kg) and allowed to cool. These "pigs" are then sold to foundries where they are remelted with coke as a fuel for the manufacture of iron castings.

Molten iron from the blast furnace must be refined because it contains

FIG. 4. Diagram of blast furnace operation. One car of the skip hoist charges raw materials into the top of the furnace while the other is being filled at the stockhouse hopper below. Molten iron is shown running into a mixer car. The iron is either taken to steel making furnaces or is allowed to solidify in molds. (Photo courtesy American Iron and Steel Institute.)

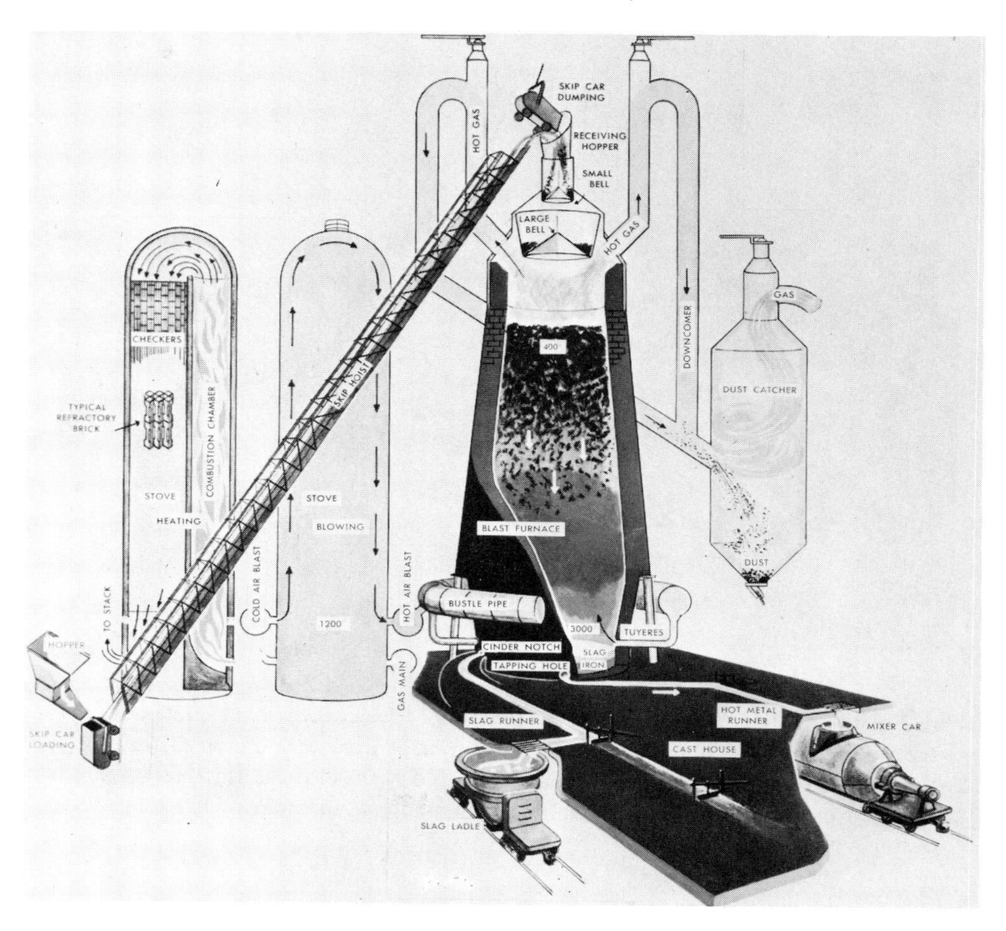

too much carbon (3.5 to 4 percent) and does not have the proper amounts of silicon and manganese. Also, the phosphorous, sulfur, and other elements must be extracted.

THE BESSEMER PROCESS

One of the oldest methods of making steel for low-carbon and mild steel products was by pouring molten pig iron directly from the blast furnace into a

FIG. 5. Modern double-strand pig casting machine. The molten iron flows from ladle into two moving chains of molds where it cools to form "pigs." The term "pig iron" comes from the original practice of letting iron flow into sand troughs and molds which resembled a sow with suckling pigs. (Photo courtesy American Iron and Steel Institute.)

pear-shaped Bessemer converter. Compressed air, blowing up through the charge, burned out the carbon, manganese, sulfur, and silicon, but not phosphorus, and most of the impurities, in a huge shower of sparks. High-quality steels, tool steels, and alloy steels were not made by this process. This method is no longer used in the United States. See Figs. 6 and 7.

THE OPEN HEARTH PROCESS

Steel can be produced in large open hearth furnaces where the flames sweep across the top of the molten steel. The charge, at 3000°F (1650°C) is held in an oval saucer-like area lined with a refractory material. The chemical composition of the refractory material used to line the furnace is extremely important. It may be acidic, basic, or neutral, and this affects the quality and type of steel produced. The charge usually consists of about 52 percent pig iron and 48 percent scrap. During the refining process, impurities are removed, the chemical content is controlled, and alloying elements such as chromium or vanadium may be added to produce alloy steels. Practically all

FIG. 6. Bessemer converter shooting flame skyward as air is blown through molten iron to burn out impurities. The flame subsides when the impurities have been burned out. At lower left are cast iron ingot molds into which steel will be poured when made. This method is no longer used in the United States. (Photo courtesy American Iron and Steel Institute.)

high-carbon and alloy spring steels were made by this process up to 1976, but very little if any tool steels were open hearth steels. In 1976, many of the large steel producing plants replaced their open hearth furnaces with basic oxygen converters to increase production and reduce costs. See Fig. 8.

THE ELECTRIC FURNACE PROCESS

In the electric furnace method, large electrodes of carbon extend downward from the roof of the furnace to within a few inches of the charge. The electric arc formed between the electrodes supplies the heat for refining the steel. The charge may consist entirely of scrap steel fortified with alloying elements such as chromium and nickel to make stainless steel or tungsten and other elements to make high-grade tool steels. Practically all stainless steels

FIG. 7. Diagram at top shows how molten iron is poured into horizontal position of Bessemer converter. Bottom diagram shows how the stream of air is forced through the bottom while the converter is in the vertical position. (Photo courtesy American Iron and Steel Institute.)

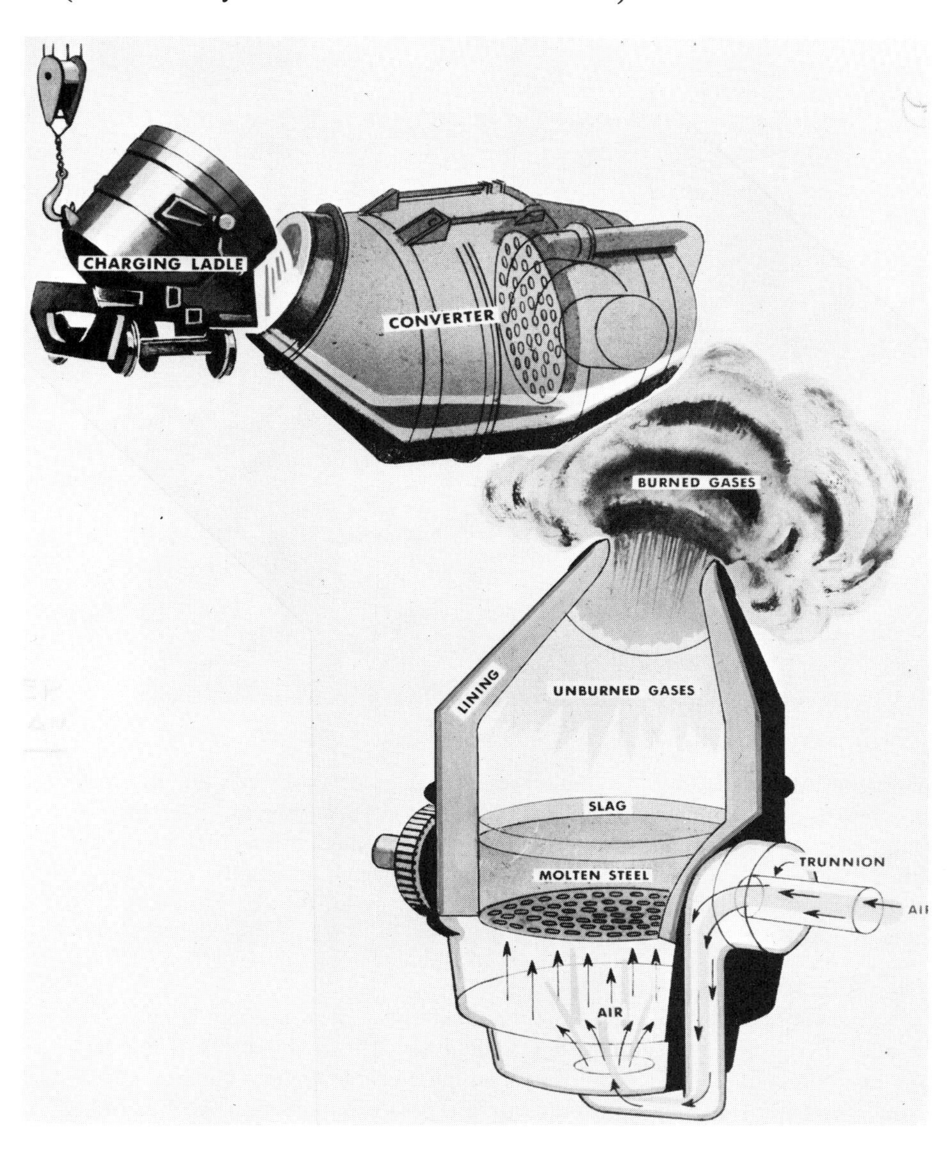

FIG. 8. Shoveling dolomite lining material against the back wall of an open hearth furnace. Furnace linings must resist the weight, heat, and chemical action of molten metal and slag. The chemical composition of refractory bricks may be acid, basic, or neutral, which affects the quality and determines the type of steel being made. Open hearth furnaces are almost obsolete in the United States. (Photo courtesy American Iron and Steel Institute.)

are made by this method. Certain types of high-grade tool steels are also made in a new type of coreless electric induction furnace operating on the principle of an electric transformer. The primary high-frequency current is in a copper coil surrounding a crucible filled with steel scrap which acts as the secondary of the transformer; this becomes so hot that melting of the charge occurs quickly. See Figs. 9, 10, and 11.

THE CRUCIBLE FURNACE PROCESS

The crucible furnace process, an old-time method of steel making, is now obsolete because of its high cost and limited output. In this process, the desired carbon content was obtained by melting a charge of pure bar iron and a fixed quantity of refined pig iron in covered pots or crucibles capable of holding about 100 lb (45 kg) of metal. Alloying materials in desired percentages were added to the melt and allowed to soak until a uniform composition was

FIG. 9. Electric furnace being tipped to pour the molten steel into a ladle. The pipe-like columns on top are adjustable carbon electrodes. (Photo courtesy American Iron and Steel Institute.)

obtained. Although this early method of steel making produced tool steels of high quality, it has been mostly replaced by the electric furnace process.

THE VACUUM MELTING PROCESS

The vacuum method of melting steel to produce a better product–stronger at high temperatures, more ductile, and with improved resistance to fatigue–is accomplished by controlling or eliminating gases contained in steel, particularly oxygen, nitrogen, and hydrogen. Vacuum induction furnaces are especially useful for making tool steels.

FIG. 10. Cutaway diagram of electric furnace shows how the electric arc supplies heat for the refining process in the manufacture of certain high-quality steels. The intense heat can be closely controlled. (Photo courtesy American Iron and Steel Institute.)

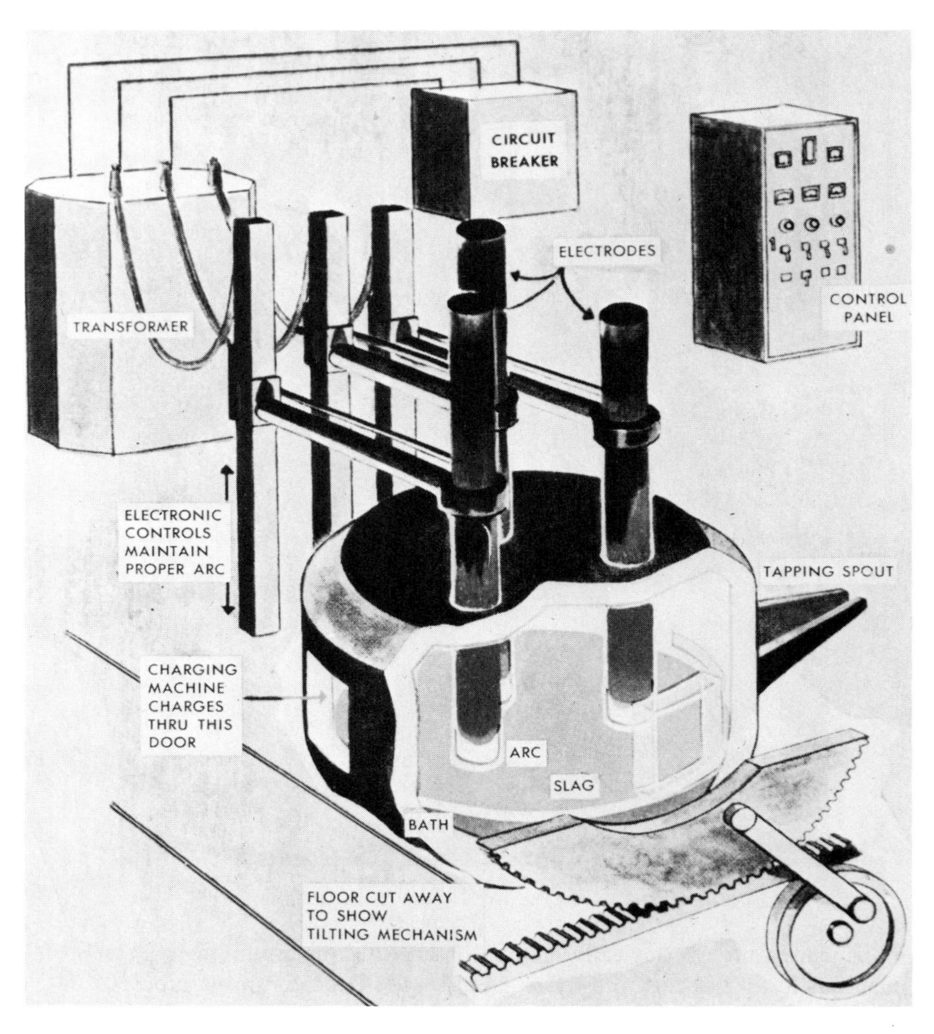

The basic unit is a stainless steel liquid-cooled vacuum tank containing a tilting high-frequency induction furnace. The furnace is surrounded by a coiled electrical conductor through which a high-frequency current is passed, thus generating a heavy secondary current in the charge. The resistance of the charge to the secondary current creates enough heat to melt it. The materials to be put into the furnace are first placed in a lock; the atmosphere in the lock is then pumped out until conditions in the lock closely resemble those in the chamber containing the furnace; then the materials are dropped into the furnace, uncontaminated by atmospheric gases.

The design makes it possible to charge, melt, pour, and remove ingots without affecting the vacuum in the melting chamber. The products of this process have proved superior for use in critical applications in aircraft engines, machine components, and for small ball bearings where quality unattainable from conventional materials is desirable.

OXYGEN CONVERTERS–BASIC OXYGEN METHOD

In 1976 many of the larger steel producing plants replaced their open hearth furnaces with oxygen converters to increase production and to reduce costs. This development replaces the customary method of blowing air through the bottom of a Bessemer converter to burn out the high percentages of several constituents, by substituting oxygen of high purity (98 percent or better). This oxygen is blown through an oxygen lance at supersonic speed onto the top of the molten metal bath (or up through it) and quickly produces steel with properties similar to the basic open hearth steel. This process can produce about three times as much steel per day as the open hearth process and at less cost for fuel. The nitrogen content of the product compares favorably with the open hearth steels, and the converters can be charged with as much as 30 percent scrap steel instead of the 7 percent limitation for conventional Bessemer converters.

The iron and steel industry is now one of the largest industrial users of gaseous oxygen. It is produced from atmospheric air by compressing, cooling, removal of moisture, removal of impurities, and separation into its components by distillation.

By enriching the blast of air in the blast furnace from its natural content of 21 percent oxygen to 30 percent production has been increased and coke consumption decreased.

In the steel making process, oxygen increases the rate of scrap melting while materially reducing heat time.

INGOTS

When a furnace is tapped, the melt flows into a large ladle with a pouring spout in the bottom. This spout is opened to permit the melt to flow into cast iron ingot molds where it is allowed to solidify. These ingots may be round, square, rectangular, hexagonal, or octagonal sections (without square corners), 6 to 36 in. (150 mm to 1 m) thick, 6 to 10 ft (2 to 3 m) high, and weighing from 6 to 20 tons. The method used to cool the ingot determines the type of steel that is produced. During the cooling operation, the metal nearest the walls of the mold cools and solidifies first, thereby causing

FIG. 11. Mixer car or "bottle car" for transporting molten iron from blast furnaces to steel making furnaces. The cars are lined with heat-resistant brick. They are somewhat like giant thermos bottles. (Photo courtesy American Iron and Steel Institute.)

a cavity due to shrinkage to form down the center of the ingot. This cavity or "pipe" would destroy the quality of the ingot if means were not taken to prevent its formation.

HOT TOP
To eliminate "piping," a hollow refractory sleeve 10 to 30 in. (250 mm to 0.75 m) high is placed on top of the ingot mold and the pouring of the ingot is stopped momentarily when the mold is filled; then the hot top is filled with molten metal. As the ingot cools, this metal flows into the center of the ingot, thus preventing the formation of a central cavity.

KILLED STEEL
The gas within a melt causes steel to be effervescent and bubble while cooling in the ingot mold. To retard this action, aluminum, ferrosilicon, or manganese is added to the furnace or to the ladle to deoxidize the steel; this is called a "killed" steel. Properly killed steel has a more uniform analysis, less segregation,

and is comparatively free from aging. Killed steel is also harder than rimmed steel of similar carbon and manganese content and usually has a lower silicon content. All steels over 0.20 percent carbon are "killed" steels. In "semi-killed" steel, incomplete deoxidiation occurs and some evolution of carbon monoxide takes place, thus tending to offset some shrinkage during solidification of the ingot. Such steels are useful for reinforcing rods used in concrete and similar applications.

RIMMED STEEL

When incomplete deoxidation is allowed, an ingot remains liquid at the top. During the cooling period a heavy "rim" of metal, of purer composition and lower in carbon content, is formed all along the sides and bottom of the ingot. This is called a "rimmed" steel. Rimmed steels are used only for low-carbon steels, usually below 0.20 percent carbon; and represents about 30 percent of steel output. This steel is made without a "hot top," but can be "capped steel."

CAPPED STEEL

To stop the rimming action after the pouring of the ingot is completed, a steel cap can be placed over the ingot mold for two or three minutes. Such molds are reduced or constricted at the top to support the cap. This is often called a "bottle top mold" because of the shape of the ingot. Rimmed steels are best for deep drawing operations, as their smooth surfaces and the uniform structure of the rim can be rolled down into smooth surface steels.

BILLETS

Ingots can be heated and then hammered, rolled, or pressed into sections about 2 to 4 in. (50 to 100 mm) square or round and several times longer than the original ingot; they are then called billets. These billets may be reheated and rolled into rods from which spring wire can be drawn.

BASIC STEEL

Basic steel is made by using limestone or other basic materials to absorb the acid phosphorous from iron containing a high percentage of this element. The furnace is protected by a lining of basic materials such as magnesite or burned dolomite. Hard-drawn steel wire is often made from basic steel.

ACID STEEL

Acid steel is made from iron containing smaller percentages of phosphorus and sulfur by omitting the limestone customarily used in making basic steel, thus making the slag acid in character. In this instance, the furnace is lined with silica sand also of acid nature. An advantage of acid steel over basic steel is that the composition and nature of the slag is more easily controlled, thereby producing a more uniform quality of clean steel.

FIG. 12. One of the earliest multiple continuous wire-drawing machines. The wire is drawn through six wire-drawing dies of gradually diminishing diameters, without annealing. Music wire, stainless steel, phosphor bronze, nickel, and other hard-drawn wires are drawn to desired diameters on such machines. Later models use water-cooling jackets and forced air for cooling and are capable of drawing wire at speeds of 2000 ft (600 m) per minute. (Photo courtesy American Iron and Steel Institute.)

WIRE MANUFACTURE
Iron ore from the earth is made into pig iron in a blast furnace, remelted, refined, alloyed, and ingots rolled on a blooming or billet mill into billets about 30 ft (9 m) long and then reheated and hot rolled to form long rods which are coiled on reels. These coils are prepared for drawing into wire by immersing in hot dilute sulfuric acid to clean the surface, then washed in water and dipped in hot borax solution, which picks up lubricant and prevents galling during the drawing operation. In some cases, a thin oxide coating may be allowed to form as a further aid to lubrication. The borax-coated coils are then flash baked and positioned to permit their ends to be welded, thereby forming rods of greater length, which can then be delivered to the wire-drawing frame. Here the welded coil is usually drawn to a convenient size such as ¼ in. (6.35 mm) diameter, in which form it is referred to as wire. The wire is then sent to the continuous wire-drawing mills for further reduction to the sizes desired. This cold drawing does not "iron" down the wire surface, as is popularly believed, but is actually a stretching operation. The grains of the steel are elongated or stretched all through the wire, and the surface becomes smooth and uniform. See Fig. 12.

SOFT PROCESS
In the soft process of wire manufacture, hot-rolled rods of a desired analysis are "patented" (so-called because this heat treatment applied to high-carbon wire received the first patent granted for such a process). Patenting is a special high-heat treatment used only on high-carbon steel wire to assure uniformity

of structure, combining high tensile strength with high ductility to prepare the wire for additional reduction. The wire is heated considerably above the upper critical temperature, sometimes as high as 1830°F (1000°C), then cooled quickly through the critical range by quenching in molten lead at a high temperature, usually about 890°F (475°C), so that martensite cannot form, thus causing a very fine equiaxed pearlitic grain structure having high strength and toughness especially suitable for reduction through the wire-drawing dies. Recently, in England, it was found that better torsional properties are obtained by heating to 1650°F (900°C), quenching in lead at 990°F (530°C), and then heating after the last pass to 600°F (315°C) for 3 min.

After patenting, the wire is drawn through a series of drafts of about 20 percent reduction in area to the desired size. The finished wire is then spheroidize annealed in a batch-type annealer in a controlled atmosphere to prevent decarburization. Such wire can easily be formed into springs and then hardened and tempered to a desired hardness usually Rockwell C42 to C46.

If the wire is not annealed as it comes from the last die, it is in a semi-hard condition and is called "untempered" wire. Such wire must also be hardened and tempered after coiling or forming into springs. Most soft-process wire, however, is hardened and tempered by the continuous method of heat treatment, thereby producing the popular oil-tempered MB and HB spring steels. In this heat-treating method, the wire is drawn through molten lead baths at a temperature above the critical range, and then hardened by immediate quenching in oil or a relatively low-temperature bath of molten lead. The hardened steel wire is then drawn through another lead bath to draw the temper of the wire to the desired hardness. Alloy steel wire in the annealed or oil-tempered condition is also made by this process. See Fig. 13.

HARD PROCESS

In the hard process method of producing wire, steel rods or wire of fairly high hardness have their tensile strength and hardness increased by the cold work performed in the process of reducing them to desired diameters by cold drawing through a series of tungsten carbide dies. A number of passes through these dies, having gradually decreasing diameters, produces wire to the desired size with very high tensile strengths. "Patented" hard-drawn wire and music wire are made by this method. Hard-drawn MB steel, music wire, "18-8" stainless steels types 302 and 304, Monel, Inconel, phosphor bronze, and spring brass are all made by variations of the hard process of wire drawing. They all obtain their high tensile strength and high hardness principally by cold drawing. See Fig. 14.

The cold working of wire in the hard process of manufacture, particularly in the cold-drawing process, induces a preferred grain orientation within the wire. The cold reduction of the body-centered cubic materials causes a desirable translation and rotation of crystal planes during plastic transformation to

FIG. 13. A single-die wire-drawing machine. Annealed rods are given large reductions of diameter by one pass through a tapered wire-drawing die in such machines. (Photo courtesy American Iron and Steel Institute.)

produce elongated gains running parallel to the length of the wire. However, the cold working of the wire, although producing the high tensile strengths and high hardnesses desired, also causes peculiar changes in the value of the modulus of elasticity—peculiar in that reductions of wire diameters from 10 to 50 percent increases the modulus values nearly 10 percent, but further reductions of diameter required for high tensile strengths suddenly reduces 90 percent of this increase. See Figs. 15 and 16.

There has in the past been much confusion in the design of springs made from wire produced by the hard process, including hard-drawn wire, music wire, and stainless steel, because the force and deflection tests of completed springs did not always agree with the results that were expected from the design formulas. The principal reason for the difference is that the modulus of elasticity varies with each size of wire, depending on the percent reduction in diameter that occurred during the wire-drawing operation. For example, music wire 0.111 in. (2.8 mm) in diameter drawn down through six drawing dies to obtain a wire diameter of 0.059 in. (1.5 mm), a reduction in diameter of 46.8 percent, will have a tensile strength of about 278 000 psi (1917 MPa), and a modulus of elasticity in tension, E, of 27 200 000 (187 537 MPa). The same

FIG. 14. Wire rod from reel in foreground being pulled through wire die near operator's right hand. Note smooth uniform wire and coil on block of machine after wire comes from die. (Photo courtesy American Iron and Steel Institute.)

FIG. 15. Wire-drawing die for making round wire. Square, hexagonal, and shaped dies are also used. (Photo courtesy American Iron and Steel Institute.)

FIG. 16. Cross section of round-hole wire-drawing die showing large entrance angle at bell mouth, reducing angle, bearing area, and small hole which produces the wire diameter. The preformed back relief after the small-diameter hole in the carbide insert reduces broken dies and scratched wire. (Photo courtesy American Iron and Steel Institute.)

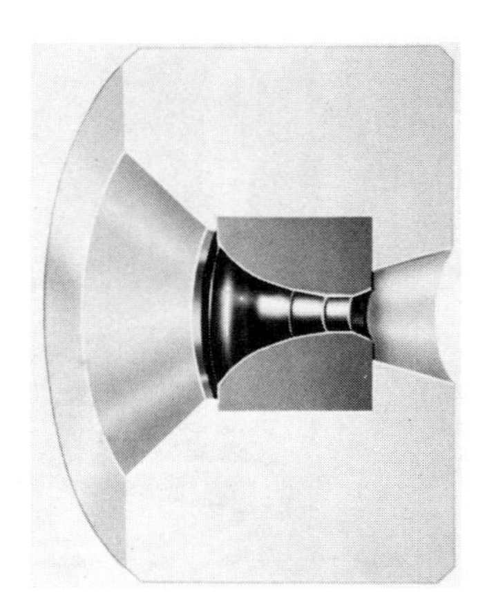

wire with further reductions of five passes to bring the wire diameter down to 0.039 in. (0.99 mm), a reduction in diameter of 64.8 percent, produces a tensile strength of 350 000 psi (2413 MPa), but here the modulus of elasticity in tension E drops to a low of 26 600 000 (183 400 MPa). A uniform relationship between the tensile strength and the elastic modulus does not exist. Also, small wire diameters drawn to high tensile strengths have a higher value for the modulus than do the larger sizes.

The steel industry can produce over 160 million miles of wire a year—enough to go around the earth more than 6500 times, and more than 160 000 different items are made from this wire—everything from paperclips and hairpins to watch springs and coat hangers. This wire is drawn through wire-drawing dies at the rate of about 1 200 ft (365 m) per minute. The dies are made from tungsten carbide, and even real diamonds to withstand the friction and forces exerted on them. Only the smaller sizes of wire are drawn through holes drilled into diamonds and sapphires. Some diamond dies have had as much as 20 miles (32 km) of wire drawn through them without appreciable wear. Most wire is round, but for some uses square, rectangular, oval, triangular, hexagonal, and other shapes are drawn or rolled.

Chemical Elements and Their Influence

Iron is the predominant element in the chemistry of steel, but is never mentioned in the analysis. It is supposed to be understood that after all alloying elements are shown in the chemical composition, the balance to make up 100 percent is iron. Alloying elements are added to spring steels during the process of melting. They frequently are combined with a certain percentage of iron to produce ferroalloys and are then thoroughly and completely mixed with the melt. These alloying elements increase the tensile strength, hardness, and toughness of steel.

Spring steels are somewhat similar to ordinary steels except that much greater care and more operations are required to produce them, and higher carbon contents are used. As a comparison of carbon contents for illustrative purposes, the range of carbon used in SAE 1020 machine steel in general use for shafts, screw machine products, and structural shapes is 0.15 to 0.25 percent, whereas the carbon range in music wire is 0.80 to 0.95 percent—an average of four to five times as much. The influence of the chemical elements found in high-carbon spring steels is briefly discussed in the following paragraphs.

CARBON

The element carbon, carefully regulated to definitely prescribed limits varying with each type of spring steel, raises the tensile strength, elastic limit, and hardness, but decreases the ductility, toughness, and shock resistance. Steels containing less than 0.40 percent carbon do not always properly respond to heat treatment for uniform hardening. In high-carbon steels the eutectic composition, at which the lowest hardening temperature is obtained, contains approximately 0.85 percent carbon. Spring steels rarely contain over 1.05 percent carbon, for larger amounts tend to cause brittleness without increasing hardness. Razors and engraving tools contain about 1.30 percent carbon, and cast iron has 2.0 to 4.0 percent carbon.

MANGANESE

The addition of manganese to the melt is done primarily to make the steel "sound" when it is first cast into an ingot. This element also makes the steel easier to roll, forge, and draw, and is as necessary as carbon because it causes the steel to harden rapidly and deeply. Because of this deep-hardening property, spring steels must be quenched in oil instead of water to avoid cracking. Without manganese, steel would harden only for a short distance from exterior surfaces and would remain soft in the central portion. Most spring steels contain from 0.60 to 1.20 percent manganese, but 0.30 percent minimum is often permitted. Music wire has 0.20 to 0.60 percent manganese. The United States imports nearly all of its manganese (95 percent) from India, South Africa, African Gold Coast, Brazil, Cuba, Russia, Korea, and Labrador. Brazil and Labrador have quantities which may not have been accurately determined, but U.S domestic resources are of such low grade that no method has yet been developed to process these ores on a truly economical basis.

SILICON

Most spring steels contain from 0.10 to 0.30 percent silicon, but additions of silicon to ordinary high-carbon steel in ranges from 1.80 to 2.20 percent raises the tensile strength without sacrificing ductility or toughness. Silicon also facilitates hot working of the steel. As an alloy, silicon is never used alone or with carbon. A deep-hardening element, usually manganese from 0.60 to 0.90 percent, is added with it, and the resulting alloy, called silico-manganese, becomes an important alloy spring steel.

PHOSPHORUS AND SULFUR

Phosphorus and sulfur are impurities and are the two most undesirable elements to have in spring steel. They reduce tensile strength, ductility, and shock resistance, and increase brittleness; for these reasons they are kept as low as possible. The maximum amount of either elements permitted in any spring steel is 0.055 percent; much smaller quantities are preferred, and the actual amount of either one found in most compositions is less than one half of 0.055 percent. In machinery steel, these elements are sometimes purposely added to make the steel more free machining—but such steels are not equal in quality to spring steels and are not customarily heat treated in the same manner as spring steels.

In addition to the elements described above, most of which have always been in steel, other alloying elements to increase the tensile strength, hardness, resistance to extreme temperatures and shock have been used only since about 1900. These elements intentionally added to steel are described below.

CHROMIUM

A slight addition of chromium, frequently called "chrome," usually in quantities from 0.80 to 1.10 percent, increases tensile strength, hardness, and toughness, reduces the need for high carbon content, increases the resistance of steel to attack by acids and alkalies, and increases steel's ability to withstand elevated temperatures. Like manganese, chromium causes a deeper penetration of hardness, but unlike manganese, it raises instead of lowers the temperature necessary for hardening. Stainless steels derive their resistance to corrosion principally by the addition of chromium in amounts varying from 12 to 20 percent. The United States imports 100 percent of its chromium from Canada, South Africa, Turkey, the Phillippines, Cuba, New Caledonia,

Russia, Rhodesia, Yugoslavia, and Pakistan. Domestic U.S. ores are of such low grade that processing them is quite difficult.

COBALT

Cobalt is another alloying element found in a wide variety of special steels, especially in high-speed tool steels, valves, drills, and other products. Its principal effect is to increase the red hardness so that cutting tools may be used at high operating speeds. Here again, the United States relies on imports for 100 percent of its requirements–obtaining the chromite ores from the Belgian Congo, Canada, Northern Rhodesia, Norway, and French Morocco. Several small domestic deposits are located in Pennsylvania, Idaho, and Missouri.

MOLYBDENUM

Molybdenum, like chromium, increases the hardness penetration and inclines the steel toward oil hardening. Molybdenum also increases the toughness of steel, the ability to withstand higher operating temperatures, and frequently is used to replace tungsten. Like tungsten, it is also used as filaments in vacuum tubes.

NICKEL

Nickel lowers the hardening temperature and tends to make a steel oil hardening rather than water hardening. Although having little effect on the hardenability of steels, it does add to the toughness and wear resistance. Large amounts, of about 7 to 8 percent, when added with 18 percent chromium, cause stainless steels to become austenitic and impervious to hardening by heating and quenching. The United States imports 74 percent of its requirements from Canada, the Scandinavian countries, England, and Europe.

TUNGSTEN

Tungsten, frequently called "wolframite" after the principal ore from which it is obtained, is added in quantities up to about 4 percent with straight carbon steels to obtain extremely hard tool steels. When 12 to 20 percent tungsten is added along with chromium, a tool steel acquires red hardness. Tool steels of the 18-4-1 variety containing 18 percent tungsten, 4 percent chromium, and 1 percent vanadium are important high-speed tool steels and can be coiled into springs for high-temperature applications. Tungsten is widely scattered over the globe. The United States produces about one third of its needs using ores from California, North Carolina, and Nevada. Ores are also imported from Korea, China, Australia, Africa, South America, and Canada.

VANADIUM

The addition of vanadium in small quantities from 0.15 to 0.25 percent raises the tensile strength, elastic limit, and toughness, and increases the ability of steel to resist impact shock. When used with chromium, the fatigue life and endurance of the steels are increased. Vanadium also helps keep the grain size desirably small and reduces the grain growth which occurs during accidental overheating. The United States imports 32 percent of its requirements from South Africa, Chile, and the USSR.

IMPORTS

The U.S. Department of Commerce predicted that in 1977 and over the next several decades, American requirements for the importation of certain selected materials would be as follows: chromium, cobalt, columbium, manganese, platinum, rubber, tantalum, and tin, 100 percent; mercury 95 percent; asbestos, bauxite, and nickel, 90 percent; zinc, 56 percent; petroleum, 38 percent; and iron ore, 30 percent.

2 High-Carbon Spring Steels

The most commonly used of all spring materials are the plain high-carbon, hard-drawn spring steels. These materials are readily available in a wide variety of sizes and are usually stocked in decimal, metric and fractional inch sizes. These materials, other than music wire, are the least expensive of all spring materials. All types of spring steels are drawn to the U.S. Steel Wire Gauge (same as Washburn and Moen Gauge) and also are frequently obtainable in half and quarter gauge sizes, even in fractions of an inch and in metric sizes. From an economic view, the high-carbon spring steels should be used in preference to other materials whenever possible.

For recommended design stresses, fatigue strength, endurance limits, and other properties, refer to the curves and tables for each material. These materials are for springs that are "cold coiled," with the material at room temperature. For minimum tensile strengths refer to Fig. 113. For Modulus of Elasticity see Fig. 112, and for curvature correction see Fig. 114.

Music Wire Spring Steel

DESCRIPTION

Music wire is the best-quality, most generally used cold-drawn high-carbon spring steel. It is considered the aristocrat of spring steels because it has a high tensile strength, high elastic limit, and can withstand high stresses under repeated loadings. This type of wire is frequently called "piano wire," but the name "music wire" is more generally acceptable and it is so called in standard specifications. The name is derived from the application of this type of wire to stringed instruments, especially the piano. The color of the wire is obtained by drawing the wire through a liquid tin solution prior to the last pass through a wire-drawing die, thus depositing a very thin coating of tin on the surface of the wire. This coating is much too thin to act as a good rust-resistant finish, but it does protect the surface for a reasonable period of time and also keeps the surface sufficiently clean so that electroplated finishes can be applied with very little acid dipping, tumbling, or abrasive cleaning.

COMMERCIAL SPECIFICATIONS

ASTM A 228 (Many government and company specifications also exist.)

MANUFACTURE

For many years, music wire of excellent quality, high tensile strength, and of the exact sizes desired was imported from Sweden. Later, Swedish steel rods were obtained and drawn into music wire by American wire mills. During World War II the importation of Swedish steels came to a stop, and it was necessary to make music wire from domestic steels. At first this was a problem and some of the American product did not have the uniformity, tensile strength, or ductility desired. However, the American mills soon developed improved techniques and now the music wire made in the United States is of good quality. Sweden is fortunate in having extremely high-grade ores, and her small iron plants can use charcoal rather than coke for fuel, resulting in cleaner iron and cleaner steel. Moreover, her rolling and wire mills are not large and produce comparatively small tonnages, so that extreme care can be more conveniently exercised in producing steel rods and wire with virtually no seams or scratches. Steel for music wire is made by the acid open hearth, basic oxygen, or electric furnace process, and the rods are drawn into wire by the "hard process" whereby the high tensile strength and hardness are obtained by patenting and cold drawing through a number of dies of diminishing diameters. The smaller sizes of wire are drawn through holes drilled into diamonds. Music wire is made to high tensile strengths only in round sections. It is rarely used in the annealed condition and is not commercially available in square or rectangular sections. Music wire can be obtained pretinned or preplated with cadmium, but plating of the finished springs is preferred when a 100-hour salt spray test is required.

APPLICATIONS

Music wire is the best, toughest, and most widely used of all spring materials for small springs. It has a high tensile strength and can withstand high stresses under repeated loading. It may be obtained readily in many gauge decimal and metric sizes, as each wire manufacturer draws it to the gauge best suited to his wire-drawing practice. The most popular sizes run from 0.004 in. (0.10 mm), which is the size employed in "E" strings of violins, to 0.125 in.

(3.17 mm) diameter, and some wire mills supply it in every three-place decimal size from 0.004 to 0.200 in. (0.10 to 5 mm) or larger, advancing in increments of 0.001 in. (0.025 mm). Although this steel wire is tough and durable, springs made from it should not be subjected to applications involving temperatures exceeding 250°F (120°C). At that temperature springs will suffer a load loss of approximately 5 percent when stressed to 90 000 psi. (620 MPa), and the load loss increases rapidly at higher temperatures. This material is used only for cold-coiled springs.

CHEMICAL COMPOSITION

Nearly all music wire contains between 0.80 and .090 percent carbon. Several commercial specifications are also available in addition to the ASTM Standard. The ASTM specification has a wide range for carbon content, but orders can specify a 10 percent range, if desired. See Fig. 17.

TENSILE STRENGTH

The tensile strength and hardness of music wire are caused by the cold work done in drawing the wire through the wire-drawing dies. The reduction of area per pass and number of passes through the dies determine the wire size and mechanical properties. Each wire manufacturer has his particular method of wire drawing, and although all methods are very similar, no two companies produce wire exactly alike, nor will two batches of wire of the same diameter and chemical composition made by any one company by the same method have exactly the same tensile strength. Some wire manufacturers make one type of music wire conforming to one set of tensile strength values. Others produce two or more tensile strength ranges, all with the same chemical composition. To draw wire to an exact tensile strength within 10 000 psi (68.9 MPa) is exceptionally difficult, especially in the fine sizes. For this reason it would be preferable to have three tensile strength ranges. In drawing a wire of 0.030 in. (0.76 mm) diameter to a desired tensile strength of 360 000 psi (2480 MPa), it might actually come out with only 340 000 psi (2345 MPa), or as high as 380 000 psi (2620 MPa). For practical purposes,

FIG. 17. Chemical Composition of Music Wire (percent)

Element	Swedish	ASTM A 228
Carbon	0.85-0.95	0.70-1.00
Manganese	0.25-0.45	0.20-0.60
Silicon	0.15-0.30	0.10-0.30
Phosphorus	0.025 max.	0.025 max.
Sulfur	0.025 max.	0.030 max.

each of the three wires has some particular advantage over the others. The wire with the lower tensile strength will have less hardness, and its greater ductility will make it especially suitable for torsion and extension springs with sharply formed radii or bends on hooks and for springs with small index ratios where the outside diameter of the spring divided by the wire diameter is less than 5. For all general springs the 360 000 psi (2480-MPa) wire is best, but for compression springs having abnormally high stresses, the 380 000 psi (2620-MPa) wire will have the longest fatigue life provided that the spring index D/d is more than 7.

As suggested in the preceding paragraph, there could be three definite ranges of desirable tensile strengths, but all three are seldom available. Spring drawings should either specify the type of music wire desired or leave it to the discretion of the spring manufacturer. Music wire should, were possible, be purchased to meet a desired tensile strength range in addition to a commercial designation or trade name.

Although not many wire mills will furnish music wire in three tensile strength ranges, it is hoped that as more spring manufacturers request it in this manner, three ranges will someday become a standard.

At the present time, the ASTM A228 specification has tensile strength recommendations that cover too wide a range for uniform spring manufacture. Army, Navy, and federal specifications are often so indefinite that inferior grades of wire frequently will meet their requirements. For these reasons, many companies have found it necessary to set their own specifications for music wire.

For important springs, it is highly desirable to state on the purchase order that "Wire must be tested for torsional properties in accordance with ASTM E 558."

SECONDARY HARDNESS

All springs made from hard-drawn spring steels, including music wire, hard-drawn steel, and hard-drawn stainless steels, acquire about 2 to 3 points of Rockwell C increase in hardness during the baking or stress-equalizing heat treatment performed after coiling. This phenomenon, called "secondary hardness," is usually beneficial because it raises the tensile strength and elastic limit of such springs.

MECHANICAL PROPERTIES OF MUSIC WIRE

Elastic Limit. Percent of tensile strength: in tension, 65 to 75 percent; in torsion, 45 to 50 percent.

Hardness, Rockwell. 42 to 46C.

Electrical Conductivity. Percent of copper, 8 to 12 percent.

Bending Properties. All diameters should be capable of being bent around an arbor equal to the wire diameter without breaking or cracking the surface. Also, a close-would extension spring 5 in. (127 mm) long cooled on an arbor 3

FIG. 18. Modulus of Elasticity

Wire diameter		Tension E		Torsion G	
(in.)	(mm)	(psi)	(MPa)	(psi)	(MPa)
Up to 0.032	Up to 0.80	29 500 000	203 400	12 000 000	82 700
0.033-0.063	0.81-1.6	29 000 000	200 000	11 850 000	81 700
0.064-0.125	1.61-3.2	28 500 000	196 500	11 750 000	81 000
Over 0.125	3.2	28 000 000	193 000	11 600 000	80 000

FIG. 19. Tolerances on Music Wire Diameters

Diameter		Tolerance (±)	
(in.)	(mm)	(in.)	(mm)
Up to 0.010	Up to 0.25	0.0002	0.005
0.011-0.028	0.26-0.71	0.0003	0.008
0.029-0.063	0.72-1.60	0.0004	0.010
0.064-0.080	1.61-2.03	0.0005	0.013
Over 0.080	Over 2.03	0.0010	0.030

FIG. 20. Tempering Time (minutes)

Wire diameter		General	Severe	High-temperature
(in.)	(mm)	service	service	service[a]
Up to 0.015	Up to 0.38	10-15	15-20	20-30
0.016-0.050	0.39-1.27	15-20	20-30	20-45
0.051-0.120	1.28-3.0	20-25	30-40	45-60
Over 0.120	Over 3.0	25-30	40-50	60-80

[a]It is not customary to use music wire at temperatures over 250°F (120°C).

to 3½ times the wire diameter should be pulled out so that it sets to about 15 in. (380 mm) long and should have a uniform pitch with no splits or fractures.

Weight. per cubic inch, 0.284 lb (0.129 kg); per cubic centimeter 7.85 g.

HEAT TREATMENT, MUSIC WIRE

Stress Equalizing. Springs made from this material can be heated to remove residual coiling stresses and to settle the structure as follows:

For general service: 400 to 420°F (200 to 215°C)

For severe service: 450 to 500°F (230 to 260°C)

For high-temperature service: 525 to 550°F (275 to 290°C)

Tempering Time. See Fig. 20.

Hard-Drawn Spring Steels

DESCRIPTION

Hard-drawn spring steels, while of lower quality than music wire or oil-tempered wire, nevertheless is an important spring steel where cost is an important factor and where long fatigue life and uniformity are of lesser importance. It is the least expensive of all spring materials. This wire is available in several grades and is used only for cold-coiled springs.

COMMERCIAL SPECIFICATIONS

For MB grade, ASTM A 227; for HB grade, ASTM A 679, and for Upholstery grades, ASTM A 407 and 417.

MANUFACTURE

Steel for hard-drawn wire is usually made from domestic ores and the wire is produced by the hard process. The tensile strengths and hardness, though lower than those of music wire, are obtained in a similar manner by cold drawing the wire through a series of dies with gradually smaller diameters. Although several grades with different tensile strengths and carbon ranges are available, only four grades are in general use in the spring industry. These are commercially known as the MB grade, HB grade, and two Upholstery grades. Other grades with a higher carbon content called Extra HB (or XHB) are occasionally used. These materials are not used in the annealed conditions except in special cases. The grades are available in different diameter ranges. Square and rectangular shapes are made when necessary, but are infrequently used. Hard-drawn wire may be obtained with special finishes including galvanized, tinned, coppered, or preplated with cadmium and zinc. Its clean, smooth surface lends itself readily to electroplating.

APPLICATIONS

MB Grade ASTM A 227 Class II. This popular grade is finding increasing usefulness for the general run of compression and extension springs due to improvements in its uniformity and quality. However, it is best suited for those applications where long life and accuracy of loads and deflections are not too important. This grade is usually obtained with the same chemical composition as the regular oil-tempered MB steel ASTM A 229, but is processed differently. It can be used in many applications where high-grade expensive steels are not needed. This material is used extensively in mechanical products, automotive equipment, hardware, and toys. Class I is generally used for torsion springs and springs requiring rather sharp bends.

HB Grade ASTM A 679. This material, having a higher carbon content than the MB grade, is used where the stress requirements are about halfway between those used for oil-tempered MB and music wire. This wire is cold drawn to tensile strengths quite similar to commercial grades of music wire. It has been referred to as a low-quality music wire, often selling at about half the price of music wire. Only a few wire mills produce this material, so it should be specified with caution and only where the quantity of springs desired merits ordering at least several hundred pounds of wire. The HB grade is satisfactory for applications where high stresses are required and where a low-cost wire is essential.

Upholstery Grade ASTM A 407. This wire, available in a small range of sizes, is, as its name implies, a low-cost wire with excellent ductility and bending properties that are especially suitable for springs used in upholstered products. This material should not be specified for important springs in machine tools or where fatigue life or accurate loads are essential, but it is a useful material for wire forms, hooks, and similar items. This wire has good ductility and can be knotted without fracture. It has a carbon content of 0.45 to 0.70 percent and is available in six different tensile strengths depending upon coiling and knotting requirements; diameters available run from 0.0348 in. (0.884 mm) to 0.162 in. (4.115 mm) only.

Upholstery Grade ASTM A 417. This grade is for zig-zag, square-formed, and no-sag types, such as are used for furniture and in the seats of automobiles. It should not be used for mechanical springs. The carbon content runs from 0.50 to 0.75 percent. The wire is available in five different tensile strengths depending on the bending and stress requirements; diameters available run from 0.0915 in. (2.324 mm) to 0.192 in. (4.877 mm) only.

CHEMICAL COMPOSITION

Although several specifications and chemical compositions for hard-drawn wire are available, the ASTM Standards are generally used. The ASTM specification is of a general nature in that a wide range of carbon is specified. However, the carbon range in the most desirable spring steels should not exceed 13 points, and small diameters of wire can have less carbon than larger sizes. For this reason, it is advisable when ordering the MB grade to specifiy a 13-point carbon range such as 0.50 to 0.63 percent for the smaller wires under 0.100 in. diameter, and a range of 0.60 to 0.73 percent for larger sizes. For the HB grade, similar ranges such as 0.70 to 0.83 and 0.80 to 0.93 percent may be specified. Manganese, too, should not exceed a 30-point range in a like manner. See Fig. 21.

MECHANICAL PROPERTIES OF HARD-DRAWN MB AND HB WIRE

Elastic Limit. Percent of tensile strength: in tension, 60 to 70 percent; in torsion, 45 to 55 percent.

FIG. 21. Chemical Composition–Hard-Drawn Steel Wire (percent)

Element	MB grade ASTM A 227	HB grade ASTM A 679
Carbon	0.45-0.85	0.65-1.00
Manganese	0.60-1.30	0.20-1.30
Silicon	0.10-0.30	0.10-0.40
Phosphorus	0.040 max.	0.040 max.
Sulfur	0.050 max.	0.050 max.

FIG. 22. Modulus of Elasticity

Wire diameter		Tension, E		Torsion, G	
(in.)	(mm)	(psi)	(MPa)	(psi)	(MPa)
Up to 0.032	Up to 0.80	28 800 000	198 600	11 700 000	80 670
0.033-0.063	0.81-1.6	28 700 000	197 900	11 600 000	80 000
0.064-0.125	1.61-3.2	18 600 000	197 200	11 500 000	79 290
Over 0.125	Over 3.2	28 500 000	196 500	11 400 000	78 600

Hardness, Rockwell. Varies considerably from 38 to 46C.

Electrical Conductivity. Percent of copper, 8 to 12 percent.

Modulus of Elasticity. See Fig. 22.

Wrap Tests. See Fig. 23.

Tolerences. See Fig. 24.

HEAT TREATMENT, HARD-DRAWN WIRE

Stress Equalizing. Springs made from hard-drawn wire can be heated to remove residual coiling stresses and to settle the structure as follows:

For general service: 420 to 450°F (215 to 230°C)

For severe service: 450 to 500°F (230 to 260°C)

For high-temperature service: 525 to 550°F (275 to 290°C)

Tempering Time. See Fig. 25.

FIG. 23. Wrap Test

Wire diameter[a]		Arbor diameter ASTM A 227		
(in.)	(mm)	Class I	Class II	ASTM A 679
Up to 0.162	Up to 4.11	Same as wire diam.	2 × wire diam.	2 × wire diam.
0.163-0.312	4.12-7.92	2 × wire diam.	4 × wire diam.	4 × wire diam.

[a]Wire should be bent around an arbor without cracking.

FIG. 24. Tolerances on Diameters (Hard-Drawn Wire)

Diameter		Tolerance (±)	
(in.)	(mm)	(in.)	(mm)
Up to 0.028	Up to 0.71	0.0008	0.020
0.029-0.075	0.72-1.90	0.001	0.030
0.076-0.375	1.91-9.53	0.002	0.050
Over 0.375	Over 9.53	0.003	0.080

FIG. 25. Tempering Time (minutes)

Wire diameter (in.)	Wire diameter (mm)	General service	Severe service	High-temperature service[a]
Up to 0.050	Up to 1.27	15-20	20-30	30-45
0.051-0.120	1.28-3.05	20-25	30-40	45-60
0.121-0.375	3.06-9.53	25-30	40-50	60-80
Over 0.375	Over 9.53	30-45	50-60	60-90

[a]It is not customary to use hard-drawn wire at temperatures over 325°F (160°C).

HEAT TREATMENT, HARD-DRAWN WIRE

Stress Equalizing. Springs made from hard-drawn wire can be heated to remove residual coiling stresses and to settle the structure as follows:

For general service: 420 to 450°F (215 to 230°C)

For severe service: 450 to 500°F (230 to 260°C)

For high-temperature service: 525 to 550°F (275 to 290°C)

Tempering Time. See Fig. 25.

Oil-Tempered Spring Steels

DESCRIPTION

Oil-tempered spring steels are good-quality, high-carbon spring steels, uniform in quality and temper, and used for a large majority of coil springs requiring wire diameters from 1/8 to 1/2 in. (3.0 to 13 mm), although other diameters are also available. Their color is usually a dull, smoky, black, mottled with gray, lacking all the lustre and sheen common to music wire. These materials are used only for cold-coiled springs.

COMMERCIAL SPECIFICATIONS

For MB grade: ASTM A 229 Class I & II and SAE 1065. For HB grade: SAE 1080.

MANUFACTURE

Oil-tempered spring steels are made by the open hearth, basic oxygen, or electric furnace process, usually from domestic iron ores. Wire to be oil tempered is first cold drawn to the desired size from annealed rods. Since the hardness and tensile strength are obtained by heat treatment, no attempt is made to cold draw the wire to high tensile strengths as is customary in making hard-drawn or music wire. After being drawn to the desired diameter, it is given a spring temper by a continuous process of heat treatment in the wire mill and shipped in large-diameter coils up to 6 ft (2 m) and larger. In this process, the wire is pulled through a furnace or lead bath where it is heated to the proper quenching temperature; then it is quenched in an oil bath and drawn through another furnace or lead bath for tempering to obtain the desired hardness. Although several ranges of carbon content are obtainable, only two are in general use for springs; these are commercially known as the MB grade and the HB grade. Other grades with higher carbon contents, called Extra HB (or XHB) are occasionally used. Of these the MB grade is used for probably over 90 percent of the applications. However, the HB grade, although more expensive than the MB grade, is often used where higher tensile strengths are required. The designations MB, HB, and XHB will gradually be replaced by standard specification numbers.

APPLICATIONS

MB Grade ASTM A 227 Class II. This oil-tempered spring steel is a general-purpose material commonly used for all types of compression, extension, and torsion springs where the cost of music wire is prohibitive and for sizes larger than those available in music wire. It is not recommended for impact or shock loading, nor for temperatures above 350°F (175°C) and is not always satisfactory for below freezing temperature applications. This grade is used extensively in many industries for springs used in machine tools and mechanical products where the rather coarse surface and slight amount of surface scale are not objectionable. Square and rectangular shapes are also available in common fractional sizes and in millimeter sizes. Annealed stock can be obtained for springs having sharp bends or for those coiled with small spring index ratios of outside diameter to wire size, such as 5 or less. The heat-treating scale naturally formed on the wire should be removed by shot blasting or tumbling before finishes are applied by electrodeposition. Class I is generally used for torsion springs and springs requiring rather sharp bends.

HB Grade SAE 1080. This material is similar to the MB grade except that the higher carbon content provides a higher tensile strength to withstand higher operating stresses. It is obtainable in the same sizes or shapes and has the same applications as the MB grade, but is not so readily available in the plants of many spring manufacturers. Usually, where this material is presently specified, it may be better to use the more expensive alloy spring steels, particularly if a long fatigue life or high endurance properties are needed. Although some spring manufacturers do not ordinarily stock this steel, others feel that it should be used much more often for highly stressed important springs. The chemical composition is quite similar to ASTM A 679.

These materials also may be obtained in the annealed condition if desired. Springs made from annealed stock require a heat treatment to harden and temper them after coiling. Extension springs made of annealed wire cannot be wound with initial tension.

CHEMICAL COMPOSITIONS

Several compositions are currently specified for these steels, but the commercial terms MB and HB alone are frequently used. The manganese content for wire diameters up to 0.192 in. (5 mm) is usually specified as 0.60 to 0.90 percent and for larger sizes 0.80 to 1.20 percent. Smaller percentages down to 0.20 percent may be used. The carbon content should not vary more than 0.13 percent in any one lot. See Fig. 26.

FIG. 26. Chemical Composition (percent)

Element	MB grade ASTM A 229	SAE 1065	HB grade SAE 1080
Carbon	0.55-0.85	0.60-0.70	0.75-0.88
Manganese	0.60-1.20	0.60-0.90	0.60-0.90
Silicon	0.10-0.35	0.15-0.30	0.15-0.30
Phosphorus	0.040 max.	0.040 max.	0.040 max.
Sulfur	0.050 max.	0.050 max.	0.050 max.

MECHANICAL PROPERTIES OF OIL-TEMPERED WIRE:
MB AND HB GRADES

Modulus of Elasticity. E in tension (for torsion springs) = 28 500 000 psi (196 500 MPa); G in torsion (for compression and extension springs) = 11 200 000 psi (77 200 MPa).

Elastic Limit. Percent of tensile strength: in tension, up to 0.25 in. (6.35 mm) = 85 to 90 percent, over 0.25 in. (6.35 mm) = 80 to 85 percent; in torsion, up to 0.25 in. (6.35 mm) = 45 to 50 percent, over 0.25 in. (6.35 mm) = 40 to 45 percent.

Hardness, Rockwell. Up to 0.125 in. (3.18 mm), 45 to 50C; 0.126 to 0.25 in. (3.19 to 6.35 mm), 42 to 48C; over 0.25 in. (6.35 mm), 40 to 45C.

Electrical Conductivity. Percent of copper = 8 to 12 percent.

Wrap Test. Wire up to 0.162 in. (4 mm) should be capable of winding on itself as an arbor, larger sizes up to 0.312 in. (8 mm) on an arbor twice the wire diameter, without splitting or breaking.

Weight. Per cubic inch = 0.284 lb (0.129 kg); per cubic centimeter = 7.85 g.

Tolerences. See Fig. 27.

HEAT-TREATMENT, OIL-TEMPERED WIRE

Hardening. Springs from annealed material can be hardened at 1475 to 1550°F (800 to 845°C). Soak light sections under ¼ in. (6.35 mm) for 4 to 8 min, larger sections 8 to 12 min, and oil quench.

Tempering. Such springs can be tempered at 600 to 700°F. (315 to 370°C) from 30 to 60 min depending upon size and hardness desired.

Stress Equalizing. Springs made from oil-tempered wire can be tempered to remove residual coiling stresses and to settle the structure as follows:

For general service: 420 to 450°F (215 to 240°C)

For severe service: 500 to 550°F (260 to 290°C)

For high-temperature use: 600 to 650°F (315 to 345°C)

FIG. 27. Tolerance on Diameters (Oil-Tempered)

Diameter		Tolerance (±)	
(in.)	(mm)	(in.)	(mm)
Up to 0.028	Up to 0.70	0.0008	0.02
0.029-0.075	0.71-1.90	0.001	0.03
0.076-0.375	1.91-9.50	0.002	0.05
Over 0.375	Over 9.50	0.003	0.08

FIG. 28. Tempering Time (minutes)

Wire diameter (in.)	(mm)	General service	Severe service	High-temperature service
Up to .050	Up to 1.25	15-20	20-30	30-45
0.051-0.120	1.25-3.00	20-25	30-40	45-60
0.121-0.375	3.01-9.50	25-30	40-50	60-80
Over 0.375	Over 9.50	30-45	50-60	60-90

Tempering Time. See Fig. 28.

Valve Spring-Quality Steels

GENERAL

Engines used in automobiles, aircraft, motorcycles, motor boats, agricultural equipment, and compressors require springs of exceptionally good quality for continued valve operation. Valve springs must have a smooth surface and be free from burrs so that small flakes or scale will not score the cylinders or valve seats. In addition, such springs are subjected to moderately high operating temperature, vibration, and surge. Only selected compositions of steel wire having uniform, smooth finish and high fatigue characteristics under suddenly applied loads should be used for valve springs. Most valve springs are coiled with reduced or variable pitches on the first two or three active coils on one end to reduce surging. This effect is explained more fully in the section on design of compression springs. The breakage of valve springs in automotive engines, so customarily encountered in the early 1920s, is now rarely found. Better selection of steel, improved design, and shotpeening have each contributed a significant part to the trouble-free performance of valve springs used in modern cars. However, a word of warning to designers of automotive engines could be included at this point: the increased duties of valve springs, imposed by higher-speed engines and higher engine temperatures caused by traffic congestion may once again create valve spring troubles. The demand for low-cost automobiles originated by intense competition among car builders may preclude the use of the more expensive grades of spring steels. Additional high-speed fatigue testing of automotive engines to determine if presently used carbon steel springs should be redesigned or replaced with other steels or alloy steels appears warranted.

In England a popular valve spring steel is a high-carbon, patented hard-drawn steel wire having a chemical composition of carbon 0.70 to 0.80, manganese 0.60 to 0.75, silicon 0.10 to 0.20, with phosphorus and sulfur kept below a maximum of 0.040 percent. The rods are first given a spheroidizing heat treatment by soaking at 1375°F (750°C) for a short period, followed by a slow cool; or normalized at 1475°F (800°C) with a short soak and then air cooled to produce a semipearlitic structure. After cleaning, the rods are lightly drawn one hole and then given a "patenting" heat which is really a "controlled isothermal transformation of austenite." This is done by austenitizing at the high temperature of 1830°F (1000°C) and quenching in a lead or salt bath to approximately 1000°F (540°C). New air quenching methods also give good results by allowing the wire to be cooled under a jet of air at 18 to 35 psi (0.124 to 0.24 MPa) pressure. The wire is then pickled to remove scale and cleaned by immersing the coils in dilute hydrochloric or sulfuric acid and water rinsed. The wire is then hard drawn in the usual manner by several passes through dies to reduce the area about 60 percent, thus obtaining the hardness and high tensile strength required. The wire is polished by passing it between twin belt polishing machines, copper coated, and then drawn one hole to finished size.

Although only two compositions of valve steels are described in this section, other high-quality steels could be used, including the chrome-silicon and chrome-moly-vanadium alloys except that their present high cost eliminates them for consideration by economy-minded engine builders.

High-Carbon Valve Spring Steel

DESCRIPTION

High-carbon valve spring steel is the highest quality of round carbon steel valve spring wire available. It is uniform in quality and temper, but available in a limited range of diameters from 0.062 to 0.250 in. (1.57 to 6.25 mm) only, although other sizes can be obtained if the quantity desired is sufficient to be of interest to the wire mills. Although the surface is supplied clean and free from scale, coatings to prevent rust during storage are frequently used; these are removed by the heat treatment or shotpeening operations. This material is used only for cold-coiled springs.

COMMERCIAL SPECIFICATIONS

ASTM A 230.

MANUFACTURE

This steel is made by the open hearth, basic oxygen, or electric furnace process, usually from domestic iron ores. It is furnished in the hardened and tempered condition similar to oil-tempered wire, although it may be obtained in the untempered condition for hardening and tempering after coiling.

APPLICATIONS

This special plain high-carbon spring steel is the most popular material in common use for automotive engine valve springs and other springs requiring

FIG. 29. Chemical Composition of Valve Spring Wire, ASTM A 230

Element	Percent
Carbon	0.60-0.75
Manganese	0.60-0.90
Silicon	0.15-0.35
Phosphorus	0.025 max.
Sulfur	0.030 max.

FIG. 30. Tolerances on Wire Diameters

Diameter		Tolerance (±)	
(in.)	(mm)	(in.)	(mm)
Up to 0.092	Up to 2.34	0.0008	0.02
0.093-0.148	2.35-3.75	0.001	0.03
0.149-0.177	3.76-4.50	0.0015	0.04
Over 0.177	Over 4.50	0.002	0.05

high fatigue properties. It is a select-quality steel having uniform properties and smooth finish especially suited for valve springs requiring long life, and is the least expensive of the high-quality steels or the alloy steels intended for such applications. It is not recommended for high-temperature applications over 350°F (175°C), and is generally used in round sections only. Some die springs have been made from this steel, and annealed wire has been rolled to rectangular shapes with round edges and furnished in the untempered condition.

MECHANICAL PROPERTIES OF HIGH-CARBON VALVE SPRING WIRE, ASTM A 230

Modulus of Elasticity. E in tension (for torsion springs) = 29 500 000 psi (203 400 MPa); G in torsion (for compression springs) = 11 200 000 psi (77 200 MPa).

Elastic Limit. Percent of tensile strength: in tension = 85 to 90 percent; in torsion = 50 to 60 percent.

Hardness, Rockwell. 44 to 48C.

Electrical Conductivity. Percent of copper = 8 to 12 percent.

FIG. 31. Tempering Time (minutes)

Wire diameter (in.)	(mm)	General service	Severe service	High-temperature service
Up to 0.128	Up to 3.25	20-25	30-40	45-60
Over 0.128	Over 3.25	25-30	40-50	60-80

Wrap Test. Wire up to 0.162 in. (4 mm) should be capable of winding on itself as an arbor, larger sizes on an arbor twice the wire diameter, without splitting or breaking.

Weight. Per cubic inch = 0.284 lb (0.129 kg); per cubic centimeter = 7.85 g.

Tolerences. See Fig. 30.

HEAT TREATMENT

Hardening. Annealed wire springs can be hardened at 1475 to 1525°F (800 to 830°C); soak for 4 to 8 min and oil quench.

Tempering. Such springs are usually tempered at 550 to 700°F (290 to 370°C) for 30 to 60 min depending upon hardness desired.

Stress Equalizing. Springs made from oil-tempered high-carbon valve steels should be tempered to remove residual coiling stresses and to settle the structure as follows:

For general service: 420 to 450°F (215 to 230°C)

For severe service: 500 to 550°F (260 to 290°C)

For high-temperature use: 600 to 650°F (315 to 345°C)

Tempering Time. See Fig. 31.

3 Alloy Spring Steels

The alloy spring steels have an important place in the field of spring materials, particularly for conditions involving high stress and where shock or impact loadings occur. Alloy spring steels can also withstand both higher- and lower-temperature applications than those of high-carbon spring steels and are obtainable in either the annealed or oil-tempered condition. Although all these materials may be obtained in round, square, and rectangular sections, none are regularly stocked by spring manufacturers in a wide variety of sizes and therefore they usually must be specially ordered. For this reason, small quantities of springs often are quite expensive and require several weeks before deliveries can be made.

All these alloys are used in springs for machines and aircraft equipment in wire sizes frequently under ¼ in. (6.0 mm) diameter and up to 0.500 in. (13 mm), but annealed bars are available in larger diameters from $^3/_8$ in. to 2 in. or larger. These larger sizes, in the annealed condition, are hot rolled into heavy springs for armament and heavy equipment. Refer to Chapter 4 for further information.

For allowable working stresses, fatigue strength, endurance limits, and other properties, refer to the charts and tables for each material. These materials are for both springs that are "cold coiled," with the material at room temperature, and for "hot-coiled" springs or for flat-leaf elliptic springs using hot-rolled bars.

Chromium-Vanadium Spring Steel Wire

DESCRIPTION

Chromium-vanadium spring steel is a good-quality, fairly high-carbon, alloy spring steel containing small amounts of chromium and vanadium to increase the hardness, tensile strength, and endurance properties. It is obtainable in round or square sections from 0.020 to 0.500 in. (0.5 to 13 mm) diameters for general use and in heavier sizes up to 2 in. (50 mm) and larger for hot-rolled springs. The surface is a mottled dull black and gray caused by a slight oxide coating formed during heat treatment.

COMMERCIAL SPECIFICATION

ASTM A 231.

MANUFACTURE

This alloy steel may be made by the open hearth, basic oxygen, or electric furnace process. The electric furnace steel has been preferred for wire, but the acid open hearth and basic oxygen process also produce good results. It may be obtained in three conditions as follows:

1. Annealed: in this condition, the steel wire is fully annealed and it must be hardened and tempered after forming to obtain the spring properties desired. Annealed wire is used for many coil springs requiring wire diameters over 3/8 in. (9.5 mm) and for those springs where the outside diameter of the spring divided by the wire size equals 5 or less. It also can be obtained in square and rectangular sections if a sufficient quantity is ordered, as such sections are not ordinarily carried in stock.
2. Untempered (Cold-Drawn Wire): in this condition, the wire is not annealed after the last pass through the wire-drawing dies and therefore has the benefit of some cold work. Wire up to 3/8 in. (9.5 mm) diameter is often obtained in this condition, as it is easier to coil more uniformly on automatic spring coilers and the cold-worked wire maintains a more even spacing or pitch of the coils. Untempered wire also requires hardening and tempering to obtain spring properties, in the same manner as annealed wire.
3. Oil tempered: wire to be oil tempered is first cold drawn to the desired size from annealed rods and then heated above the transformation temperature, quenched in oil, and then tempered in a manner quite similar to that of regular oil-tempered carbon steel wire. Controlled temperatures and extended tempering times are used to produce the desired hardness and tensile strength.

APPLICATIONS

Chrome-vanadium wire is a popular alloy spring steel for applications involving higher stresses than can be used with the high-carbon spring steels and for springs subjected to impact or shock, such as in pneumatic hammers. This material is also used for moderately elevated temperature applications up to 425°F (220°C). Many types and sizes of die springs are made from this material in round, square, and rectangular sections with round edges. The rounded-edge section is made by rolling round wire between the flattening rolls of small rolling mills.

CHEMICAL COMPOSITION

Although several compositions are available for this alloy, the two most generally known, produced more often in larger quantities, and recommended for general applications are ASTM A 231 and SAE 6150, and they have the same chemical composition. Vanadium in all compositions is desired in amounts of about 0.18 percent, although a 0.15 minimum is specified. See Fig. 32.

MECHANICAL PROPERTIES OF CHROMIUM-VANADIUM WIRE, ASTM A 231

Modulus of Elasticity. E in tension (for torsion springs) = 29 500 000 psi (203 400 MPa); G in torsion (for compression and extension springs) = 11 200 000 psi (77 200 MPa).

Elastic Limit. Percent of tensile strength: in tension = 88 to 93 percent; in torsion = 65 to 75 percent.

Hardness, Rockwell. 45 to 50C.

Electrical Conductivity. Percent of copper = 8 to 12 percent.

Wrap Test. Wire up to 0.162 in. (4 mm) should be capable of winding on

FIG. 32. Chemical Composition of Chromium-Vanadium[a] (percent)

Element	ASTM A 231 & SAE 6150
Carbon	0.48-0.53
Manganese	0.70-0.90
Chromium	0.80-1.10
Vanadium	0.15 min.
Silicon	0.20-0.35
Phosphorus	0.040 max.
Sulfur	0.040 max.

[a]This alloy, when made in electric furnaces, generally has reduced phosphorus and sulfur of 0.025 percent maximum.

FIG. 33. Tolerance on Diameter

Diameter		Tolerance (±)	
(in.)	(mm)	(in.)	(mm)
Up to 0.028	Up to 0.70	0.0008	0.02
0.029-0.072	0.71-1.80	0.001	0.03
0.073-0.375	1.81-9.50	0.002	0.05
Over 0.375	Over 9.50	0.003	0.08

FIG. 34. Tempering Time (minutes)

Wire Diameter		General	Severe	High-temperature
(in.)	(mm)	service	service	service
Up to 0.050	Up to 1.25	15-20	20-30	30-45
0.051-0.120	1.26-3.00	20-25	30-40	45-60
0.121-0.375	3.01-9.50	25-30	40-50	60-80
Over 0.375	Over 9.50	30-45	50-60	60-90

itself as an arbor, larger sizes up to 0.312 in. (8 mm) on an arbor twice the wire diameter, without splitting or breaking.

Weight. Per cubic inch = 0.284 lb (0.129 kg); per cubic centimetre = 7.85 g.

Tolerances. See Fig. 33.

HEAT TREATMENT, CHROMIUM-VANADIUM, ASTM A 231

Hardening. For springs made from annealed material, heat springs to 1600 to 1650°F (870 to 900°C), soak light sections under 3/8 in. (9.5 mm) for 8 to 10 min, larger sections 15 to 20 min, and oil quench.

Tempering. Such springs are usually tempered at 750 to 950°F. (400 to 500°C) for ¾ to 1½ hours, depending upon size and hardness desired.

Stress Equalizing. Springs made from oil-tempered chrome-vanadium wire can be tempered to remove residual coiling stresses and to settle the structure as follows:

For general service: 450 to 500°F (240 to 260°C)

For severe service: 550 to 600°F (290 to 315°C)

For high-temperature service: 650 to 700°F (345 to 370°C

Tempering Time. See Fig. 34.

Chromium-Vanadium Valve Spring Steel

DESCRIPTION
Chromium-vanadium valve spring steel is one of the highest-quality alloy steels intended for the manufacture of valve springs, and can be obtained in a variety of sizes from 0.020 in. to 0.500 in. (0.5 to 12 mm) diameters. In appearance it resembles high-carbon valve spring steel. This material is used only for cold-coiled springs. It is often called "chrome"-vanadium.

COMMERCIAL SPECIFICATION
ASTM A 232.

MANUFACTURE
This steel is made by the open hearth, basic oxygen, or electric furnace process, usually from domestic ores. This alloy steel is generally furnished in the hardened and tempered condition as in the case of oil-tempered wire, although it can be obtained in the annealed and also in the untempered condition for hardening and tempering after coiling, if desired.

APPLICATIONS
This special alloy steel is intended specifically for the manufacture of valve springs requiring high fatigue properties, especially when used at moderately elevated temperatures up to about 425°F (220°C) and for applications where plain high-carbon valve spring steels are not satisfactory. This material has been used for valve springs in engines for aircraft, racing cars, and speed boats. This material also has higher hardenability and greater resistance to shock loading than the plain carbon steels.

CHEMICAL COMPOSITION
See Fig. 35.

FIG. 35. Chemical Composition of Chromium-Vanadium, ASTM A 232

Element	Percent
Carbon	0.48-0.53
Manganese	0.70-0.90
Silicon	0.12-0.30
Chromium	0.80-1.10
Vanadium	0.15 min.
Phosphorus	0.020 max.
Sulfur	0.035 max.

MECHANICAL PROPERTIES OF CHROMIUM-VANADIUM VALVE SPRING WIRE, ASTM A 232

Modulus of Elasticity. E in tension (for torsion springs) = 29 500 000 (203 400 MPa); G in torsion (for compression springs) = 11 200 000 (77 200 MPa).

Elastic Limit. Percent of tensile strength: in tension = 88 to 93 percent; in torsion = 65 to 75 percent.

Hardness, Rockwell. 46 to 51C.

Electrical Conductivity. Percent of copper = 8 to 12 percent.

Wrap Test. Wire up to 0.162 in. (4 mm) should be capable of winding on itself as an arbor, larger sizes up to 0.312 in. (8 mm) on an arbor twice the wire diameter without splitting or breaking.

Weight. Per cubic inch = 0.284 lb (0.129 kg); per cubic centimetre - 7.85 g.

Tolerances. See Fig. 36.

HEAT TREATMENT, CHROMIUM-VANADIUM WIRE, ASTM A 232

Hardening. Annealed wire springs can be hardened at 1600 to 1650°F (870 to 900°C), soaked for 15 min, and oil quenched.

FIG. 36. Tolerances on Wire Diameters

Diameter		Tolerance (±)	
(in.)	(mm)	(in.)	(mm)
Up to 0.075	Up to 1.9	0.0008	0.02
0.076-0.148	1.91-3.75	0.001	0.03
0.149-0.375	3.76-9.5	0.0015	0.04
Over 0.375	Over 9.5	0.002	0.05

Fig. 37. Tempering Time (minutes)

Wire Diameter		General	Severe	High-temper-
(in.)	(mm)	service	service	ature service
Up to 0.050	Up to 1.27	15 to 20	20 to 30	30 to 45
0.051 to 0.120	1.28 to 3.0	20 to 25	30 to 40	45 to 60
Over 0.120	Over 3.0	25 to 30	40 to 50	60 to 80

Tempering. Such springs are usually tempered at 750 to 900°F (400 to 480°C) from ¾ to 1 hour depending upon the hardness desired.

Stress Equalizing. Springs made from oil-tempered wire should be tempered to remove residual coiling stresses and to settle the structure as follows:

For general service: 450 to 500°F (230 to 260°C)

For severe service: 550 to 600°F (290 to 315°C)

For high-temperature use: 650 to 700°F (345 to 370°C)

Tempering Time. See Fig. 37.

Chromium-Silicon Alloy Spring Steel Wire

DESCRIPTION

Chromium-silicon alloy spring steel is one of the newer types of alloy spring steels; it was originally developed in England for recoil springs in antiaircraft guns and for control springs in British torpedos. The alloy has a composition somewhat similar to silicon-manganese spring steel SAE 9262 except that it has a slightly smaller amount of silicon and slightly more chromium. These small differences, however, are especially significant, as they induce a deeper, more uniform hardening and higher mechanical properties. Round sections are most generally used, but square and rectangular sections can be obtained on special order. Diameters from 0.032 to 0.437 in. (0.80 to 11 mm) are obtainable. The oil-tempered wire has a surface appearance similar to chromium-vanadium. Hard-drawn wire has a surface quite similar to hard-drawn MB wire.

COMMERCIAL SPECIFICATION

ASTM A 401.

MANUFACTURE

This alloy steel may be made by the basic or acid, open hearth, basic oxygen, or electric furnace process. Although it can be furnished in the annealed condition, it is most generally used in the spheroidized annealed and cold-drawn condition (untempered), for hardening and tempering after forming, and is often used in the oil-tempered condition, just as in the case of chromium-vanadium and other alloy steels. Wire is obtainable and often used in the hard-drawn condition, similar to music wire, usually in sizes from 0.020 to 0.207 in. (0.5 to 5 mm). Hard-drawn wire is easier to coil, can be plated without acid dipping, is free of decarburization, has a smooth surface free of scale, but is not available in diameters over 0.207 in. (5 mm).

APPLICATIONS

Chromium-silicon is a special alloy spring steel wire especially suited for highly

FIG. 38. Chemical Composition of Chromium-Silicon, ASTM A 401

Element	Percent
Carbon	0.51-0.59
Manganese	0.60-0.80
Chromium	0.60-0.80
Silicon	1.20-1.60
Phosphorus	0.035 max.
Sulfur	0.040 max.

stressed springs subjected to shock or impact loading such as recoil springs in antiaircraft guns and for moderately elevated temperatures up to 475°F (245°C). This material is capable of being heat treated to higher hardnesses without losing ductility than most other alloy steels, thus obtaining higher tensile strengths; a Rockwell hardness of C50 to C53 is quite common. This alloy has been used for cold-coiled springs, but sizes over 3/8 in. (9.5 mm) can be hot rolled. Wire in the oil-tempered condition with a smooth surface can be obtained and used for valve springs.

CHEMICAL COMPOSITION

See Fig. 38.

MECHANICAL PROPERTIES OF CHROMIUM-SILICON OIL-TEMPERED ALLOY STEEL WIRE, ASTM A 401

Modulus of Elasticity. E in tension (for torsion springs) = 29 500 000 psi (203 400 MPa); G in torsion (for compression and extension springs) = 11 200 000 psi (77 200 MPa).

Elastic Limit. Percent of tensile strength: in tension = 88 to 93 percent; in torsion = 65 to 75 percent.

Hardness, Rockwell. 50 to 53C (oil-tempered).

Electrical Conductivity. Percent of copper = 8 to 12 percent.

Wrap Test. All types of this alloy steel wire up to 0.162 in. (4 mm) should be capable of being wound around an arbor equal to the wire diameter, larger sizes on an arbor equal to twice the wire diameter, without cracking or breaking.

Weight. Per cubic inch = 0.284 lb (0.129 kg); per cubic centimetre = 7.85 g.

Tolerances. See Fig. 39.

FIG. 39. Tolerance on Diameter (Oil Tempered)[a]

Diameter (in.)	Diameter (mm)	Tolerance (±) (in.)	Tolerance (±) (mm)
Up to 0.072	Up to 1.8	0.001	0.030
Over 0.072	Over 1.8	0.002	0.050

[a]Hard-drawn wires are available at one half these tolerances.

HEAT TREATMENT, CHROMIUM-SILICON ALLOY STEEL WIRE

Hardening. Springs from annealed material can be hardened at 1475 to 1525°F (800 to 830°C); soak light sections under ¼ in. (6 mm) for 8 to 10 min, larger sections 10 to 20 min, and oil quench.

Tempering. Springs can be tempered at 700 to 850°F (370 to 455°C) from 30 to 60 min depending upon size and hardness desired.

Stress Equalizing. Springs can be tempered to remove residual coiling stresses and to settle the structure as follows:

For general service: 500 to 550°F (260 to 290°C)

For severe service: 600 to 650°F (315 to 340°C)

For high-temperature use: 700 to 750°F (370 to 400°C)

Special Note. Springs made from this material should be heated as soon as possible after coiling; a delay of 3 to 4 hours may be too long. The high residual stresses caused by coiling this high-hardness wire combined with unusually high elastic limit and unique grain structure can cause the springs to crack or even break unless heated quickly. Under no circumstances should unheated springs be left overnight! Some wire mills recommend heating 700 to 750°F (370 to 400°C) for 30 min, but the temperatures listed above are often used.

Tempering Time. See Fig. 40.

FIG. 40. Tempering Time (minutes)

Wire Diameter (in.)	Wire Diameter (mm)	General service	Severe service	High-temperature service
Up to 0.050	Up to 1.27	15-20	20-30	30-45
0.051-0.120	1.28-3.0	20-25	30-40	45-60
0.121-0.375	3.1-9.5	25-30	40-50	60-80
Over 0.375	Over 9.5	30-45	50-60	60-90

4 Hot-Rolled Bars for Hot-Coiled Springs

General

The large majority of springs used in industry, especially those having wire sizes under 3/8 in. (9.5 mm), are "cold coiled" with the material at room temperature. If the spring index D/d is above 6, cold coiling can be accomplished with springs having wire diameters up to 5/8 in. (16 mm).

For smaller indexes, and for heavier wire or bars, the customary practice is to heat the bars to a distinct red heat, often above the hardening temperature, and "hot coil" the springs. Steel is quite easy to coil or form while red hot. Important considerations in the selection of one coiling process in preference to another are (1) size of wire or bar, (2) spring index, (3) equipment available, and (4) type of material used. Certain materials, particularly those which cannot be hardened by heat treatment, are not hot coiled to make springs; such materials, among others, are the copper-base alloys, brass, phosphor-bronze, and beryllium-copper; also, certain types of nickel-base alloys including Monel, Inconel, and the austenitic stainless steels of the "18-8" or 300 series. The description of the materials that follow are for springs that are "hot coiled" with the material red hot. Bars may be obtained with a wide variety of finishes such as cold-drawn, turned and polished, and centerless ground, as described in ASTM A 331.

These materials are usually supplied in the form of annealed bars, but the smaller sizes up to 5/8 in. (16 mm) can be furnished in the oil-tempered condition for cold coiling if desired.

MECHANICAL PROPERTIES OF ALL HOT-ROLLED SPRING STEEL BARS

For design purposes, most of the mechanical properties are quite similar. The modulus of elasticity, E in tension, as used for torsion springs, is 29 000 000 psi (200 000 MPa) and G in bending as used for compression and extension springs is 10 750 000 psi (74 100 MPa). These moderately reduced values allow for some slight decarburization and coarse surface.

The elastic limit, as a percentage of the ultimate tensile strength, for hot-rolled carbon steel bars in tension is 65 to 75 percent, and in torsion is 50 to 60 percent. For all alloy steel bars, the values are somewhat higher; in tension they are 78 to 85 percent, and in torsion equal 60 to 70 percent.

The Rockwell hardness of the carbon steel bars after hardening and tempering should be 40 to 44C and the alloy steel bars 45 to 50C.

The tensile strength of the carbon steel bars varies from 175 000 to 195 000 psi (1200 to 1345 MPa) and the alloy steel bars from 180 000 to 200 000 psi (1240 to 1380 MPa). Bars with slightly higher values are also obtainable.

Weight for all types per cubic inch is 0.284 lb (0.129 kg) and per cubic centimeter is 7.85 g.

The tolerances on diameters are in an extremely wide range, depending principally upon the type of finish and size of bars. Permissible variations for all types of hot-rolled bars are covered in 31 tables listed in ASTM A 29 and are too complex to include here.

Hot-Rolled Carbon Steel Bars (ASTM A 68 and SAE 1095)

Hot-rolled carbon steel bars are available in round and square sections up to 2 in. (50 mm) or larger, and in rectangular sections with rounded edges up to 6 in. (150 mm) wide in a large variety of thicknesses up to 1 in. (25 mm).

APPLICATIONS

This plain high-carbon spring steel is the most widely used of all materials for "hot-coiled" springs because of its low cost and general availability. It has been used for springs in railway cars, trucks, buses, automobiles, ships, and as buffer and safety springs for elevators. In heavy sections this steel has rather poor hardenability and lowered mechanical properties, particularly to resistance against shock loading, overloads, and settling in service.

HOT COILING

Heat bars to 1600°F (870°C) for hot coiling and air cool springs.

HARDENING

Reheat springs to 1550°F (845°C), soak light sections under ¼ in. (6 mm) diam. for 20 min, larger sections up to 40 min, and quench in moderately agitated warm quenching oil not exceeding 150°F (65°C).

TEMPERING
Remove springs from oil while still quite hot (200 to 300°F) (90 to 150°C) and temper immediately at 850 to 950°F (450 to 510°C) for ¾ to 1½ hours depending upon hardness desired.

Hot-Rolled Alloy Steel Bars

Hot-rolled alloy steel bars are especially useful for springs subjected to shock or suddenly applied loads and have been used continuously at elevated temperatures up to 800°F (425°C) and up to 950°F (510°C) intermittently. A brief description of some of the various types that are used follows.

Hot-Rolled Chromium Alloy Steel Bars. ASTM A 331, grade 5160 (SAE 5160). These bars are usually furnished in round and square sections in the annealed or oil-tempered condition. This material has been used extensively for knee-action coil springs in automobiles and similar applications. The ASTM grade designations are the same as the AISI and SAE designations. SAE 5150 and SAE 5155 have slightly lowered percentages of carbon, but are also used.

Hot-Rolled Silicon-Manganese Alloy Steel Bars. ASTM A 331, grade 9260 (SAE 9260). Bars are furnished in round, square, and rectangular sections in the annealed or oil-tempered condition in a wide variety of sizes. This material, often used as a substitute for other alloy steels, has been used in many railway applications, but is now seldom used in the United States. In England it has been used quite extensively, where it has always been made under unusually strict conditions to avoid the seams so often found in it. SAE 9255 is also used.

Hot-Rolled Chromium-Molybdenum Alloy Steel Bars. ASTM A 331, grades 4150 (SAE 4150) and SAE 4161. These bars have been used for railway and U.S. Army equipment. They are obtainable in round, square, and rectangular shapes in a wide variety of sizes.

Hot-Rolled Nickel-Chromium-Molybdenum Alloy Steel Bars. ASTM A 331, grades 8645 (SAE 8645), SAE 8655, and SAE 8660. These bars, available in round sections (also in square and rectangular sections), are obtainable up to 2.5 in. (65 mm) in diameter, in a wide variety of sizes and finishes. SAE 8660 is especially recommended for all sizes over 1 in. (25 mm).

MANUFACTURE
This steel may be made by the open hearth, basic oxygen, or electric furnace process and is usually furnished in annealed, hot-rolled bars for hardening and tempering after forming.

APPLICATIONS
This alloy spring steel has become quite popular ever since it was introduced

during World War II as a National Emergency Steel. Although brought forth as a substitute or alternate for chromium-vanadium and other alloy steels, this new alloy has proved to have excellent mechanical properties, is easy to use, and is well adapted for production of both light and heavy hot-coiled springs. The surface characteristics are also excellent, and the alloy has the high hardenability properties so necessary for hot-coiled springs. The mechanical properties of the different alloys are so nearly alike that for practical purposes no differences need be specified. This material is now used extensively for hot-rolled springs in automobiles, trucks, buses, railroad cars, and U.S. Army equipment and is fast replacing chromium-vanadium in many applications. It is also less expensive than chromium-vanadium and has a favorable record with respect to cleanliness or lack of nonmetallic inclusions. More consideration is also being given to the possibilities of producing this material in the oil-tempered condition, in the form of wire, up to 3/8 in. (9.5 mm) in diameter.

HOT COILING
Heat bars to 1650°F (900°C) for hot coiling and air cool springs.

HARDENING
Reheat springs to 1600°F (870°C), soak light sections under ¼ in. (16 mm) diameter for 20 min, larger sections up to 40 min, and quench in moderately agitated warm quenching oil not exceeding 150°F (65°C).

TEMPERING
Remove springs from oil while still quite hot (200 to 300°F; 95 to 150°C) and temper immediately at 850 to 950°F (450 to 510°C) for ¾ to 1½ hours depending upon hardness desired.

Tool Steels for Springs

GENERAL
Tool steels of the oil-hardening type occasionally have been coiled satisfactorily into springs for special applications. Often, however, early breakage occurs due to incorrect heat treatment. Toolmakers familiar with the high hardness of Rockwell C58 to 62, or higher, customarily used on cutting tools, punches, and dies, often made springs with hardnesses far too high and caused brittleness. The high cost of tool steels and the short lengths in which they are available preclude their use for most production items. Some sizes can be obtained in standard lengths of 3 ft (1 m) and 12 ft (4 m). Tool steels are supplied in the annealed condition, with accurately ground surfaces, and are stocked in steel warehouses usually as high-carbon drill rods with a carbon content of 0.95 to 1.10 percent. Other analyses are obtainable directly from the steel producer and from some warehouses.

TYPES OF TOOL STEEL

Although many types of tool steels are available, best results for springs are obtained with the oil-hardening types. All tool steels are high-quality special steels and have individual brand names to distinguish them from ordinary steels. The oil-hardening types used for springs are supplied by more than 50 different steel companies. The popular high-speed tool steels have been used satisfactorily for springs in high-temperature applications up to 775°F (415°C) with torsional stresses up to 70 000 psi (485 MPa).

RECOMMENDED DESIGN STRESSES

Use the same curves as for oil tempered chromium-silicon ASTM A 401 projected to larger diameters.

MECHANICAL PROPERTIES

The tensile strength, elastic limit, and other properties of these tool steels are not sufficiently established in all sizes and conditions to justify complete tabular data. However, these properties are at least equal to, and probably higher than, those shown for the regular alloy steels such as chromium-vanadium and chromium-silicon.

HEAT TREATMENT

The oil-hardening carbon tool steels can be hardened at temperatures of 1420 to 1450°F (770 to 790°C) depending on size, soaked 5 min, and quenched in oil until the bath temperature is reached. Draw or temper immediately at 650°F (345°C) for 30 to 60 min to obtain a hardness of Rockwell 50 to 54C.

The 18-4-1 types should be heated slowly and uniformly to 1550 to 1600°F (845 to 870°C) and transferred to a superheating furnace, heated to 2340 to 2390°F (1280 to 1310°C), and quenched in oil to 200°F (95°C). Draw or temper immediately at 1250°F (675°C) for 1 to 2 hours to obtain a hardness of Rockwell 50 to 54C.

5 Flat High-Carbon Spring Steel Strip

High-Carbon Steel Strip

The mainsprings of clocks, coiled from blue tempered and polished flat spring steel strip, so familiar to all, actually are made from one of the best high-carbon spring steels produced. Several types of steel strip, in a variety of chemical compositions, are made for specific applications such as band saws, piston rings, cutlery, surgical instruments, razor blades, and flexible steel rules. However, more than 95 percent of the flat spring clips and similar parts used in industry are made from only two types of steel. These are the 0.70 to 0.80 percent carbon and the 0.90 to 1.03 percent carbon steels. These two steels are called cold-rolled spring steel, and both are available in a variety of tempers, finishes, and edge conditions; for this reason, both types may be described simultaneously. Special applications representing nearly 5 percent of the flat springs used in industry are made from either a lower-carbon steel of 0.60 to 0.70 percent carbon, especially recommended for deep-drawn and severely formed parts, or from a very high-carbon steel having 1.25 to 1.32 percent carbon called "barcoid" steel, which is used for band saws, cutlery, scraper blades, and other parts requiring extreme hardness and wear resistance.

CHEMICAL COMPOSITION

See Fig. 41.

MANUFACTURE

These steels are made by the open hearth, basic oxygen, electric arc, or

FIG. 41. Chemical Composition, Flat High-Carbon Spring Steel Strip, ASTM A 682

Element	SAE 1074	SAE 1095	SAE 1064	Barcoid
Carbon	0.70-0.80	0.90-1.03	0.60-0.70	1.25-1.32
Manganese	0.50-0.80	0.30-0.50	0.40-0.60	0.10-0.25
Silicon	0.15-0.30	0.15-0.30	0.15-0.30	0.15-0.30
Phosphorus	0.040 max.	0.040 max.	0.040 max.	0.040 max.
Sulfur	0.050 max.	0.050 max.	0.050 max.	0.050 max.

electric induction furnace under rigidly controlled conditions and using selected ores. Great care is exercised in melting and casting the ingots, cropping the billets to eliminate any possibility of pipes, seams, segregations, or other mechanical flaws, with careful inspection exercized in the rolling operations. A superior hot-finished raw material of forging quality is used to produce hot-rolled rods which later are cold rolled into strip. These hot-rolled rods are cleaned by hot pickling followed by a baking operation and then annealed to refine the structure. The iron carbide (cementite) in the hot-rolled steel is pearlitic in structure and the annealing spheroidizes the carbides, thus making the steel more suitable for cold working. The surface finish of the cold-rolled steel depends upon the finish of the rolls through which the steel passes, and for this reason a smooth finish on the rolls is extremely important. Process annealing between rolling is often necessary to soften the steel and to make it more malleable and ductile so that additional cold rolling can be done.

APPLICATIONS

ASTM A 682, SAE 1074 (0.70 to 0.80 percent Carbon). This is the most popular flat cold-rolled spring steel. It is readily available in a wide variety of thicknesses from 0.002 to 0.093 in. (0.050 to 2.362 mm), in the annealed condition, in various hard-rolled tempers, and in the hardened and tempered condition. Annealed sizes may be obtained in thicknesses up to about ¼ in. (6.35 mm). This material is frequently available in warehouses and is a material commonly formed by four-slide machines and presses. The annealed stock is readily processed and easily hardened and tempered. However, thin sections must be carefully hardened to prevent excess decarburization. This material is widely used for spring clips, flat springs, and spiral springs.

ASTM A 682, SAE 1095 (0.090 to 1.03 percent Carbon). This popular type, used principally for clock and motor springs, can withstand higher stresses than the SAE 1074 type. Its principal use is in the tempered polished and colored condition in sizes from 0.002 to 0.062 in. thick (0.050 to 1.58 mm), but thicknesses up to 0.093 in. (2.36 mm) are obtainable. It too can be furnished in both the annealed and hard-rolled conditions. Spiral and clock springs made from hardened and tempered stock usually have their ends annealed for bending and piercing operations.

ASTM A 682, SAE 1064, Deep Drawn Steel (0.60 to 0.70 percent Carbon). This material is the most ductile of the regular spring steels and is especially useful in the annealed condition for deep-drawn or severely formed parts. After hardening and tempering, the parts will have quite good spring properties. SAE 1050 is also used for these purposes.

Barcoid Steel (1.25 to 1.32 percent Carbon). This composition is useful for parts requiring extreme hardness and high wear resistance. This steel, lacking the ductility of the other spring steels, is fabricated in the annealed condition and then hardened and tempered. It is especially suited for band saw blades, cutlery, and scraper blades.

RECOMMENDED DESIGN STRESSES

For SAE 1095 oil-hardened and tempered flat spring steel with rounded or smooth edges and hardened to Rockwell C46 or higher, the curves for torsion springs for oil-tempered MB steel wire ASTM A 229, Fig. 85, may be used. For lower hardnesses and for SAE 1074 the curve for torsion springs for high-tensile hard-drawn HB steel wire ASTM A 679, Fig. 83, may be used.

Cold-Rolled Steel Strip, Not Hardened and Tempered

TEMPER

High-carbon spring steel strip may be obtained in several conditions of temper or hardness, produced solely by cold rolling. The bending abilities of these tempers vary considerably and all bends described are over a radius equal to the thickness of the stock.

ANNEALED

Temper No. 5 and 4. After cold rolling to the thickness required, the steel is annealed to make it dead soft, pliable, and ductile. In this condition, called Temper No. 5, it is suitable for severe forming operations and the material can be bent flat on itself either across or with the grain. It is also sufficiently ductile for rather deep-drawing operations. The maximum hardness recommended for the 0.70 to 0.80 percent carbon steel is Rockwell B83. For the 0.90 to 1.05 carbon steel, the maximum recommended is Rockwell B86. Most annealed commercial strip has a hardness range of Rockwell B75 to 85, which is satisfactory for most commercial requirements. Hardening and tempering to a desired hardness is accomplished after forming. If the steel is given a very light pass between rolls after final annealing to improve the finish, it is labeled Temper No. 4, although some companies use these two temper numbers interchangeably.

FIG. 42. Cold rolling annealed flat steel to reduce thickness of stock to desired dimensions. (Photo courtesy American Iron and Steel Institute.)

Temper No. 3. In this condition, also known as soft rolled, or one-pass rolled, the cold-rolled steel is especially suitable for blanking and severe forming, and still retains some stiffness and resistance to accidental deformation. The steel need not be hardened and tempered after forming, although this frequently is performed. Quarter-hard material will bend flat on itself across the grain, or diagonally and to a right angle with the grain without fracture. The final pass through the rolls may reduce the thickness as little as 0.001 in (0.025 mm), but increases the hardness of the annealed stock at least 3 points on the Rockwell B scale. The hardness recommended for the 0.70 to 0.80 percent carbon steel is Rockwell B87 to 93. For the 0.90 to 1.05 percent and higher carbon steels, the range is Rockwell B89 to 95. See Fig. 42.

HALF HARD

Temper No. 2. In this temper, also known as medium hard, the cold-rolled steel is intended to cover applications where a reasonable amount of forming ability is required and where some slight spring properties are needed. The material will bend to a right angle across the grain, but will not bend along the grain without fracture unless a bend radius equal to at least several times the stock thickness is used. The hardness recommended for all compositions is Rockwell B91 to 97.

FULL HARD

Temper No. 1. In this condition, also known as hard rolled, the cold-rolled steel in all compositions have a minimum hardness of Rockwell B98 although a hardness

FIG. 43. Continuous hardening, quenching, and tempering six cold-rolled annealed flat strips to obtain blue tempered spring steel. (Photo courtesy American Iron and Steel Institute.)

range of Rockwell B95 to 105 is considered satisfactory. This temper is recommended for blanking of parts requiring limited spring properties and when little or no bending is to be done. This condition is usually considered the top hardness range produced by cold rolling without heat treatment. Although limited forming operations can be made on such parts, they are rarely bent after blanking. See Fig. 43.

Full-hard-rolled 0.70 to 0.80 percent carbon steel strip can also be obtained with a Rockwell hardness of C28 and a tensile strength of approximately 129 000 psi (890 MPa) and the 0.90 to 1.03 percent carbon to a Rockwell hardness of C33 and a tensile strength of 148 000 psi (1020 MPa) in some warehouses.

Cold-Rolled Spring Steel Strip, Hardened and Tempered

Annealed cold-rolled spring steel strip can be hardened and tempered at the mill to produce tempered spring steels with exceptionally high mechanical properties. The spheroidized annealed cold-rolled steel is hardened and tempered by a continuous process wherein long strips of steel are uncoiled from reels and drawn through a long hardening oven at temperatures above the hardening temperature, usually 1450 to 1540°F (790 to 840°C). They are then quickly cooled in a quenching bath of oil or lead at a temperature of about 620°F (325°C) and additional cooling takes place in air. The strips are then pulled through a long tempering oven at about 770°F (410°C). These ovens may be 70 ft (21 m) long, and from 4 to 12 strands can be heat treated simultaneously. Since hardening and tempering depend on a proper combination of temperature and time, both must be carefully controlled to obtain a uniform product. The temperatures, quenching medium, and speed of strip through the process are governed by the chemical composition, thickness of stock, number of strands being treated, and

FIG. 44. Batch annealing coils of cold-rolled strip to soften them so that further reduction by cold rolling can be done. (Photo courtesy American Iron and Steel Institute.

the hardness desired. Heavy sections take longer than thinner sections. The hardening and tempering is performed in either direct or indirect fired muffle ovens or in baths of lead or salt. For certain alloys, a blast of air may serve as the quenching media, and very thin sections may be quenched between water-cooled steel plates. See Fig. 44.

HARDNESS AND TENSILE STRENGTH

Hardened and tempered spring steel strip as used in clock springs can be obtained, if necessary, in three temper or hardness ranges. A few steel mills apply temper numbers to these ranges such as T1 for the highest hardness range, T2 for the medium, and T3 for the low hardness range. Temper numbers for hardened and tempered spring steels are not nationally recognized; if used, the hardness range or the tensile strength desired should be specified after the temper number.

The T1 temper, with the highest hardness range, has the best spring properties, but this temper is not recommended for bending or forming properties, except possibly for some moderate forming operations.

The T2 temper, with the medium hardness range, is for tempered spring steel having reasonable forming ability combined with good spring properties.

The T3 temper, with the low hardness range, is especially suitable where tempered spring steel is required for blanking purposes. This hardness is satisfactory for parts where moderately good spring properties are needed.

An average Rockwell hardness range for each of the three tempers and the average tensile strength at these hardnesses are shown in Fig. 45 in tabular form.

Most warehouse stocks of tempered strip are supplied to one hardness range: for 0.70 to 0.80 percent carbon the range is Rockwell C46 to 49, and for the 0.90 to 1.03 percent carbon the range is Rockwell C48 to 51.

EDGES

The shape and condition of the thin edges of flat strip are important in terms of the ultimate use of the material and should be carefully considered before ordering. For important springs formed from strip without blanking and where fatigue life must be considered, a round edge is necessary because a slit edge has burrs which become stress raisers and cause early breakage. Spiral clock and motor springs should always be made from strip with a round or smooth square edge. However, if a spring clip is to be blanked from strip, a less expensive slit edge with burrs is satisfactory.

ASTM Standard A 682 specifies the type of edges briefly and assigns numbers to them. Hardened and tempered strip is practically always furnished with a no. 1 edge. Annealed strip can be obtained with any type of edge, but edges 1, 3, and 5 are most commonly used.

No. 1 Edge. This is a nicely finished, smooth edge without burrs, either round or square, and is necessary when small tolerances on width are required; it is produced by filing, by cold drawing, or by cold rolling. This edge is always supplied on pretempered clock spring steel. In filing, the strip is drawn between a number of

FIG. 45. Hardness and Average Tensile Strength of Cold-Rolled Spring Steel, Hardened and Tempered

Thickness[a] in. (mm)	Temper no.	Rockwell hardness	Tensile strength, 1000 psi (MPa)	
			SAE 1074	SAE 1095
0.005 (0.125)	T1	87-89, 15N	305 (2100)	314 (2165)
	T2	84-86, 15N	300 (2070)	298 (2055)
	T3	81-83, 15N	250 (1725)	260 (1795)
0.010 (0.25)	T1	87-89, 15N	287 (1980)	292 (2015)
	T2	84-86, 15N	278 (1915)	276 (1900)
	T3	81-83, 15N	232 (1600)	246 (1695)
0.015 (0.38)	T1	71-74, 30N	270 (1860)	273 (1880)
	T2	66-69, 30N	262 (1800)	261 (1800)
	T3	61-64, 30N	220 (1500)	234 (1615)
0.020 (0.50)	T1	70-73, 30N	258 (1780)	266 (1835)
	T2	65-68, 30N	250 (1725)	249 (1715)
	T3	61-64, 30N	212 (1460)	226 (1560)
0.025 (0.65)	T1	76-78A	252 (1735)	258 (1780)
	T2	73-75A	242 (1670)	240 (1655)
	T3	70-72A	205 (1415)	218 (1500)
0.030 (0.75)	T1	76-78A	244 (1680)	252 (1735)
	T2	73-75A	236 (1625)	234 (1615)
	T3	70-72A	202 (1390)	212 (1460)
0.040 (1.00)	T1	48-51C	234 (1615)	242 (1670)
	T2	44-47C	226 (1560)	223 (1540)
	T3	39-42C	193 (1330)	203 (1400)
0.050 (1.25)	T1	47-50C	225 (1550)	234 (1615)
	T2	43-46C	216 (1490)	215 (1480)
	T3	39-42C	185 (1275)	196 (1350)
0.063 (1.60)	T1	46-49C	214 (1475)	224 (1545)
	T2	43-46C	203 (1400)	207 (1430)
	T3	39-42C	177 (1220)	187 (1290)

[a]Intermediate sizes can be interpolated. Metric sizes are shown in parentheses, rounded off.
[b]Average tensile strengths shown, can vary ± 10 percent.

stationary inclined files set at angles so that all burrs are removed and the edges become rounded. Square or special-shaped edges can be produced by varying the settling of the files.

No. 2 Edge. This is a natural, round edge resulting from the hot-rolling operation, without additional processing. It is not often used because stock widths rarely coincide with purchasers' requirements, and width tolerances are quite large.

No. 3 Edge. This edge is approximately square and is the natural slit edge with burrs produced by either single strip shearing or multiple shearing of several strands from a wide strip. This edge is not filed, but is suitable for strip used for blanking purposes.

No. 4 Edge. This is a rounded edge produced by edge rolling. It is usually made by rolling a no. 2 or no. 3 edge and is useful when an approximately round edge is desired and when a perfectly smooth edge is not needed.

No. 5 Edge. This is a square edge produced by slitting and filed sufficiently to remove the burrs. This edge can also be made by rolling. This edge is not of a quality equivalent to a smooth no. 1 edge.

No. 6 Edge. This, too, is an approximately square edge, but it is produced by edge rolling only and all the characteristics of the hot-rolled or slit edge may not be entirely removed. This edge can be specified when width, tolerance, and finish are not so exacting as for a no. 1 edge.

EDGE LIMITATIONS

Practical limitations to the above edge descriptions, based upon cross-section sizes of the strip and upon manufacturing equipment, should be considered as follows: for edges no. 1 and no. 5, the maximum thickness suitable for filing is approximately 0.070 in. (1.78 mm). The maximum thickness suitable for slitting is approximately 0.130 in. (3.3 mm). The maximum width suitable for edge rolling is 2½ in. (63.5 mm).

Thin sections cannot have their edges rolled due to buckling of the section. Heavy sections such as $^3/_8 \times ^5/_{32}$ in. (9.5 X 4 mm) can be cold-drawn instead of cold-rolled and some sections may be processed by a combination of cold-drawing and cold-rolling.

FINISH AND COLOR

The finish and color of hardened and tempered strip is not always a good indication of the quality, hardness, or temper. However, the finish and color are important and should be specified when ordering. Full agreement on exact terminology for each finish has not been reached, but all mills understand the meaning of the following generally accepted terms.

Black Tempered. This is a dark oxidized surface condition formed during the hardening operation in a muffle-type oven or between the hardening and oil-quenching operations. It is used mostly for heavy sections from about $^1/_{32}$ in. to $^1/_4$ in. (0.78 to 6.35 mm) thick, and some heat treating scale may be expected.

Blue Tempered. This popular scaleless blue oxide finish is obtained by protecting the strip from air as it passes through the hardening oven and into the quench bath. Quenching in lead instead of oil is used and helps to prevent the formation of scale. The shade of blue cannot be accurately controlled nor uniformly distributed over the surface. Standard thicknesses run from 0.005 to 0.091 in. (0.127 to 2.3 mm).

Tempered and Polished. The above finishes may be removed and the strip made into a bright clean steel by polishing. This is accomplished by pulling the strand between high-speed wire brushes or cloth buffing wheels charged with abrasives.

Tempered, Polished, and Colored. The smooth, uniform straw or blue oxide color, long established as a standard characteristic of clock spring steel, is obtained by pulling strands of bright, clean tempered and polished strip through a lead bath and controlling the temperature and rate of cooling as it passes into the air. These finishes have no relationship to the Rockwell hardness or temper and are only mildly corrosion resistant.

TOLERANCES

Although ASTM A 682 specifies commonly accepted industry standards, some mills still maintain their own tolerances based upon the limitations of their equipment and methods of production. Closer tolerances are commonly negotiated between purchaser and producer in accordance with the special equipment, inspection, and processing methods required. Complete tables of tolerances are available from most mills, and although they are quite similar in many respects, some variations will be observed. ASTM A 682 specifies nine comprehensive tables, which are much too complex to include in this handbook.

Heat Treatment of Steel Strip

HARDENING

Springs made from annealed or untempered strip can be hardened at 1450 to 1500°F (790 to 815°C), soaked for a short time, about 1 to 3 min, and quenched in warm agitated oil. Care must be taken to avoid burning or decarburizing the steel.

TEMPERING

Drawing the temper at 600 to 700°F (315 to 370°C) to obtain the hardnesses required is done by varying the temperature and length of time at heat.

STRESS EQUALIZING

Springs made from hardened and tempered strip can have their residual forming stresses removed by heating from 400 to 600°F (200 to 315°C) for 15 to 30 min depending upon the thickness of the stock.

6 Stainless Steel Wire and Strip

General

Stainless steel is an alloy somewhat similar to ordinary steels except that it contains from 12 to 20 percent chromium—the only alloying element, out of some 40 possible elements intentionally added to steel, which produces a condition approaching complete resistance to atmospheric corrosion. Nickel also may be added in amounts up to 10 percent, in addition to chromium, to improve corrosion resistance and to alter certain properties. Only seven types of stainless steel are in general use for springs, but several other types are occasionally used.

CORROSION RESISTANCE

Except for the precious or noble metals, stainless steel stands alone as the one metal that has the ability to completely resist atmospheric corrosion. Marine atmospheres, however, will occasionally cause injury because of the presence of chlorine ion; also, sulfurous industrial atmospheres may discolor some types of stainless steel. The industrial reagents to which properly prepared surfaces of stainless steel are immune or highly resistant include alcohol, alkaline solutions, ammonia, atmosphere, crude oil, fruit and vegetable juices, gasoline, mercury, mine water, perspiration, sea water, steam, soap and sugar solutions, and many mineral solids and organic chemicals. Comprehensive tests to determine the resistance of each type of stainless steel to hundreds of different specific solutions are constantly being made. Results of these tests and up-to-date information regarding resistance can be obtained, when desired, from producers of stainless steel.

FILM THEORY

The corrosion resistance of stainless steel is best explained by the "film thoery." Although a controversial issue, substantial evidence indicates that an extremely thin, invisible protective oxide film is formed on the surface of the steel due to a chemical reaction of its elements. If this film is destroyed by machining, pickling, or by heat treatment, several hours may be required before the film is naturally restored. During this short period, the material may corrode, and moist fingerprints left on clean surfaces of strip will remain permanently and mar the appearance. Also, minute particles or ordinary steel or foreign matter touching the clean stainless surface may rust and give the erroneous impression that the stainless steel is rusting.

PASSIVATING

"Passivating" or "immunizing" is a process which in addition to dissolving foreign particles that might be adhering to the stainless steel also accelerates restoration of the protective oxide film. This is done by dipping the stainless steel parts in a passivating solution consisting of 20 percent (by volume) of nitric acid in water. The solution is most effective when warmed to 140°F (60°C). The parts should be immersed for 10 min and then rinsed in water. A 5-min dip in a 50-50 solution is also used and is especially effective, particularly for strip made from the straight chromium heat-treatable 400 series such as types 416, 420, and others. Other solutions containing varying percentages of nitric acid (from 20 to 40 percent) are also used. Passivating springs made from round wire is usually unnecessary, although many specifications require it to be done. Many spring companies ignore passivating treatments for "18-8" (300 series) stainless steel springs made from round wire, as very little or no benefit is obtained. However, if such springs are shotpeened, passivating is beneficial as it removes small particles of steel shot that may adhere to the surface of the springs. These finally divided steel particles, picked up during the shotpeening operation, will quickly rust and pit the surface of the stainless steel springs if not removed immediately after shotpeening.

MANUFACTURING

Stainless steel to be used for springs is usually made by the electric furnace process under rigidly controlled conditions, and wire is thinly coated with lead or copper to provide lubrication when used with automatic spring coilers.* Other coatings, including stearate of lime, zinc, cadmium, lead, and tin, or combinations of two or more of such materials, are also used. Because of these coatings, spring wire does not have the highly polished

* ECOLOGY COATING, NOW BEING USED IS NON-METALLIC, BIO-DIGRADABLE, AND NON-CONTAMINATING

appearance that is generally associated with stainless steel, but this does not materially affect its corrosion resistance. Such coatings are not required for springs coiled over arbors nor for wire forms or flat strip. For many applications these coatings can remain on the wire; however, where the wire contacts foods or sensitive chemicals the coatings can be removed by dipping the springs for 5 min in a standard 20 percent nitric acid passivating tank at room temperature. Stronger acid solutions heated to 180°F (82°C) will quickly remove all coatings.

GROUPS

The American Iron and Steel Institute has assigned type numbers of all stainless steel chemical compositions. These type numbers are universally recognized and accepted by all producers. The 300 series such as 302, 316, and the like, is assigned to the chromium-nickel compositions popularly known as "18-8," indicating 18 percent chromium and 8 percent nickel. The 400 series, such as 414 and 420, is assigned to the straight chromium heat-treatable grades. Technical societies such as ASTM and others have also prepared complete specifications covering some compositions, mechanical properties, and other characteristics.

Austenitic Group 300 Series

The 300-numbered series of stainless steels, including the 18-8 types, comprise the austenitic group 300 series. They are the chromium-nickel compositions and only three types (302, 304, and 316) are in general use for springs, although a few other types (such as 301) are occasionally used. These types cannot be hardened by heat treatment. They acquire their hardness and high tensile strength by the cold work resulting from the process of reducing rods to smaller diameters by cold drawing through hardened steel dies in a manner similar to music wire. Flat strip is cold rolled in the same way as cold-rolled steel strip.

Nonmagnetic properties are present only when these alloys are in the fully annealed condition. Slight magnetism is acquired by realigning the structure during the cold-working process of wire drawing and in cold rolling. Some of these magnetic properties are reduced by the low-temperature thermal treatment used to relieve the residual stresses caused by the coiling process.

Sizes most commonly used are from 0.005 to 3/16 in. (4.75 mm) diameters in spring or full-hard temper. Larger sizes are available in ½ or ¾ hard temper with correspondingly lower tensile strengths. For sizes about 3/16 in. (4.75 mm) diameter, it may be preferable to use the 17-7 PH type or a straight chromium, martensitic, heat-treatable type, to acquire the high tensile properties often needed for springs.

Intergranular corrosion may occur only in the austenitic steels in this group. When these steels are heated in an oven or as in welding or brazing, in the temperature range between 800 and 1400°F (425 and 760°C), a tendency of carbon to precipitate to the grain boundaries occurs. This forms a path for corrosion between the grains, and disintegration of the metal may soon follow. If it is necessary to heat these compositions through this range, it should be done as quickly as possible. The addition of columbium or titanium to these compositions minimizes this danger, as these elements have a strong affinity for carbon and form carbides which remain uniformly dispersed throughout the composition. Water quenching from these temperatures also reduces carbon precipitation.

APPLICATIONS

ASTM A 313, Type 302. This is the most popular 18-8 composition of all the stainless spring steels because it is readily available and has high tensile strength in wire diameters up to 3/16 in. (4.75 mm) combined with uniform properties. The wire is cold drawn to high tensile strengths but cannot be hardened by heat treatment. Coil springs made from this material are suitable for below-freezing temperatures and for elevated temperatures up to 550°F (290°C) with less than 5 percent loss of load provided that torsional stresses are kept below 55 000 psi (380 MPa). This material is also available in hard-rolled flat strip, but square and rectangular wire are seldom used. Type 302 is especially useful for highly stressed compression springs.

ASTM A 313, Type 304. This material is quite similar to type 302, but usually has about 5 percent lower tensile strength and slightly lower hardness; for these reasons it has better bending properties. This 18-8 alloy also has a lower carbon content than type 302 and therefore is easier to pull through wire-drawing dies. It is used frequently as an alternative to type 302 in a most satisfactory manner.

ASTM A 313, Type 316. This composition is not a true 18-8 composition, although it frequently is called an 18-8 alloy. It is also called 18-12-2, signifying 18 percent chromium, 12 percent nickel, and 2 percent molybdenum. This modified composition is more resistant to corrosion than types 302 and 304 in the presence of chlorides, phosphates, sulfates, and other salts, and also in the presence of reducing acids such as sulfuric, sulfurous, acetic, and phosphoric acids. For this reason it is preferred by the Armed Forces, particularly for aeronautical supplies. This type was originally developed for use in the chemical, paper, and textile industries and has found many other commercial applications even though its tensile strength may be 10 to 15 percent lower than type 302.

Martensitic Group 400 Series

The Martensitic group 400 series strainless steels are the straight chromium heat-treatable types that are usually formed in the annealed condition and then hardened and tempered in a manner similar to carbon steels. Such alloys have little or no nickel and are as magnetic as regular carbon steels. These steels should be clean hardened or pickled, and best corrosion resistance is obtained with clean, polished surfaces. Brittleness and inability to withstand shock or impact loading at low temperatures make these materials unsatisfactory for below-freezing applications. They are generally recommended for springs having wire diameters above 3/16 in. (4.75 mm) and for smaller wire diameters where the spring index (mean coil diameter divided by the wire diameter) is less than 4. These types are subject to grain growth and embrittlement during heat treatment unless precautions are taken.

APPLICATIONS

Type 414. This material is obtainable in the hard-drawn condition in sizes up to 3/16 in. (4.75 mm) diameter, but the tensile strength is 15 percent lower than type 302. For this reason, this alloy is usually formed in the annealed condition and then hardened and tempered to obtain tensile strengths ranging from 175 000 to 225 000 psi (1200 to 1550 MPa). Several applications for this magnetic type exist in the field where hard-rolled strip for stampings and flat springs are required.

Type 420. This composition is the most frequently used type for large diameters above 3/16 in. (4.75 mm), although it may also be used in smaller sizes, such as the 0.057 in. (1.45 mm) diameter wire originally used in the recoil springs for Garand rifles. It is generally formed in the annealed condition and then hardened and tempered after forming. This type does not have stainless properties until after it is hardened. Clean, bright surfaces provide best corrosion resistance and all heat-treating scale should be removed whenever possible.

Type 431. New processing methods have made this type one of the best to use for highly stressed springs. Wire is hardened and tempered at the mill and then cold drawn. This combination produces bright, clean wire with tensile strengths nearly as high as music wire, but its corrosion resistance is not quite equal to type 302. This wire has been used for flexible watchbands where the band is made by winding the wire into flat, rectangular, tight-wound coils. The hardened wire is magnetic and is obtainable in diameters from 0.050 to 0.312 in. (1.25 to 8 mm) and in bars and strip.

CHEMICAL COMPOSITION

See Fig. 46.

HARDNESS, TENSILE, AND YIELD STRENGTHS

See Figs. 47 and 48.

FIG. 46. Chemical Composition of Stainless Steel (percent) ASTM A 313

	Chromium-nickel austenitic				Chromium martensitic, hardenable		
Element	Type 302 18-8	Type 304 18-8	Type 316 18-12-2	Type 17-7 PH	Type 414	Type 420	Type 431
Carbon	0.08-0.15	0.08 max.	0.08 max.	0.09 max.	0.08-0.15	0.30-0.40	0.20 max.
Chromium	17.00-19.00	18.00-20.00	16.00-18.00	16.00-18.00	11.50-13.50	12.00-14.00	15.00-17.00
Nickel	8.00-10.00	8.00-12.00	10.00-14.00	6.50-7.75	1.25-2.50	–	1.25-2.50
Manganese	2.00 max.	2.00 max.	2.00 max.	1.00 max.	1.00 max.	1.00 max.	1.00 max.
Silicon	1.00 max.	1.00 max.	1.00 max.	1.00 max.	1.00 max.	1.00 max.	1.00 max.
Phosphorus	0.045 max.	0.045 max.	0.045 max.	0.040 max.	0.040 max.	0.040 max.	0.040 max.
Sulfur	0.030 max.	0.030 max.	0.030 max.	0.030 max.	0.030 max.	0.030 max.	0.030 max.
Other	–	–	Moly. 2.00-3.00	Alum. 0.75-1.50	–	–	–

FIG. 47. Properties of Stainless Steel Strip Types 302, 304, 316 (ASTM A 313)

Cold Rolled	Hardness: Reduction of area (%)	Hardness: Rockwell Scale	Tensile minimum (psi)	Tensile minimum (MPa)	Yield strength minimum[a] (psi)	Yield strength minimum[a] (MPa)
¼ hard	20	B85-100	130 000	896	100 000	690
½ hard	40	C22-27	145 000	1 000	130 000	896
¾ hard	60	C27-36	170 000	1 172	150 000	1 034
Full hard	80	C38-44	185 000	1 276	175 000	1 207

[a]The yield strength is taken at 0.20 percent offset. The elastic limit in tension is 65 to 75 percent of the tensile. Full-hard temper for the 300 series is usually used for spring clips where sharp bends are not required. For sharp bends, use ¾ hard temper; ¼ and ½ hard are useful only where sharp bends and limited spring properties are needed.

FIG. 48. Properties of Stainless Steel Strip Type 420

Cold Rolled	Hardness: Reduction of area (%)	Hardness: Rockwell Scale	Tensile minimum (psi)	Tensile minimum (MPa)	Yield strength minimum[a] (psi)	Yield strength minimum[a] (MPa)
¼ hard	20	B95-100	120 000	827	90 000	620
½ hard	40	C23-28	140 000	965	100 000	690
¾ hard	60	C28-32	150 000	1 034	130 000	896
Full hard[b]	80	C42-52	220 000	1 517	200 000	1 380

[a]The yield strength is taken at 0.20 percent offset. The elastic limit in tension is 65 to 75 percent of the tensile.
[b]These properties apply after hardening and tempering. For type 414, the tensile and yield strengths should be reduced about 15 percent.

MECHANICAL PROPERTIES OF STAINLESS STEEL STRIP, TYPE 420

See Fig. 48.

Precipitation-Hardening Group 17-7 PH

The precipitation-hardening group 17-7 PH (ASTM A 313) are usually placed in the category of the austenitic types; however, their unusual properties warrant a

FIG. 49. Mechanical Properties of Stainless Steel Wire, ASTM A 313 (after stress equalizing or hardening)

	Chromium-nickel		Chromium martensitic		
Property	302, 304, 316	17-7 PH	Type 414	Type 420	Type 431
Elastic limit in tension, percent of a tensile	65-75%	75-80%	65-70%	65-75%	72-76%
Elastic limit in torsion, percent of tensile[a]	45-55%	55-60%	42-55%	45-55%	50-55%
Modulus of elasticity in tension, E	28 000 000[b] 193 000 MPa	29 500 000 203 400 MPa	29 000 000 200 000 MPa	29 000 000 200 000 MPa	30 000 000 206 000 MPa
Modulus of elasticity in torsion, G	10 000 000 68 950 MPa	11 000 000 75 840 MPa	11 200 000 77 200 MPa	11 200 000 77 200 MPa	11 500 000 79 300 MPa
Hardness, Rockwell	C42-47	C47-50	C43-48	C46-51	C47-51
Elongation in 2 in. (50 mm)	2-5%	2-4%	4-6%	4-6%	3-5%
Weight per cubic inch	0.288 lb 0.131 kg	0.277 lb 0.126 kg	0.280 lb 0.127 kg	0.280 lb 0.127 kg	0.280 lb 0.127 kg
Weight per cubic centimetre	7.97 g	7.67 g	7.75 g	7.75 g	7.75 g

[a]The term "elastic limit" is taken as a percentage of the ultimate tensile strength for convenience purposes. Some materials do not have a true elastic limit, but terms such as "proportional limit," "proof stress at 0.01 percent offset," "yield strength," etc., often add to the confusion in a designer's mind. All he seeks is the maximum stress that should not be exceeded before plastic flow causes an undesirable amount of permanent set.

[b]If not fully hard, this value may drop to 27 000 000 (186 000 MPa).

separate description. Each supplier furnishes the material with a slightly different composition, but all use a small percentage of aluminum as the precipitation hardening element. The name 17-7 PH signifies 17 percent chromium, 7 percent nickel; the PH indicates precipitation hardening. This material has a useful combination of spring properties and excellent resistance to both heat and corrosion.

APPLICATIONS AND HEAT TREATMENT OF 17-7 PH

Annealed. Both wire and strip are obtainable in a soft annealed condition (Rockwell B78 to B92) so that severe forming and deep-drawing operations can be accomplished. The parts are then given a two-stage heat treatment after forming. The first stage is a solution heat treatment at 1400°F (760°C) for 1½ hours, followed by cooling in air or quenching in water. Then the parts are precipitation hardened at 850 to 950°F (455 to 510°C) for 30 to 90 min to produce a hardness of Rockwell C40 to C45.

Cold Worked. In this condition, both wire and strip are cold drawn or cold rolled to a high tensile strength and high hardness (Rockwell C40 to C44) at the mill. This severe cold reduction transforms the structure so that it is martensitic without going through the first stage of solution heat treatment. In this condition the material may be coiled or formed similarly to the 18-8 grades. Spring wire and strip that will receive only limited forming are usually obtained in this condition. After forming or coiling, the parts are precipitation hardened at 850 to 950°F (455 to 510°C) for 30 to 90 min–900°F (480°C) for 1 hour is usually preferred. Springs do not alter or change position of hooks during this simple heat treatment, so allowance for distortion is not necessary. Hardness after heat treatment ranges from Rockwell C47 to C50. Wire can be obtained in sizes from 0.005 to 7/16 in. (0.13 to 11 mm) diameters.

Precipitation Hardened. It is also possible to obtain this material precipitation hardened at the mill. This is called "extra-hard temper." In this condition, only limited forming can be accomplished, but straightened and cut wire and rods occasionally find usefulness for specific applications.

Workability of Stainless Steel

COILING

All of the stainless steel spring materials can be coiled on automatic spring coilers provided that a proper coating of lead, copper, or other lubricant is on the wire. Such coatings prevent the coiling tool from seizing or galling and are not required for wire that is coiled over an arbor or formed on four-slide machines. Strip also need not be coated, but a lubricant is helpful when forming, cutting, blanking, or drawing with tools used in presses.

WRAP TEST
Spring wire in the hardened condition up to 0.162 in. (4 mm) diameter should be capable of being wound on itself as an arbor without breaking or cracking the surface. Larger diameters should wind on an arbor having a diameter twice the wire diameter without breaking or cracking the surface.

HEAT TREATMENT
Heat treating stainless steels does not differ much in general from the methods used in the heat treatment of other steels. The conventional type of gas-fired or electric furnace may be used for hardening, although hardening may also be done in salt baths and induction ovens. In gas-fired furnaces care should be taken that the material is not exposed to the direct flame from the burners, thus avoiding decarburization. In salt bath furnaces the commonly used types of salts are satisfactory, but in no case should cyanide by used as this would cause carburization of the stainless steel, thereby impairing its corrosion resistance. Atmospheric-type furnaces using manufactured atmospheres for ordinary carbon steels are not often employed for heat treating stainless steel, as these atmospheres produce a scale that is difficult to remove from the stainless steel. It is possible, however, to use dry cracked ammonia or dry hydrogen.

ANNEALING
Type 414 can be process annealed in preparation for severe forming operations by heating to 1250°F (675°C) for 4 hours and air cooling. This type does not fully anneal by slow cooling methods.

Type 420 is process annealed at 1420°F (770°C) for 3 hours. Full annealing is acquired by heating to 1650°F (900°C) for 1 hour and reducing the temperature 30 to 50°F (0 to 10°C) per hour until 1100°F (595°C) is reached and then cooling in air, oil, or water.

Types 302, 304, and 316 can be annealed by heating to 1850°F (1010°C) for 15 to 30 min. These types should be water quenched after heating to keep the carbides in solution, thus preventing intergranular corrosion and embrittlement. Water quenching will not reharden these types.

HARDENING
Types 414 and 420 stainless steels are in general use for springs and are capable of being hardened by an oil-quenching heat treatment. These types are rather sluggish in hardening and should be soaked slightly longer at the hardening temperature than other types of steels. It is good practice to soak types 414 and 420 at least 20 min per inch (25 mm) of thickness after they have reached the proper hardening temperature. Both these steels should be quenched in warm oil, up to 250°F (120°C). Type 414 is best hardened at 1825 to 1850°F (995 to 1010°C) hardening temperature.

Type 420 is more difficult to heat treat properly than most of the other stainless steels. This material should be put into a hardening oven having a temperature of approximately 1000°F (540°C) and the temperature should be raised slowly to 1450°F (790°C) and then brought up to the proper hardening temperature between 1825 to 1850°F (995 and 1010°C). The material should always be quenched in warm oil and then stress relieved or tempered immediately following the quenching process. Under no conditions, if hardening cracks are to be avoided, should the material be allowed to remain fully hardened for any length of time. The Rockwell hardness after quenching and in the hardened condition is C50 to C54. After quenching, the material should be drawn at a temperature of 450 to 700°F (230 to 370°C) in order to obtain the proper hardness for springs. Rockwell C46 to C51 should be the final hardness.

The chromium-nickel austenitic 18-8 types cannot be hardened by heat treatment.

CLEANING

Heat-treating scale can be removed after tempering by immersing the parts in a solution of 50 percent muriatic (hydrochloric) acid and 50 percent water. The solution should be heated to 140 to 150°F (60 to 65°C). The temperature range is important. The springs should then be washed in clean running water and dipped in a 20 percent nitric acid solution, either warm or cold, and rinsed in hot water. Reheating to 375 to 400°F (190 to 205°C) for 30 to 60 min before dipping in the nitric acid solution will relieve hydrogen embrittlement caused by acid dipping.

Electropolishing to clean and brighten the surface is the reverse of electroplating. Standard electroplating lead-lined tanks are generally used and the work is made anode instead of cathode. A thin layer, 0.0005 to 0.001 in. (0.013 to 0.025 mm) in thickness, is removed from the surface of the steel in about 5 to 10 min, leaving a clean, bright surface.

Shotblasting followed by passivating will clean the surface faster and more easily than most other methods.

A bright finish is produced by tumbling with no. 1 burnishing balls in soft water and white soap for several hours.

Greases and oils are speedily removed in vapor degreasers using trichlorethylene. Organic solvents such as kerosene, gasoline, and carbon tetrachloride are useful for washing the steel and for removing surface dirt.

WELDING

The chromium-nickel, 18-8, stainless steels are most suitable for welding. They produce strong, tough welds. The high heat, however, may cause precipitation of carbide to the grain boundaries, resulting in intergranular corrosion, electrolytic attack, and physical embrittlement, unless proper precautions are taken. The stabilized grades of 18-8 containing columbium and titanium are recommended when welding is required, since these elements retard intergranular corrosion.

Straight chromium stainless steels also may be welded, but the welding heat is above the heat-treating temperature used for hardening, and when the weld cools, the weld and the parent metal becomes air hardened and brittle. The metal should be annealed after welding to remove embrittlement to relieve weld stresses and restore uniformity and corrosion resistance. A passivating treatment should be given.

Oxyacetylene welding with uncoated rods having the same composition as the parent metal is used mainly for light gauges up to 0.037 in. (1 mm) diameter or thickness.

Electric are welding with flux-coated rods is used primarily for medium and heavier gauges, usually above 0.037 in. (1 mm). The rod or electrode should be connected to the positive terminal, and the work to the negative terminal.

Electric resistance welding, including spot, seam, butt, and flash welding, produce uniform, high-quality welds.

Atomic hydrogen welding may be used. Hammer, forge, or open fire welding are not recommended.

Welding springs is generally not recommended because the welding heat might destroy the spring temper of the material. An occasional spot of weld, or tacking the end coil to a plate, is about all the welding that should be done to springs.

BRAZING

Although brazing stainless steel can be done without difficulty, this process of joining is not endorsed except where stainless is joined with brass, bronze, or other nonferrous metals. The high melting point of brazing alloys causes carbide precipitation and intergranular corrosion on the 18-8 grades. The high brazing temperature also causes the straight chromium steels to air harden and become brittle unless an annealing treatment is given. Where brazing is necessary, Tobin bronze with a suitable flux is recommended.

SOLDERING

Thoroughly cleaned surfaces can be soft soldered quite well with solder composed of 50 percent tin and 50 percent lead, provided that a suitable flux is used. The strong acid fluxes should be thoroughly washed off to prevent local corrosion and pitting.

Silver solders are used extensively with excellent results, particularly with 18-8 grades, provided that care is used to retard intergranular embrittlement and carbide precipitation that frequently occurs when 18-8 is heated between 800 and 1400°F (425 and 760°C). Silver solders commonly used are those that flow between 1200 and 1600°F (650 and 870°C). When used with straight chromium stainless steels, the silver solder should flow freely at temperatures below 1450°F (790°C) and thus prevent air hardening of the parent metal.

HYDROGEN EMBRITTLEMENT

Practically all steels become embrittled by contact with an acid such as muriatic or sulfuric acid. The adsorption of hydrogen by the steel results in interference with the slip that ordinarily occurs between the grains when a member is subjected to deflection. This embrittlement is more pronounced on the high-carbon and hardened steels and to a lesser amount on stainless steels. The brittleness can be removed—provided that it is done before serious damage occurs to the grain structure—by heating the steel to a temperature of 420°F (215°C) for 1 to 2 hours. This should be done as soon as practicable after the steel has been in contact with such acids. Nitric acid in the passivating process does not cause embrittlement.

SHOTPEENING

The process of shotpeening is one of the greatest advances in the spring industry because it increases the fatigue life of highly stressed springs. This process, frequently called shotblasting, consists of peppering the surface of springs with small round steel shot about 0.015 in. (0.40 mm) or larger in diameter. The shot, hurled with tremendous force by compressed air or by centrifugal force, not only cleans the surface and hammers smooth the minute surface irregularities, but also develops compressive stresses, called prestresses, in the outer surface layers of the springs. The working tensile stress will then be the tensile stress developed by the force less the compressive prestress, thus reducing the working stress and thereby lengthening the life of the springs. An increase in life from 3 to 10 times is not unusual.

Some authorities maintain that shotpeening should not be done to stainless steel because minute particles of the steel shot might adhere and cause spots of rust. The additional benefit of longer spring life, however, warrants the use of shotpeening and the extra operation of cleaning and passivating to prevent rusting. Shot made from stainless steel has been used successfully, but it is not always readily available. Highly stressed springs of stainless steel that have been shotpeened are now being used successfully.

PACK HARDENING

Although it is possible to obtain a higher surface hardness on stainless steel by pack hardening or carburizing, this process is not generally endorsed because the carbon added to the surface destroys most of the stainless properties of the steel. Pack hardening, particularly on the nonhardenable types of stainless steel, should be avoided. Slight success has been obtained by using nitrogen as the hardening element, but even nitride hardening will cause some reduction in corrosion-resisting properties.

7 Metallurgy of Copper-Base Alloys

Copper was the first metal to be worked and used by man. Metallurgists have for many years directed their efforts to finding some element or combination of elements which, when added to copper, would produce an alloy of high tensile strength. This objective has been reached in varying degrees by several alloys used for springs, the most common of which are spring brass, phosphor-bronze, and beryllium-copper. These three copper-base alloys and others are described in detail in this section.

History

Centuries ago, copper was used for spears and knives by a half-savage tribe on the island of Cyprus near Greece. They called it "cyprian," and the ancient Romans later called it "cyprium," later contracted to "cuprum," but through the years the name gradually changed to "copper." Copper is one of the most widely distributed of all metals and was used by early races of mankind in every corner of the world, including the races of Asia Minor, the Chinese, the early Europeans, the Incas of Peru, and the American Indians. Many Egyptian relics in our museums dating back to over 1500 B.C. are made of copper. These relics are still in a good state of preservation after 3500 years due to their excellent resistance to corrosion. Copper is the basic metal for making brass and bronze. The Egyptians cast bronze to make the great doors of the massive temple of Karnak, and Benvenuto Cellini cast many statues and other works of art in bronze. Leonardo da Vinci used bronze in several of his inventions, including guns. Caesar's legions used bronze as a major item in their implements of war, and cannons were cast from both brass and bronze by the Florentines as early as 1326. Brass cannon were used effectively against the British by the American Colonists during their War for Independence, and were used on the copper-sheathed historic ships "Constitution" and "Constellation." In the Armoury of the Tower of London is a gun of cast bronze made in 1469 and another cast in 1530; both are still in excellent condition and capable of being used. The Dead Sea Scrolls were written on sheets of copper hammered flat before the birth of Christ and are still legible.

The most famous of all statues, considered one of the wonders of the modern world, is the Statue of Liberty in New York Harbor. This statue, made of copper, was designed by Auguste Bartholdi and took 20 years to create. It was presented as a gift from the people of France to the people of the United States, and arrived in the United States on June 19, 1885. The statue is 152 ft (46 m) high and is constructed of 300 separate pieces of sheet copper 3/32 in. (2.38 mm) thick, fitted together like a huge jigsaw puzzle, and riveted.

Paul Revere, famous for his participation in the American Revolution, founded the Revere Copper and Brass Company in Massachusetts in 1801. The first manufacturer to produce a high-strength phosphor-bronze alloy was the Phosphor Bronze Smelting Company, founded in 1874.

Mining

Copper is now mined in nearly every country in Europe, in Siberia, Africa, Japan, Australia, Mexico, South America, and Canada, but the chief non-American mines are in Chile, Peru, Rhodesia, Spain, and the Belgian Congo. About one third of the world's total output of copper is produced in the United States. The states containing most mines are Arizona, Utah, Montana, Nevada, Michigan, New Mexico, Tennessee, and California.

The ores are mined by underground shaft-mining methods in Michigan, but open-cut mining is used in many areas where the ore deposits occur near the earth's surface, such as in Utah, Arizona, Montana, New Mexico, and elsewhere. The largest open-cut mine in the United States is in Bingham Canyon, Utah. Here an entire mountain half a mile high and two and one half miles long is being eaten away by electric shovels. Another large mountain is being consumed in Butte, Montana.

Ores

COPPER ORES

About 70 percent of U.S. copper is obtained from sulfide ores from Utah and Montana. The principal sulfide ores are cholcocites, but others are also mined.

About 5 percent of the domestic output is "lake" copper of very pure concentration, averaging above 99.9 percent copper in the metallic state. This free or native copper is found scattered through beds of amygdaloidal rocks, of lava origin, and conglomerate rocks in the Upper Michigan peninsula. These rocks contain about 1 percent pure copper, and the only impurities are slight traces of arsenic, silver, iron, and nickel.

REDUCTION OF ORES

After ore leaves the mine it is treated in various ways to separate the copper from other minerals. The copper content of ores is not high; it is often as low as 2 to 3 percent and rarely exceeds 10 to 15 percent. The method of smelting varies with each type of ore. Lake or native copper is concentrated by being crushed and ground to fine particles, washed and heated in a furnace using limestone as a flux, and the copper melted and cast into ingots.

The oxide ores are usually mixed with sulfide ores for smelting. This extraction process is complicated and consists of concentration as described above for low-grade ores, or flotation for the great majority of ores. In the flotation process the finely ground ore is oiled with water in the presence of less than 1 percent of coal tar and creosote for selective wetting purposes. The compound is then frothed so that the oiled mineral particles are buoyed up, leaving the waste silicate matter or gangue to settle out. The froth is then removed and the metallic ore recovered and roasted to expel the volatile oxides of arsenic and antimony and to oxidize the major portion of the sulfur.

The smelting process refines the product from the roasting furnace in a reverberatory furnace not unlike an open hearth furnace used for steel making and the sulfide mixture is now called "matte." A converter, somewhat similar to the Bessemer converter, is then used to blow air through the melt to burn out the sulfur and ferrous sulfide and to oxidize the copper sulfide. The yield, called "blister copper," is 99 percent pure and must be further refined to reduce the minute quantities of gold, silver, and other impurities until it is 99.95 percent pure. This refining is done by the electrolytic method or by fire refining in furnaces. In the electrolytic process, the copper is suspended as anodes in long tanks containing the electrolyte, a solution of copper sulfate and sulfuric acid. The copper flows from the anode to the cathode to become a practically pure copper of 99.955 percent purity. Copper is used in the spring industry for the manufacture of parts used in switches, in components for electronic equipment, in television and radar parts, and for alloying to produce stronger metals.

Alloying

Pure copper is so soft and ductile that it is necessary to add other elements to it to obtain the strength and resilience needed for many industrial applications. These alloying elements, however, always considerably reduce the electrical conductivity and have a marked effect upon other properties. If the electrical conductivity of copper is taken as 100 percent, a slight addition of 5 percent phosphor-tin to produce the popular phosphor-bronze spring alloy reduces the conductivity to only 18 percent. Even spring brass consisting of 70 percent copper and 30 percent zinc has only 27 percent conductivity. This value, however, is still two to three times that of steel.

Brass is an alloy of copper and zinc. Bronze is an alloy of copper and tin, although other alloys are used, such as in manganese-bronze, silicon-bronze, and aluminum-bronze.

EFFECTS OF ALLOYING ELEMENTS

Zinc is added to copper, in amounts up to 42 percent, to form brass and to increase the tensile strength and corrosion resistance; spring brass contains about 30 percent zinc. The United States imports more than 50 percent of its zinc from Canada, Mexico, Peru, and Australia.

Tin is added to copper, in amounts up to 11 percent, to form bronze and to make a strong, ductile alloy, especially resistant to sea water.

Lead is added to copper in amounts up to 1 percent and to brass in amounts up to 4 percent, as it greatly improves machinability. However, it has a harmful effect on ductility, deep drawing, and cold-working properties. The United States imports 26 percent of its needs from Canada, Australia, Peru, and Mexico.

Aluminum is added to brass, in amounts of 2 percent, to increase the mechanical properties. It is also added to copper, in amounts of 5 to 10 percent, to form a series of aluminum bronzes. The United States imports 96 percent of its requirements from Jamaica, Surinam, Canada, and Australia.

Silicon is added to copper, in amounts of about 3 percent, to form a silicon bronze alloy especially useful for marine hardware. It also increases the strength and toughness and is added to the aluminum bronzes for these same reasons along with its ability to improve corrosion resistance and machinability.

Iron is added to copper, in small amounts up to 3 percent, as a strengthening agent, along with aluminum, in certain aluminum-bronzes and with manganese in the manganese-bronzes. It also retards grain growth during annealing, but it reduces ductility and drawing capacity.

Phosphorus is added to copper, in small amounts up to 0.35 percent, along with tin to form phosphor-bronze high-tensile-strength alloys. It acts as a deoxidizer and improves spring properties.

Nickel is added to copper to produce cupro-nickel alloys. When added to brass it whitens the alloy to a silver color and strengthens it. These alloys are called nickel-silver, and the composition having 18% nickel has excellent spring properties for flat strip used for switch contact fingers.

Beryllium is added to copper, usually in amounts up to 2 percent, as an alloying constituent to produce a precipitation or age-hardening composition especially useful for springs.

Manganese is added to copper alloys containing nickel, to act primarily as a desulfurizing and degassifying agent, and to form manganese-bronze alloys.

Chromium is added to copper, principally to aid the properties desired when producing resistance-welding electrodes.

Antimony can be added to aluminum-brasses as an inhibitor against dezincification, but it also increases brittleness.

Tellurium or Selenium, in amounts up to 1 percent, also may be added to copper-base alloys to improve machinability.

Wire Manufacture

The manufacture of nonferrous wire is not unlike the method for producing steel wire. Bars of the desired alloy are heated to 1750°F (955°C) and passed through a series of graded grooves in rolling mills. These grooves gradually become smaller and are alternately square and oval to "knead" the material, thus breaking up the cast structure and replacing it with a uniform fine-grain structure. These rods then have more uniform mechanical properties than the cast bars from which they were made. The rods are then quenched in water, as quenching does not harden the material as in the case of steel, but does accelerate cooling and retard oxidation. Scale and oxide are removed by a hot water rinse to prepare the bars for further processing.

The bars are cold drawn through wire-drawing dies to reduce diameters, increase lengths, and obtain the high tensile strengths desired in a manner similar to drawing steel wire, except that larger reductions of area per pass are permissible. This reduction of area through the dies also determines the gauge size of the wire.

Gauges

All copper-base alloys for springs are drawn and rolled to the American Wire Gauge (same as Brown & Sharpe gauge) and to metric sizes. On drawings and bills of material, the gauge number and its decimal gauge thickness or metric size should be specified whenever possible. The most popular sizes in round wire are from 1/32 to 3/16 in. (0.78 to 4.75 mm) diameter, but smaller diameters ranging to 0.005 in. (0.127 mm) and larger diameters ranging to ½ in. (12.7 mm) are available. Intermediate and half gauges are sometimes available.

Strip with rounded edges is used for frequently deflected flat springs, because slit edges have a tendency to tear under flexure and cause early breakage. Slit edge strip, however, is customarily used for parts that are to be stamped or blanked.

Square and rectangular sections of wire are infrequently used; they are not generally recommended because of the lack of availability with good spring properties although round wire rolled to a rectangular shape having round edges may be obtained on order.

Strip may be obtained in the American Wire Gauge, and is also obtainable in many even three-place-decimal thicknesses, such as 0.018, 0.020, and 0.022 (0.45, 0.50, and 0.55 mm), and in many odd three-place-decimal figures and in metric sizes as well. Some mills also supply nonferrous materials in the Birmingham or Stub's gauge, but stock in this gauge with good spring properties is not always readily available.

Hardness and Temper

Wire made from spring brass, phosphor-bronze, and many other nonferrous materials receives its high tensile strength and hardness by the cold work done in reducing the rods or wire to smaller diameters by cold drawing through hard dies. Flat strip acquires its high tensile strength and hardness by cold rolling.

The amount of hardness acquired in cold working is frequently called the "temper" or hardness number. The "temper number" is governed by and depends upon the reduction of area from the annealed to the final state; it is equal to the number of American Wire Gauge numbers through which the material is drawn or rolled from the annealed to the final state without intermediate annealing. The American Wire Gauge is used because the gauge numbers have a definite dimensional relationship to each other. In this system the thickness dimension is reduced by approximately 50 percent for every six gauge numbers; for example, no. 0 is 0.3248 in. (8.25 mm), no. 6 is 0.1620 in. (4.12 mm), and no. 12 is 0.0808 in. (2.06 mm).

The grain size of the material in the annealed condition before drawing or rolling must be accurately controlled to acquire uniformity of temper.

Although both wire and strip are frequently purchased by temper designations, such as the numbers of hardness, they are also obtainable in hardness measurements. The standard hardnesses listed have been universally adopted. In specifying the hardness, it should be remembered that the popular Rockwell scales should be used with caution for strip less than 0.020 in. (0.50 mm) thick. A superficial tester for such thin strip is desirable.

In the copper-base alloys, hardness readings are seldom taken on round wire smaller than $^3/_{16}$ in. (4.75 mm) diameter. A reading taken on smaller diameters may read from 2 to 5 points on the low side, depending on the diameter of the wire; the smaller the diameter, the larger is the error. As a general guide, the corrected Rockwell hardness readings carefully taken on round surfaces should run as follows: spring brass wire, 89 to 95B; phosphor-bronze wire, 90 to 97B; beryllium-copper wire drawn 4 nos. hard, not precipitation hardened, 93 to 100B, and for precipitation-hardened wire, 37 to 41C. Recommended hardness ranges for spring brass strip, phosphor-bronze strip, and beryllium-copper strip are listed in tables in ASTM Standards.

Tolerances

For wire, use ASTM B 250; for strip, use ASTM B 248.

The tolerance tables are numerous and far too complex to list in this handbook.

Corrosion Resistance

Copper-base alloys have long been known for their excellent resistance to atmospheric corrosion and to attack by both fresh water and sea water. Spring brass, phosphor-bronze, and beryllium-copper have somewhat similar characteristics in this respect but, of course, some variation exists in their ability to resist attack by certain chemicals. The difference in this property depends upon the composition of the respective alloys.

Corrosion of the copper-base alloys may result in the green color of the carbonate, as usually observed on bronze statues and on copper roofs, the red of their oxides or the black of their sulfides.

For practical purposes, the corrosion resistance of phosphor-bronze and beryllium-copper may be considered the same as that of pure copper. These materials can be used in alcohol, acetic acid, beer, mild brine solutions, cold dilute hydrochloric and sulfuric acids for short periods of time, gasoline and hydrocarbons, organic salts and acids, phosphoric acid, sea air, and phenol. Extreme conditions of temperature, concentration, impurities, and aeration will seriously affect the corrosion resistance of these alloys.

Copper-base alloys are not recommended for use in contact with the following reagents: ammonia, ammonium hydroxide, cyanide, ferric or mercuric salts, oxidizing acids, sulfur or chlorine gas, but may be used in water treated with chlorine. Tin coatings or nickel plating are employed to protect the base alloys from some types of corrosion. Spring brass is seldom used successfully in direct contact with active chemical substances.

Stress-Corrosion Cracking

An unusual form of failure, called "season cracking," is occasionally observed in some of the copper-base alloys. Phosphor-bronze and beryllium-copper are practically immune to season-cracking or stress-corrison failures when properly processed. Season cracking is most prevalent among the brass alloys that are low in copper and high in zinc content. The following combination of conditions may cause such cracks to occur:

1. High internal stresses in the material resulting from cold drawing or cold rolling plus additional stresses caused by forming operations and stresses introduced by external loads.
2. Small traces of ammonia in air. Ammonia is very detrimental to yellow brass. Cracks may be caused to occur in this metal even by the small amounts of ammonia normally present in damp atmospheres such as are usually found in cellars.
3. A composition susceptible to such failures.
4. The length of time under which any or all of the first three conditions exist.

Susceptibility to stress-corrison cracking can be reduced in several ways:

1. Avoid storing the material in damp cellars or outdoor locations.
2. Avoid areas where ammonia may be present.
3. Avoid using susceptible alloy (high in zinc) compositions.
4. Use a low temperature (325 to 375°F (165 to 190°C), 30 to 60 min) stress relief anneal after forming.

Workability of Copper-Base Alloys

WELDING

All the common welding methods, with proper care, can be used in joining the copper-base alloys. Cleanliness of the surfaces to be welded is essential to assure that no oxide on the surfaces will interfere with fusion.

Brass is welded best with oxyacetylene welding, but carbon-arc and resistance or spot welding also can be employed. Phosphor-bronze can be welded with carbon-arc or metal-arc methods quite successfully; oxyacetylene welding is not recommended, but resistance or spot welding is frequently employed. Beryllium-copper is readily welded by the carbon-arc method; gas-fusion and metallic-arc welding are not recommended, but resistance or spot welding can be used.

Welding operations on springs must be done with extreme care because the welding heat will destroy the spring temper of the material at the weld. An occasional spot of weld, or tacking the end coil of a spring to a plate, is about all the welding that is usually done to springs. A broken spring should not be welded, as it will not withstand repeated deflections without breaking.

BRAZING

Spring brass and phosphor-bronze can be brazed, but no thoroughly satisfactory technique has been developed for brazing thin sections of beryllium-copper. Silver alloy brazing is best for brass. Phosphor-bronze alloys are readily brazed with torches, in furnaces, and by electrical resistance. Silver brazing alloys and copper-zinc alloys are used with a borax or boracic-acid flux. The brazing heat, however, will cause all copper-base alloys to lose their spring temper near the brazing area unless special care is taken.

SOLDERING

All the copper-base alloys can be soft soldered satisfactorily, because the low melting temperatures used for relatively short periods do not destroy the spring temper. Surfaces to be soldered should be carefully cleaned either by chemical or mechanical means such as scraping or scouring. Customary fluxes or commercial soldering compounds can be used. A solder containing 90 percent tin and 10 percent zinc is recommended for strong joints. Silver soldering, frequently called brazing, can also be done with the low-melting-point grades. Extreme care must always be taken to avoid losing the spring temper when soldering the copper-base alloys.

STAMPING

In strip form, the copper-base alloys are especially suitable for stamping, blanking, and bending operations. Formed parts are easily bent to intricate shapes. They are used in switches, fuse plugs, receptacles, and in other electrical apparatus.

CLEANING

Greases, oils, lubricants, and dirt can be cleaned from copper-base alloys by immersing the pieces in hot alkaline solutions or by washing with kerosene, gasoline, or other solvents. Vapor degreasing can also be used. Tarnish and corrosion effects can be removed by pickling for 1 to 7 min in a sulfuric acid pickle consisting of four parts water and one half to one part sulfuric acid at a temperature from 100 to 160°F (38 to 70°C). This solution is also frequently used at room temperatures.

A glossy finish can be obtained by immersing the parts in a bright dip solution consisting of one part water, four parts nitric acid, eight parts sulfuric acid, and with 1 oz of hydrochloric acid for each 5 gal of solution. This bright dip can be used at room temperature, with a soaking time range of 5 sec to 5 min. Many other solutions are also used for bright dipping.

It is important to rinse the parts well immediately after all pickling operations. Also, the immersion time in acids should be kept as short as possible to reduce acid attack on the parent metal. Too long a period in acid will reduce the size of the cross section, thus resulting in lower permissible loads and in an increase in expected deflection.

These materials are for springs that are cold coiled, with the material at room temperature.

ELASTIC LIMIT

Copper-base alloys do not have a true yield point as does steel, but the apparent elastic limit listed under mechanical properties has more general usefulness to designers than proportional limit or proof stress.

Spring Brass Wire, ASTM B 134; Strip, ASTM B 36, Alloy 260

Spring brass commonly used in the spring industry is a nonferrous, corrosion-resistant, nonmagnetic copper-base alloy containing approximately 70 percent copper and 30 percent zinc. Small amounts of other metals such as lead and tin are sometimes added to improve certain properties. This alloy, which is in the group called yellow brass or high brass, is frequently called cartridge brass because in the annealed condition the strip is used extensively for deep-drawn cases of rifle cartridges.

Spring brass is made from a wrought metal. The alloying metals in its composition are first melted and then cast in the form of bars; then processed into wire by cold drawing and into strip by cold rolling. The course crystalline structure of the cast metal is broken down during these cold-working processes; the grains are also elongated in the direction of drawing or rolling so that the metal takes on directional properties, which should be considered when parts are to be fabricated from strip by bending or forming operations.

Spring brass wire acquires improved tensile strength, hardness, and spring properties by the cold working inherent in the process of reducing rods to smaller diameters by cold drawing through hard dies. Flat strip acquires similar properties by cold rolling. The amount of reduction in cross-sectional area to obtain proper spring properties is from 75 to 85 percent for wire and from 50 to 60 percent for strip.

APPLICATIONS

Spring brass wire made in accordance with ASTM B 134 alloy no. 260 and spring brass strip to ASTM B 36 alloy no. 260 are often used in the extra hard (6 nos. hard) or in the spring hard (8 nos. hard) tempers. These materials, although having lower spring qualities than some of the other nonferrous spring alloys, are frequently used for severely cold-formed parts such as wire forms or flat stampings that will not be subjected to high stresses in operation.

Spring brass is also the least expensive of the copper-base spring materials and is sometimes used where its color harmonizes with surrounding parts. Its high electrical conductivity, combined with easy manufacturing workability and low cost, make it an excellent material for many electrical components where repeated flexure under high stress or at high temperatures is

FIG. 50. Chemical Composition, Spring Brass Wire, ASTM B 134, and Strip, ASTM B 36, Alloy No. 260

Element	Percent
Copper	68.5-71.5
Iron	0.05 max.
Lead	0.07 max.
Zinc	Remainder (approx. 30%)

FIG. 51. Tensile Strength of Spring Brass Wire, ASTM B 134, Alloy No. 260, Wire Diameters 0.020 to 0.250 in. (0.50 to 6.35 mm)

Drawn temper	Numbers of hardness	Reduction by drawing (%)	Minimum Tensile Strength[a] (psi)	(MPa)
Hard	4	60.5	102 000	700
Extra hard	6	75.0	115 000	790
Spring hard	8	84.4	120 000	830

[a]The maximum tensile strength runs about 10 000 to 15 000 psi (70 to 100 MPa) higher than the minimum.

not a function of the part. The alloy is readily plated, welded, brazed, and soldered. Brass springs are not recommended for use in temperatures much above 150°F, 200°F maximum (66°C, 90°C maximum), but they are especially suitable for below-freezing-temperature applications.

CHEMICAL COMPOSITION

See Fig. 50.

MECHANICAL PROPERTIES OF SPRING BRASS WIRE AND STRIP

Modulus of Elasticity, Wire and Strip. E in tension (for torsion and flat springs) = 15 000 000 psi (103 400 MPa); G in torsion (for compression and extension springs) = 5 000 000 psi (34 500 MPa).

Elastic Limit. Percent of tensile strength; in tension = 75 to 80 percent; in torsion = 45 to 50 percent.

FIG. 52. Tensile Strength of Spring Brass Strip, ASTM B 36, Alloy No. 260

Rolled temper	Numbers of hardness	Reduction by rolling (%)	Minimum tensile strength[a] (psi)	(MPa)
Hard	4	37.10	71 000	490
Extra hard	6	50.15	83 000	570
Spring hard	8	60.50	91 000	625
Extra hard	10	68.65	95 000	655

[a]The maximum tensile strength runs about 10 000 to 15 000 psi (70 to 100 MPa) higher than the minimum.

Electrical Conductivity. Percent of copper = 28 percent.

Tensile Strength. See Figs. 51 and 52.

Weight. Per cubic inch = 0.308 lb = 0.140 kg; per cubic centimeter = 8.53 g = 0.300 lb.

HEAT TREATMENT OF SPRING BRASS

Spring brass cannot be hardened by heat treatment. It acquires high hardness and high tensile strength solely by cold-working operations such as cold drawing or cold rolling.

Spring brass is not usually annealed because annealing destroys the spring properties. It softens rapidly at temperatures from 500 to 575°F (260 to 300°C). Also, high annealing temperatures cause a large grain structure. Low annealing temperatures produce the more desirable fine grain structure.

After coiling or forming it is highly desirable to stress relieve springs made from spring brass. Such treatment relieves the residual stresses caused by cold working and also reduces the tendency to season cracking. Immersion in boiling water for 1 hour is sufficient for many parts that will be subjected to low stress and where operating conditions are not too severe. By heating from 325 to 375°F (160 to 190°C) for 30 to 60 min, highly stressed springs may have their spring properties and fatigue life increased.

Phosphor-Bronze Wire and Strip Wire: ASTM B 159, Alloy 510, and ASTM B 103, Alloy 524*

The phosphor-bronze material used for springs is also a nonferrous, corrosion resistant, nonmagnetic copper-base alloy. It differs from brass in that it contains tin and phosphorus and very little or no zinc.

*Alloy 510 is commercially called alloy A; alloy 521 is commercially called alloy C; and alloy 524 is commercially called alloy D.

This alloy has a high percentage of copper (90 to 96 percent) to which has been added about 4 to 10 percent tin and a small amount of phosphorus. The small amount of phosphorus and tin are combined to form a compound of "phosphor tin" which is added to the melt to rid the metal of oxides. The compound rises to the top of the melt and is then removed. A slight amount of phosphorus may remain in the melt without harm to the metal and guarantees that complete deoxidation has taken place.

This alloy is also a wrought metal, and like brass it is cast into bars, given a homogenizing heat treatment to refine the grain structure, then hot rolled into rods and cold drawn or cold rolled into wire and strip. The cold work elongates and further refines the grain structure and produces directional properties that are so important in parts stamped or formed from strip.

Phosphor-bronze wire acquires its high tensile strength and hardness solely by cold drawing through hard dies. Strip acquires similar properties by cold rolling. Wire is reduced in area from 75 to 85 percent and strip from 50 to 60 percent to obtain the required spring properties.

NEW-TYPE PHOSPHOR-BRONZE

Newly processed alloys having the same chemical compositions as ASTM B 159 alloys no. 510, 521, and 524 (same as grades A, C, and D) but with superfine grain structures produced by special heat treatments are now produced by several companies and have excellent spring properties. This material has an unusually high endurance limit and a long fatigue life. The wire mills modestly claim a 30 percent better endurance limit, but fatigue tests made by the author indicate that this percentage can be easily doubled, and at quite high operating stresses as follows:

Compression springs of superfine grain structure were stressed from 27 000 to 64 400 psi (187 to 444 MPa), did not break and showed very little loss of load and no appreciable permanent set after 1 000 000 deflections. These springs were made from alloy 510, 0.093 in. (2.36 mm) diam. phosphor-bronze wire having a tensile strength of 128 200 psi (884 MPa), a yield strength at 0.2 percent offset of 122 500 psi (845 MPa), a proportional elastic limit of 85 600 psi (590 MPa), and a Rockwell hardness of B99.

This new superfine-grain phosphor-bronze is available in diameters up to $^{3}/_{16}$ in. (4.75 mm) and in strip up to $^{1}/_{16}$ in. (1.6 mm) thick, at nearly the same price as the regular types of phosphor-bronze. Also, its forming and coiling properties are better than the regular grades. For this reason, this new material is especially useful for formed parts, contact fingers, and similar parts.

APPLICATIONS

Phosphor-bronze spring wire made in accordance with ASTM B 159, alloy 510 (grade A) and spring strip to ASTM B 103 alloys 510 and 521 (grades A and C) are usually used in the extra hard (6 nos. hard) or in the spring hard (8 nos. hard) tempers. Extra spring hard is not always available.

This material is by far the most commonly used nonferrous alloy for coil springs, wire forms, and in flat strip for stampings and blanking to produce parts with good spring properties. It is used extensively in switches as contact fingers to carry electrical current, and because of its resistance to the effects of arcing it is used in circuit breakers, controllers, and signalling devices. Its excellent spring properties also make it especially useful where a contact or stamped part is subject to flexure.

Although this material costs only a little more than brass, it may be stressed 30 to 50 percent higher and has longer fatigue life. This alloy is readily plated, welded, brazed, and soldered and, like brass, is especially suitable for subzero temperatures. Also, it may be used satisfactorily in temperatures up to 150°F (66°C) with average stresses and up to 212°F (100°C) with low stresses.

CHEMICAL COMPOSITION

See Fig. 53.

MECHANICAL PROPERTIES OF PHOSPHOR-BRONZE WIRE AND STRIP

Modulus of Elasticity. E in tension, wire = 15 000 000 psi (103 400 MPa), strip = 16 000 000 psi (110 300 MPa); G in torsion, wire = 6 000 000 psi (41 370 MPa), strip = 6 250 000 psi (43 100 MPa).

Elastic Limit. Percent of tensile strength: in tension = 75 to 80 percent; in torsion = 45 to 50 percent.

FIG. 53. Chemical Composition of Phosphor-Bronze (percent)

	Wire	Strip	
Element	ASTM B 159 Alloy 510 (Grade A)	ASTM B 103 Alloy 521 (Grade C)	ASTM B 103 Alloy 524 (Grade D)
Tin	4.2 to 5.8	7.0 to 9.0	9.0 to 11.0
Phosphorus	0.03 to 0.35	0.03 to 0.35	0.03 to 0.35
Iron max.	0.10	0.10	0.10
Lead max.	0.05	0.05	0.05
Zinc max.	0.30	0.20	0.20
(Copper + tin + phosphorous, min.)	99.5	99.5	99.5

Other alloys are also used for wire and strip, but are not so readily available.

FIG. 54. Tensile Strength of Phosphor-Bronze Spring Wire, ASTM B 159, Alloy 510 (Grade A)

Wire diameter		Tensile	
(in.)	(mm)	(psi)	(MPa)
Up to 0.025	Up to 0.64	145 000	1000
Over 0.025 to 0.0625	Over 0.64 to 1.60	135 000	930
Over 0.0625 to 0.125	Over 1.60 to 3.25	130 000	890
Over 0.125 to 0.250	Over 3.25 to 6.25	125 000	850
Over 0.250 to 0.375	Over 6.25 to 9.50	120 000	830
Over 0.375	Over 9.50	105 000	720

FIG. 55. Hardness and Tensile Strength of Spring Strip, ASTM B 103, Alloy 521 (Grade C)

Thickness		Extra Hard		Spring Hard	
(in.)	(mm)	Rockwell min.	Tensile min.	Rockwell min.	Tensile min.
0.010-0.29	0.25-0.75	77 (30-T)	97 000 psi	78 (30-T)	105 000
0.20-0.039	0.50-1.00	93B	(670 MPa)	95B	(720 MPa)
Over 0.029	Over 0.75	78 (30-T)	(670 MPa)	79 (30-T)	(720 MPa)
Over 0.039	Over 1.00	95B	(670 MPa)	97B	(720 MPa)

Electrical Conductivity. Percent of copper, wire = 17 percent; strip = 13 percent.

Tensile Strength. See Figs. 54 and 55.

Weight. Per cubic inch: alloy 510 (grade A) = 0.320 lb (0.145 kg) = 8.86 g/cubic centimetre; alloys 521 and 524 (grades C and D) = 0.318 lb (0.144 kg) = 8.80 g/cubic centimetre.

HEAT TREATMENT

Process annealing phosphor-bronze spring material is seldom done because annealing destroys the spring properties; however, the material may be annealed by heating from 800 to 1200°F (425 to 650°C) depending upon the amount of anneal required.

Hardening cannot be accomplished by heat treatment. Phosphor-bronze acquires hardness and high tensile strength solely by cold drawing or cold rolling.

Stress relieving after coiling or forming phosphor-bronze springs is highly desirable for both coil springs and formed parts subject to flexure. The heat relieves the residual stresses caused by coiling and forming, thereby improving the spring properties and increasing the fatigue life. Immersion in boiling water for 1 hour is sufficient for many designs where the stress is not too severe. Highly stressed springs requiring long life should be heated from 325 to 375°F (165 to 190°C) for 30 to 60 min. A precipitation of tin compound will sometimes cause a silvery surface appearance, but this does not impair the spring properties.

Beryllium-Copper Wire, ASTM B 197 and Strip, ASTM B 194, Alloys 170 and 172 *

Beryllium oxide, identified in 1797 and first isolated as a metal in 1827, was added to copper in 1926 and a beryllium-copper alloy that could be hardened by heat treatment was made. This alloy was called beryllium-bronze, but this name is seldom used today. The first American commercial producer of this alloy started its operations late in 1930; before this date, however, some American laboratories had acquired a considerable amount of data through research.

Beryllium is an element and is in fourth place in the periodic classification. In a natural state is is extremely hard, being capable of cutting glass, yet it is one of the lightest metals known (35 percent lighter than aluminum) and is practically as light as magnesium.

ORES

Beryllium is found in about 30 complex minerals predominating in silicates, and the ores are found in practically every country of the world. However, the only commercially important source is beryl, a beryllium aluminum silicate containing about 3.6 to 4.5 percent beryllium. Beryl is found in opaque crystals in many colors including light green, light gray, yellow, blue, and white, and in its rarer forms beryl occurs as gems such as the aquamarine and the emerald. The United States has been importing up to 90 percent of its beryl. The largest ore reserves are in Brazil, but recent discoveries in North Carolina indicate ore reserves sufficient for U.S. needs, at the present rate of use, for several decades. However, the use of beryllium for atomic energy purposes may cause a rapid dimunition of reserves. Beryllium is one of the most effective materials for slowing down neutrons and has only a slight tendency to absorb them, therefore its use as a moderating material for thermal energy nuclear reactors may be expanded.

When beryllium is added in small amounts to copper, it affects the

*Alloy 170 is commercially called alloy 165; alloy 172 is commercially called alloy 25.

copper in a manner similar to the effect that small amounts of carbon have on steel. It makes the copper alloy amenable to hardening by heat treatment and raises the hardness and tensile strength.

USE

Beryllium-copper is the only popular copper-base, nonferrous alloy in general use for springs that can be hardened by heat treatment. The heat-treatable characteristics of this corrosion-resistant alloy containing approximately 98 percent copper and 2 percent beryllium were still in the laboratory stage as late as 1930. In the following decade, desirable compositions and proper heat treatments were discovered. New compositions are still being experimented with; one composition made available in 1948 contains only 1.7 percent beryllium. The electrical conductivity and allowable stress values of beryllium-copper are nearly twice that of phosphor-bronze, thus making it an excellent material for electrical components.

Beryllium-copper is obtainable and used in several different conditions depending upon the design and application. In the heat-treatable condition the material may be obtained annealed fully soft for deep drawing and for unusually severe forming operations; it is then precipitation hardened after forming. In this soft condition, however, it is seldom used for springs.

The material is more often used cold drawn or cold rolled and is ordered from the mill to a degree of hardness such as ¼ hard, ½ hard, ¾ hard, or hard. Most wire and strip are obtained 3 to 4 numbers hard for ordinary coiling and stamping and then precipitation hardened by the spring manufacturer after forming.

Beryllium-copper is also available in the hardened or pretempered condition such as hard drawn and heat treated. In this condition the material is usually obtained in straight lengths or in large-diameter coils and is used for parts requiring little or small forming operations or for coil springs with large index, D/d above 6. It may be used without further heat treatment, although a light heat of 300 to 350°F (150 to 175°C) for 15 to 30 min is often used to reduce the residual coiling stresses. Some mills recommend a higher heat of 400 to 500°F (200 to 260°C) for 30 min.

Another available condition is called "overaged and drawn," (mill hardened). The material in this condition has been given a special overaging heat treatment followed by additional cold work at the mill. This material may be coiled into springs or formed into shapes with large bending radii and used without a hardening heat treatment, but a low-temperature heat to relieve residual stresses caused by coiling or bending is recommended. In this condition, the material has a tensile strength approximately 90 percent as high as material that is precipitation hardened after forming, but its usefulness is curtailed principally to springs with large indexes D/d usually over 8 and to parts that require limited forming operations during fabrication.

For general purposes, beryllium-copper, ASTM B-194, alloy no. 170, is available in strip form only and does not necessarily require heat treatment. This material in the mill-hardened condition combines good formability with high spring properties. This composition is also obtainable solution annealed or in the solution-annealed and then cold-rolled condition. In the mill-hardened condition, this alloy has many uses in industry, principally because of the ease with which it may be formed; also, the precipitation-hardening heat treatment is not necessary.

APPLICATIONS

Beryllium-copper spring wire made in accordance with ASTM B 197, alloy 172 (formerly called 25) and spring strip to ASTM B 194 (alloys 170 and 172) are used extensively in electrical switches, relays, circuit breakers, fuse clips, and as brushholder springs. These alloys have demonstrated remarkable abilities and have justified their place as highly useful spring materials.

This material is especially suitable for carrying electrical current; however, its field of application is not limited to electrical apparatus. Its properties of low drift and low hysteresis are especially useful for springs used in measuring instruments, and also for diaphragms and bellows.

The phenomenon of hysteresis prohibits the use of some materials for delicate and sensitive control units. Hysteresis may be briefly described as the difference in the load for each increment of deflection when compared with the load at the same point when the spring is slowly released. For example, a compression spring with set removed may exert a load of 10 lb (44.48 N) when compressed half way from free to solid, but on the return stroke, at the halfway point the load exerted is slightly less and may read 9.95 lb (44.26 N), the difference in loads (0.05 lb; 0.22 N) is called hysteresis. Other types of measurement, such as in percentage or in inches (or millimetres), are sometimes used. Although the difference may be slight, it will affect recording instruments using high ratios on the dimensions of lever arms or indicating pointers.

Extension springs made from beryllium-copper do not usually have initial tension holding the coils together, because the material is obtained hard drawn and then precipitation hardened after forming; some initial tension, however, may be coiled into springs made from material obtained in the pre-tempered condition.

Strip stock is especially useful for stampings, because this material does not have directional properties of high magnitude such as is always found with brass or phosphor-bronze. Bends may be made to any angle without regard for direction of grain. Beryllium-copper, like most copper-base alloys, may be used at below-freezing temperatures. It is also useful at temperatures higher than those permissible for the other copper-base alloys; up to temperatures of 300°F (150°C) it continues to have good spring properties.

COMPOSITIONS

ASTM B 197, alloy no. 172, is the standard grade used for the majority of both wire and strip. This is the most useful alloy where high tensile strengths are required.

ASTM B 194, alloy no. 170, is for strip only. This composition has less beryllium and therefore is less expensive than alloy no. 172. Its electrical conductivity is 20 to 25 percent, the same as for alloy no. 172.

Alloy no. 10 also is for strip only. This composition has about one-third as much beryllium as does alloy no. 170 and therefore is less expensive. This composition has quite low tensile strength, but it has twice the amount of electrical conductivity and therefore is especially useful for current carrying switch parts and contact fingers having low design stresses.

TEMPER DESIGNATIONS

Some mills use the following code (particularly for strip) to denote tempers.

A = solution annealed and quenched
AT = condition A, precipitation hardened
½ H = solution annealed, quenched, and cold rolled, 20.7 percent area reduction
½ HT = condition ½ H precipitation hardened
H = solution annealed, quenched, and cold rolled, 37.1 percent area reduction
HT = condition H, precipitation hardened

WIRE, TEMPER

Only alloy composition no. 172 is used extensively for wire. Although several tempers are available and used, the most popular for spring purposes are as follows:

Wire, Pretempered—In this temper condition, the no. 172 alloy wire is cold drawn and precipitation hardened at the mill to a hardness of Rockwell C36 to C40 and a tensile strength of 180 000 to 210 000 (1240 to 1448 MPa) for wire diameters up to 0.040 in. (1 mm) and 170 000 to 200 000 (1170 to 1380 MPa) for larger sizes. In this condition, the wire is suitable for the great majority of general springs, especially compression springs, as no heat treatment is required if the stresses are low. For springs with average stresses, the springs can be heated from 300 to 400°F (150 to 200°C) or 400 to 500°F (200 to 260°C) for 30 min for highly stressed springs to reduce the residual coiling stresses, thereby improving the quality of the springs and reducing their tendency to take a permanent set. Wire in this condition should not be subjected to sharp bends, but small wire diameters up to 0.072 in. (1.8 mm) are ductile enough to be wrapped around their own diameter.

Wire, Hard Drawn, ¾ Hard—This type of temper condition permits the alloy no. 172 wire to be quite sharply coiled and is especially useful for extension and torsion springs that require rather sharp bends and small radii that cannot be made with pretempered wire. Springs made from this temper wire must be precipitation hardened after coiling–usually at 600°F (315°C) for 1 to 2 hours. After heat treatment the hardness is Rockwell C39 to C43 and the tensile strength is 190 000 to 230 000 psi (1310 to 1585 MPa), which averages about 25 percent higher than the pretempered wire. Additional bending or forming operations should not be made after heat treatment.

Wire, Annealed–Although the alloy no. 172 wire may be obtained solution annealed or dead soft, it is seldom used in this temper except for exceptionally severe forming operations. After forming, the springs are precipitation hardened at 600°F (315°C) for 2 to 3 hours to acquire a hardness of Rockwell C36 to C40 and a tensile strength of 165 000 to 180 000 psi (1138 to 1240 MPa).

Strip, Temper–Strip, unlike wire, is available in three different chemical compositions, alloys nos. 170, 172, and 10, in a variety of tempers. The highest mechanical properties are found in alloy no. 172, with the lower-cost alloy no. 170 a close runner-up. Alloy no. 10, however, has low tensile strengths and is used principally in applications where its higher electrical conductivity is desirable. Most strip is available in four tempers, but the ¼ hard temper is most popular for diaphragms and where sharp bends and deep drawing are required. The ½ hard temper is widely used for springs and where higher strengths are needed, provided that liberal radii on bends equal to three times the stock thickness or more and only slight drawing operations are performed. The hard temper produces maximum mechanical properties, but is recommended principally for flat parts or those that have bend radii equal to four to seven or more times the stock thickness and where only limited forming requirements are needed. The dead soft solution annealed temper is seldom required, but it can be obtained for those applications requiring deep drawn or severe forming operations. Inasmuch as some mills have limited fabricating equipment or specialize on only a few conditions of temper, it is always well to check with suppliers before standardizing on a specification.

CHEMICAL COMPOSITION.
See Fig. 56.

MECHANICAL PROPERTIES OF BERYLLIUM-COPPER WIRE AND STRIP
See Figs. 57 through 62.

Weight. Alloy 172 (any condition) = 0.297 lb/in.^{3} = 8.22 g/cm^3; alloy 170 (any condition) = 0.296 lb/in.^{3}; alloy no. 10 (any condition) = 0.316 lb/in.^{3}; = 8.75 g/cm^3.

FIG. 56. Chemical Composition of Beryllium-Copper (percent)

Element	Wire ASTM B 197 (Alloy 172)	Strip ASTM B 194 (Alloy 172)	Strip ASTM B 194 (Alloy 170)	Strip Alloy No. 10
Beryllium	1.80-2.00	1.80-2.05	1.60-1.79	0.40-0.70
Additive elements:				Co 2.35-2.70
nickel or cobalt or both	0.20 min.	0.20 min.	0.20 max.	Ir, Si, Al, Ni at 0.10 max. for each
Nickel plus cobalt plus iron	0.60 max.	0.60 max.	0.60 max.	
Copper plus beryllium plus additive elements	99.50 min.	99.50 min.	99.5 min.	99.5 min.

FIG. 57. Modulus of Elasticity

Alloy	Condition	E (psi)	E (MPa)	G (psi)	G(MPa)
170, 172	Hard, unheated	17 000 000	117 200	6 500 000	44 815
170, 172	Hard, heated	19 000 000	131 000	7 300 000	50 330
No. 10	½ hard, heated	18 000 000	124 000	7 000 000	48 265

FIG. 58. Elastic Limit (percent of tensile)

Alloy	Condition	In tension	In torsion
170, 172	Hard, unheated	70%	50%
170, 172	Hard, heated	75%	50-55%
No. 10	½ hard, heated	65%	45-50%

FIG. 59. Electrical Conductivity

Alloy	Condition	Percent of copper
170, 172	Hard, unheated	15-18%
170, 172	Hard, heated	22-27%
No. 10	½ hard, heated	45-48%

FIG. 60. Tensile Strength (minimum) Wire Alloy 172

Condition	psi	MPa
Pretempered	175 000	1207
¾ hard, unheated	130 000	896
¾ hard, heated	190 000	1310

FIG. 61. Strip

Tensile strength (minimum)				Rockwell hardness		
Alloy	Condition	psi	MPa	(minimum) and scale		
170	½ hard, unheated	85 000	590	88 (B)	74 (30T)	88 (15T)
170	½ hard, heated	170 000	1 170	37 (C)	57 (30N)	78.5 (15N)
170	Hard, unheated	100 000	690	96 (B)	79 (30T)	91 (15T)
170	Hard, heated	180 000	1 240	39 (C)	59 (30N)	79.5 (15N)
172	½ hard, unheated	85 000	590	88 (B)	74 (30T)	88 (15T)
172	½ hard, heated	185 000	1 280	39 (C)	59 (30N)	79.5 (15N)
172	Hard, unheated	100 000	690	96 (B)	79 (30T)	91 (15T)
172	Hard, heated	190 000	1 310	40 (C)	60 (30N)	80 (15N)

FIG. 62. Rockwell Scales for Various Strip Thicknesses

		Cold worked	
in.	mm	Unheated	Heated
0.015-0.020	0.381-0.508	15T scale	15N scale
0.020-0.032	0.508-0.813	30T scale	30 N scale
0.032 and up	0.813 and up	B scale	C scale

HEAT TREATMENT

Beryllium-copper can be hardened by a two-stage heat treatment. The first stage, called a supersaturating or solution treatment, is customarily provided by the producer at the mill prior to cold working. The second stage, called precipitation hardening (also called age or dispersion hardening), is performed by the spring manufacturer after coiling or forming.

The supersaturating treatment is actually an annealing heat treatment. It also causes the beryllium hardening element or compound to go into a solid solution with the copper. At room temperature less than ½ percent of beryllium remains in solution with copper, but by heating to 1400°F (760°C) (plus or minus 2 percent, the solubility of beryllium in copper increases to about 2 percent. Rapidly quenching in water from 1400°C) keeps the beryllium in a supersaturated solid solution, in which condition the material is soft and ductile and may be drawn, rolled, or formed into wire or strip for further processing; it is also in condition for precipitation hardening.

Parts requiring reworking can be reannealed by heating to 1450°F (788°C) for 30 min, or until the load at heat is thoroughly soaked, and then quenched in water. Heating to higher temperatures such as 1470°F (800°C) may damage the material beyond reclamation.

The precipitation hardening treatment used after parts are formed or coiled adds additional desirable spring properties to the material by increasing the tensile strength and hardness. This increase in properties is accomplished according to some authorities by throwing out from the solution a high beryllium precipitate known as gamma phase, causing a keying action to take place on the slip planes. Recrystallization of the cold-worked structure takes place, thereby relieving internal stresses and increasing stability and freedom from drift.

To acquire the absolute maximum hardness is quite difficult because materials produced by different manufacturers require slightly different temperatures and time at heat, thus necessitating experimental trial heats for each composition. However, for the great majority of spring applications, peak hardness is not entirely necessary and a close approach to it will suffice.

For commerical results, precipitation hardening of all compositions can be accomplished by heating to 600°F (315°C) plus or minus 5°F (3°C) for 2 hours, and cooling in air or quenching in water. For best results, the time at 600°F (315°C) can be varied as follows:

For ¾ hard wire, 1 h

For ½ hard wire, 1½ h

For ¼ hard wire, 2 h

For annealed wire, 3 h

A temperature of 650°F (345°C) for 12 to 20 min is also used. To obtain the highest amount of electrical conductivity (about 30 to 34 percent of copper), it is necessary to keep the material at heat for longer periods, sometimes as long as 7 hours. The time at heat stated is for muffle-type furnaces and may be reduced 25 percent if heating takes place in a salt bath. Annealed material requires a longer time at heat than does the cold worked material.

FIG. 63. Summary of Applications Data for Copper-Base Alloys

Material	Specific	General
Spring brass	Lowest cost copper-base spring alloy. Best for electrical conductivity where repeated flexure is not required. Excellent bending, drawing, and forming properties and especially useful for flat or bent stampings. Adaptable for color harmony. Satisfactory in temperatures up to 175°F (80°C). Not hardenable by heat treatment. Do not use for high stresses or where long fatigue life is required.	These three copper-base alloys are especially suitable for springs used in electrical apparatus and for carrying current. They all have excellent corrosion-resistance properties and are suitable for subzero-temperature applications. Easily formed and blanked, they are extensively used for stampings and bent shapes. They are nonmagnetic and are readily welded, brazed, soldered, and electroplated. They are used extensively in round wire and flat strip form. Wire is readily available in sizes from 0.005 to 3/16 in. (0.13 to 4.75 mm) and in fractional sizes up to ½ in. (13 mm). Strip is available in many gauge sizes. Always use the American Wire Gauge (same as Brown & Sharpe gauge) and state the size or decimals of an inch or use SI metric sizes.
Phosphor-bronze	Most extensively used copper-base spring alloy. Good electrical conductivity combined with ability to withstand repeated flexure. Good bending and forming properties. Satisfactory in temperatures up to 225°F (105°C). Not hardenable by heat treatment. Most sizes are readily available.	
Beryllium-copper	High electrical conducitivity combined with ability to withstand high stresses with long fatigue life. Excellent forming properties. Small drift and low hysteresis. Especially useful for measuring instruments, switches, and regulators. Does not have cold-rolling directional properties. Generally formed annealed and then precipitation hardened. May be used in temperatures up to 300°F (150°C). Highest cost of the copper-base alloys.	

(continued)

FIG. 63 (continued)

Material	Specific	General
Copper-manganese Nickel-silver Cupro-nickel	Silvery-white color for harmonizing with surroundings. Frequently used for switch contact fingers.	These materials are special alloys useful principally as alternates for the more popular copper-base alloys. All have good stamping and forming qualities with excellent corrosion-resistance properties. Investigate cost, availability, and properties of each composition before specifying.
Silicon-bronze Aluminum-bronze Copper-silicon Manganese-bronze Tin-brass	Especially high electrical conductivity in some compositions.	

Keeping the material at a temperature of 750°F (400°C) for longer than 30 min will cause softness, but a short-time high-temperature precipitation heat treatment is frequently employed with good results. This treatment is usually done in salt pots at a temperature of 700 to 720°F (370 to 380°C) for 6 to 8 min.

Distortion during this second stage of heat treatment is quite noticeable. Clamping to hold formed parts to required shapes is desirable. Fixture heat treating is commonly employed using ordinary cold-rolled steel brackets to hold formed parts. Coil springs tend to reduce in diameter and twist during heat treatment and should, where possible, be supported on rods. A reduction in diameter causes an increase in load or rate and should be allowed for by the spring maker, except when supporting rods are used during heat treatment.

Summary of Applications Data for Copper-Base Alloys

A summary of these data is given in Fig. 63.

Other Copper-Base Alloys

Some materials are occasionally used as alternatives for the more popular copper-base alloys or for some special-purpose application. Such materials are not readily available and their specific mechanical properties, cost, and availability in sizes required should be investigated before specifying their use.

COPPER-MANGANESE

This is a copper, nickel, manganese alloy resembling cupro-nickel. The nominal chemical composition is copper, 60 percent; nickel, 20 percent; and manganese, 20 percent. This alloy can be precipitation hardened to give an ultimate tensile strength of 200 000 to 225 000 psi (1380 to 1550 MPa) with a yield strength of 170 000 psi (1170 MPa) and a proportional limit of 120 000 psi (827 MPa). It is not recommended for carrying electrical current because its conductivity compared with copper is only 3 percent. The modulus of elasticity in tension is 21 000 000 psi (144 800 MPa) and 7 100 000 psi (48 950 MPa) in torsion.

NICKEL-SILVER
Wire ASTM B 206, Alloy 770; strip, ASTM B 122, alloy 770. This material comes in a wide variety of chemical compositions, but the most commonly used spring alloy nominally contains copper, 55 percent; zinc, 27 percent; and nickel, 18 percent. Cold-drawn strip spring hard has an ultimate tensile strength of 108 000 to 123 000 (745 to 848 MPa) with an apparent elastic limit of 90 000 psi (620 MPa). The alloy is used principally in electrical apparatus and optical goods because of its silvery white color and good corrosion-resistance properties. The modulus of elasticity in tension is 16 000 000 psi (110 300 MPa) and 6 000 000 psi (41 370 MPa) in torsion.

SILICON-BRONZE
Many commercial designations are given to copper-base alloys high in silicon content. Such alloys contain about 96 percent copper and 3 percent silicon and may contain 1 percent manganese. Cold-rolled strip has a tensile strength of 110 000 to 120 000 psi (758 to 827 MPa) with an apparent elastic limit of 100 000 psi (690 MPa). These alloys are quite similar to phosphor-bronze and in some applications are interchangeable. The modulus of elasticity is 15 000 000 psi (103 420 MPa) in tension and 5 000 000 psi (34 475 MPa) in torsion.

ALUMINUM-BRONZE
Copper-base alloys containing 91 percent copper, 7 percent aluminum, 1 percent iron, and 1 percent manganese have been cold drawn in wire form to ultimate tensile strengths of 150 000 to 160 000 psi (1034 to 1100 MPa). This material will, if it becomes more readily available, find many opportunities to serve as an alternative to other copper-base alloys. The modulus of elasticity in tension is 16 500 000 psi (113 760 MPa) and in torsion is 6 500 000 psi (44 815 MPa). A similar alloy containing only 3.5 percent aluminum and 1 percent silicon is sometimes used.

TIN BRASSES
Brass alloys containing from 85 to 95 percent copper, 0.5 to 2.0 percent tin, and balance of zinc are recommended as substitutes for phosphor-bronze where the maximum mechanical properties of the latter are not necessary. Their physical constants and mechanical properties are slightly better than ordinary brasses; cost is the same as for brasses of the same zinc content.

OTHERS

In addition, other copper-base alloys including copper-silicon, manganese-bronze, cupro-nickel, and some age-hardening alloys with high electrical conductivity are useful for special purposes where initial cost is not a principal factor.

The general and specific information covering the commercially standard copper-base spring materials should enable engineers and designers to specify a correct composition for practically any spring application requiring the use of copper-base alloys. The data concerning specifications, physical constants, mechanical properties, workability, and resistance to corrosion should aid considerably in making the proper selection of material.

Since new compositions with improved characteristics are occasionally discovered, it is strongly recommended that product engineers and designers leave the choice of the specific composition or exact specification directly to a competent spring consultant or to an experienced spring manufacturer.

To assure optimum service from materials, correct design for long life, and adequate performance requirements needed for the job, state the application and the operating conditions under which the spring must work, and specify the material simply—without regard to an exact chemical composition—using terms such as spring brass, phosphor-bronze, and beryllium-copper.

Progressive spring manufacturers and spring consultants give to designers the best material and spring design for the application requirements. Also, they are able to give advice regarding suitability of material, applicable stresses, load and dimensional tolerances, and to recommend simplified designs to obtain economical manufacturing operations.

8 Metallurgy of Nickel-Base Alloys

Nickel is one of the basic elements of the universe. It has a white metallic color somewhat like silver, and it imparts this color to the alloys in which it is used. It is a strong, tough metal having resistance to high temperatures and corrosion. Nickel is the third most magnetic element after iron and cobalt.

History

Large deposits of nickel-bearing copper ores were found in Saxony early in the eighteenth century. In 1751 the element was isolated and named nickel. Early in the nineteenth century the element was produced in the pure form and an accurate account of its properties was determined.

Prior to the Christian Era, coins containing nickel, probably as a natural alloy with copper, were used in Asia Minor. The Chinese, in the seventeenth century, used an alloy of nickel, copper, and zinc called "packfong" or "white copper"; and the Germans in 1824 duplicated the Chinese alloy and called it "German silver," although it does not contain silver. In 1975 production of nickel began in New Caledonia, and in 1883 large deposits were found in Sudbury, Ontario, Canada.

Nickel ore is widely distributed, being found on every continent, but deposits rich enough to be mined are found principally in Canada, Cuba, New Caledonia, Germany, Greece, Norway, the United States, and the USSR. Over 90 percent of the world's annual production comes from Canada. The only nickel produced in the United States is a relatively small quantity, in the form of nickel salts which are recovered during the electrolytic refining of blister copper. Large amounts of nickel ore are known to occur in the USSR, but the quantity refined is not usually included in world production figures.

Nickel Ores

There are three classes of nickel ores as follows:

SULFIDE ORES

Sulfide ores furnish the largest commercial sources of nickel, and the ore supply in Canada furnishes practically all of the nickel used in the United States. Pentlandite is a mineral having 18 to 40 percent nickel and is most commonly used. Polydymite mineral, having about 59 percent nickel, is also found in Canada, but is not common.

NICKEL SILICATE ORES

Nickel silicate ores, low in iron, are second in importance to pentlandite. Garnierite, a green mineral with 15 to 33 percent nickel, is found in large deposits in New Caledonia. Significant deposits are also found in Brazil, Indonesia, Venezuela, and the United States.

NICKELEFEROUS IRON ORES

Nickeleferous iron ores constitute the largest known potential source of nickel in the world. Large quantities of nickeleferous iron laterites have been found in Cuba. Large deposits, measurable in hundreds of millions of tons, are found in the Philippine Islands, Celebes, and Borneo. Other deposits are known to exist in Greece, Japan, Puerto Rico, and the USSR. The reduction of these ores is difficult, expensive, and not always considered commercially economical. These ores contain about 40 percent iron and 1 to 1.5 percent nickel.

Mining

In Canada, the open pit method is used for surface ore, and short shafts are sunk with connections made at different levels after the surface ore is removed. Shrinkage mining is used extensively, and some "cut-and-fill" and "blast-hole" mining is also done.

Reduction of Ores

The smelting process varies with the nature of the ores, but the general principles in the early stages are practically identical. The practice is quite similar to that described for sulfur-bearing copper ores; the ores are roasted to reduce the sulfur content, then smelted to separate the gangue, and the resulting sulfide matte is Bessemerized.

There are several methods for extracting the nickel from the copper-nickel matte as follows:

ORFORD PROCESS

The matte with coke and sodium sulfate is charged into a blast furnace and the reducing action of the carbon converts it into sodium sulfide. The mass is then poured into pots and cooled; during cooling, two well-defined layers occur. The upper layer contains sodium and copper sulfides, and this mass is "blown" in a Bessemer converter to produce blister copper. The lower layer, after proper treatment, is roasted to produce a black nickel oxide which is reduced in an open hearth furnace to metallic nickel.

A better and more economical process is covered by U.S. patents 2,419,973 and 2,425,760. Other processes are the Falconbridge Hybinette process, the New Caledonia process, the Mond process, and modifications of these processes.

MOND PROCESS

In the Mond process, the lower layers or "nickel bottoms" produced by the Orford process are ground very fine, calcined to convert the sulfides to oxides, and leached with sulfuric acid. The nickel is recovered by heating the mass to 350°C, further treated in reduction towers 40 ft (12 m) high, cooled and volatilized, and small pellets of nickel formed.

MONEL PROCESS

The production of Monel is based on the direct smelting and refining of a special magnetic cut of ore which is low in precious metals and contains about two parts of nickel to one part of copper. Smelting is done in blast furnaces to produce a low-grade matte which is blown in Bessemer converters. The matte is then crushed, ground, roasted, and reduced in acid open hearth furnaces, and further refined in arc-type electric furnaces to produce ingots of Monel which contain about 67 percent nickel and 28 percent copper.

Another process, first described in 1958, uses a direct electrolysis of nickel matte, an artificial sulfide, thus eliminating a metal anode. In this process, nickel sulfide of low copper content from a Bessemer converter or other source is cast directly into sulfide anodes and electrolyzed to produce high-quality nickel. This process also permits, for the first time in nickel refining, the commercial recovery of sulfur and selenium in addition to cobalt.

Alloying

Commercially pure A nickel is 99.4 percent pure nickel with minor impurities consisting of traces of copper, iron, manganese, and carbon. When nickel is added to carbon steels it acts as a ferrite strengthener, contributes to toughness, and aids heat treatment by lowering the critical cooling rate, thus permitting greater depths of hardening.

The impurities that are soluble in solid nickel usually increase the strength and hardness and decrease the thermal expansion and thermal conductivity. Individual effects follow:

Carbon increases hardness, strength, and ease of hot working, but increases the difficulty of cold working by increasing the tendency to work harden.

Cobalt raises the magnetic transformation temperature and the electrical resistivity, but has no other appreciable effects on nickel.

Copper is miscible with nickel in both the solid and liquid states in all proportions. The amount commonly found in nickel is usually less than 0.10 percent, and in this amount it has no effect upon the properties of the metal.

Iron is always present in small amounts in commercial nickel, usually in amounts less than 1 percent, and has no appreciable effect.

Manganese in small amounts is usually added to combine with the sulfur normally present, thereby improving the malleability and surface quality.

Oxygen lowers the melting point of nickel—for example, the addition of 0.22 percent oxygen lowers the melting point of the eutectic 20°C below the melting point of nickel.

Sulfur also lowers the melting point of nickel. The sulfide particles in the body of the grains reduces ductility, malleability, and other mechanical properties.

Aluminum, titanium, columbium, and tantalum are added in small amounts to cause nickel alloys to age harden.

Wire Manufacture

Wire is made by the same methods used for making steel wire, but carbon-steel dies should not be used because of the danger of galling. It is customary to place the nickel-copper matte in a reverbatory furnace, then the mass of nickel-copper oxide is alloyed in an electric furnace with ferrochromium, electrolytic nickel, or refined copper to obtain the composition desired and then poured into ingots. The ingots are milled, chipped, and hammered into billets, which then go to the bar and rod mill for rolling into square and round bars. The square bars are flattened and rolled into strip at a strip mill, and the round bars are hot rolled into rods, then either hot or cold drawn through carbide dies to produce the desired wire diameters.

Gauges

All nickel-base alloys for springs are drawn and rolled to the American Wire Gauge (same as Brown & Sharpe gauge). However, other gauges, intermediate gauge sizes, common fractional, decimal sizes, and metric sizes are obtainable if the quantity of material ordered is sufficiently large to justify making special sizes. Both round wire and strip are obtainable, but the range of available sizes varies depending upon the type of alloy.

Hardness and Temper

Certain types of nickel-alloys such as Monel 400 and Inconel 600 obtain high tensile strength solely by cold drawing or rolling and cannot be hardened by heat treatment. Other types, such as Monel K-500 and Inconel X-750, are often coiled or formed in the soft- or hard-rolled condition and then hardened by a long age-hardening treatment. Hardness and tempers are described under a "Mechanical Properties" for each alloy.

Corrosion Resistance

All the nickel-base alloys are especially characterized by their high resistance to all kinds of corroding substances and for their ability to withstand both elevated and below-freezing-temperature applications. Also, the nonmagnetic characteristics of some alloys are important for gyroscopes and indicating instruments. They are highly resistant to the corrosive media found in most industries.

In general, these alloys are immune to fresh water and inside and outside atmospheres except for sulfurous conditions, which cause a brownish to green tarnish without destruction to the alloy. Quiet sea water will cause some pitting, but agitated or flowing currents of sea water have less effect.

ALKALIS

Nickel-base alloys are practically resistant to the alkalis and have outstanding resistance to caustic soda. However, ammonium hydroxide concentrations over 1 percent will cause nickel to corrode, although Monel and Monel K can withstand concentrations up to 3 percent, and Inconel and Inconel X resist all concentrations.

GASES

Inconel is resistant to steam, air, and carbon dioxide, but wet chlorine, bromine, and sulfur dioxide will attack the nickels, Monels, and Inconels. The nickels and Inconels resist mixtures of nitrogen, hydrogen, and ammonia, but gases of sulfur compounds will attack all the nickel-base alloys at elevated temperatures.

SALTS

All the nickel-base alloys are highly resistant to neutral or alkaline salts, but they are attacked by oxidizing acid salts, ferric, cupric, and mercuric chlorides. Inconel is not affected by hydrogen peroxide solutions and is less affected than the Monels and nickels to hypochlorites.

ACIDS

The nickel and Monel alloys are attacked by oxidizing acids such as nitric acid, but Inconel is resistant at moderate temperatures. Sulfurous acid solutions, especially when hot, also attack these alloys. Nickel alloys are not affected by neutral and alkaline organic compounds, but organic acids cause moderate corrosion. The Monel alloys have good resistance to organic acids, neutral and alkaline organic compounds, as well as fruit and other food acids. Nickel and Monel alloys resist air-free sulfuric acid, and Inconel has fair resistance to sulfuric and hydrochloric acids at room temperature, but is not used for hot or concentrated hydrochloric acid.

Workability of Nickel-Base Alloys

WELDING
The usual joining processes for steel can also be used for the nickel-base alloys, with slight modifications. The corrosive surroundings in which the parts are used will often determine the joining method, although the thickness of the material, strength required, and design also are factors to be considered. The welding of spring temper materials, however, must be done with extreme caution; otherwise the spring quality or temper of the material will be destroyed. Satisfactory joints are produced by oxyacetylene gas welding, electric arc welding, and by various types of resistance welding such as spot, seam, and flash welding. These materials can be spot welded to themselves, to each other, and also to brass, bronze, steel, and stainless steels. Welding rods should be of the same composition as the parent metal. Other types of welding such as oxyhydrogen, hammer welding, carbon arc, and atomic hydrogen are not recommended for spring-quality materials.

BRAZING
All nickel-base alloys can be brazed, "silver soldered," or "hard soldered" in a satisfactory manner with strong joints by using silver-brazing alloys which flow at temperatures under 1400°F (760°C), such as the 50 to 60 percent silver grades. Commercial fluxes for such brazing alloys are readily available.

SOLDERING
Soft solder using either 50-50 or 60-40 tin-lead compositions with fluxes commonly used for copper make fairly good joints on Monel and nickel, but stronger fluxes are needed with the Inconels due to their chromium oxide film. Soft solders should not be used for high-strength joints nor where they will be affected by corrosion or high temperatures.

STAMPING
All the nickel-base alloys described can be stamped, blanked, or bent into various shapes. The heat-treatable alloys can be severely formed in the soft condition and then precipitation hardened after forming.

CLEANING
Oils, lubricants, and greases should be removed by standard cleaning methods

in degreasers, washers, or alkali cleaning tanks before pickling is done. Flash pickling or bright dipping for the Monels can be done in a pickling bath of 1 gal (9 litres) of water, 1 gal (4 litres) of 38° Bē nitric acid, and ½ lb of sodium chloride. A 5-sec dip in such a bath at 70 to 100°F (21 to 38°C) is usually satisfactory. The Inconels should not be flash pickled, due to their tightly adhering chromium oxide film, until after descaling. Heavy scale can be removed by sand blasting or by a descaling pickle.

Oxides can be removed from Inconel and Inconel-X by the following solution: water, 1 gal.; nitric acid (38° Bē); 1 gal; hydrofluoric acid (40 percent 1¼ pints); immersion time 15 to 90 min.

The Monels occasionally will precipitate copper to the surface during pickling, but this can be removed by a 1-min dip in an ammonia solution containing approximately 1 pint (0.5 litre) of commercial aqua to 1 gal (4 litres) of water, and then rinsing in water.

Monel 400: Spring Wire and Strip*

DESCRIPTION

Monel 400 is a solid solution, nearly nonmagnetic, nickel-base alloy, roughly two-thirds nickel and one-third copper, possessing a good combination of reasonably high tensile strength and good resistance to corrosion. It is available as wire or strip and is the least expensive of the nickel-base alloys, but it can be used only for cold-formed springs as it cannot be hardened by heat treatment.

SPECIFICATIONS

Wire: AMS 4730C, federal QQ-N-00281c, class A; strip: AMS 4544B, federal QQ-N-00281c, class A.

MANUFACTURE

Rods are cold rolled or cold drawn to desired sizes and shapes by the hard-forming process in a manner quite similar to regular steels. The hardness and tensile strength are obtained solely by cold rolling or cold drawing. It is available in the full hard condition for spring use up to about ¼ in. (6 mm) diameter, but larger sizes are obtainable at reduced hardness and lowered tensile strength. Square and shaped sections are seldom used for springs, but flat cold-rolled strip is often used. Rods are reduced from 33 to 75 percent, with the larger reductions being used for the smaller wire sizes. Spring wire for use on automatic spring coilers seldom requires metallic coating, as the organic lubricant used for wire drawing is usually sufficient. However, some oil dropped onto the wire prior to the coiling point is helpful. All

*All types of Monel, Inconel, Duranickel, and Ni Span C described are registered trademarks of International Nickel Company, Inc.

FIG. 64. Chemical Composition (percent)

Element	Wire AMS 4730C	Strip AMS 4544B
Nickel	63-70	63-70
Copper	Remainder	Remainder
Iron, max.	2.50	2.50
Silicon, max.	0.50	0.50
Manganese, max.	2.00	1.25
Aluminum, max.	–	0.50
Carbon, max.	0.30	0.30
Sulfur, max.	0.024	0.024

lubricants should be removed before heating, usually by acid dipping, to prevent scale formation.

APPLICATIONS

Although Monel is the least expensive and has the lowest tensile strength of the nickel-base alloys, it is especially useful due to its corrosion-resistant properties in sea water and because it is nearly nonmagnetic. It can be stressed slightly higher than phosphor-bronze and nearly as high as beryllium-copper. It can be used in temperatures between -100 and +425°F (-73 and 220°C) at average design stresses, and is often used in equipment having contact with food and beverages.

CHEMICAL COMPOSITION

See Fig. 64.

MECHANICAL PROPERTIES OF MONEL 400 SPRING QUALITY WIRE AND STRIP

Modulus of Elasticity. E in tension (for torsion springs) = 26 000 000 psi (179 265 MPa); G in torsion (for compression and extension springs) = 9 500 000 psi (65 500 MPa).

Tensile Strength. See Figs. 65 and 66.

Elastic Limit. Percent of tensile, wire: tension = 65 to 70 percent (for torsion springs); torsion = 38 to 42 percent (for compression and extension springs).

Electrical Conductivity. Percent of copper = 3.6 percent.

Weight. Per cubic inch = 0.319 lb (0.145 kg); per cubic centimetre = 8.83 g.

FIG. 65. Tensile Strength (average): Spring Wire Rockwell 97B (min.)

in.	mm	psi	MPa
Up to 0.028	Up to 0.71	165 000	1 140
Over 0.028 to 0.057	Over 0.71 to 1.45	160 000	1 100
Over 0.057 to 0.114	Over 1.45 to 2.90	150 000	1 035
Over 0.114 to 0.229	Over 2.90 to 5.80	145 000	1 000
Over 0.229 to 0.313	Over 5.80 to 8.00	140 000	965
Over 0.313 to 0.375	Over 8.00 to 9.50	135 000	930
Over 0.375	Over 9.50	130 000	895

FIG. 66. Tensile Strength (average) and Elastic Limit of Strip

Hardness,	Tensile Strength		Elastic Limit (tension)	
Rockwell	(psi)	(MPa)	(psi)	(MPa)
¾ hard, 90B	97 000	670	43 650	300
Full hard, 95B	114 000	785	57 000	390
Spring hard, 100B	132 000	910	79 200	545

FIG. 67. Time at Heat (minutes)

Size		Service		
(in.)	(mm)	Average	Severe	High temperature
Up to 0.025	Up to 0.65	30-45	50-60	1-2 h
Over 0.025	Over 0.65	45-60	60-80	2-3 h

Heat Treatment. Monel cannot be hardened by heat treatment, but should be stress equalized after coiling to remove residual stresses:

For average service: 450 to 500°F (232 to 260°C)

For severe service: 550 to 600°F (288 to 316°C)

For high-temperature service: 650 to 700°F (343 to 371°C).

For time at heat, see Fig. 67.

Monel K-500: Spring Wire and Strip

DESCRIPTION
Monel K-500 is quite similar to Monel 400 and also is composed of about two-thirds nickel and one-third copper except that the addition of small amounts of aluminum and titanium transposes it to the precipitation-hardening group of alloys. It is often used in sizes larger than is customarily used for Monel 400, as it may be formed in the soft or nearly hard condition and then heat treated. It is nonmagnetic.

SPECIFICATIONS
Wire: MIL-W-4471, comp. A; QQ-N-00286c, class A; strip: QQ-N-281a, class B.

MANUFACTURE
Rods are rolled or drawn to required sizes or shapes and supplied soft or in the fairly hard condition in a manner quite similar to annealed spring steels.

APPLICATIONS
This corrosion-resistant alloy is nonmagnetic and obtainable in a wide variety of sizes usually up to about ½ in. (13 mm) diameter for cold-wound springs and in much larger sizes for hot-wound springs. It can be easily formed in the softened condition (or in fairly hard condition), can withstand sharp bends, punching, and piercing, and then be hardened by a long age-hardening heat treatment to tensile strengths and hardnesses above Monel and nearly as high as stainless steel. It can be used at temperatures from -100 to +450°F (-73 to 175°C) at average design stresses usually below 50 000 psi (344.7 MPa).

CHEMICAL COMPOSITION
See Fig. 68.

MECHANICAL PROPERTIES OF MONEL K-500
SPRING WIRE AND STRIP (After Precipitation Hardening)

Modulus of Elasticity. E in tension (for torsion springs) = 26 000 000 psi (179 265 MPa); G in torsion (for compression and extension springs) = 9 500 000 (65 500 MPa).

Tensile Strength and Elastic Limit. See Fig. 69.

Elastic Limit. Percent of tensile: tension = 65 to 70 percent (for torsion springs); torsion = 38 to 42 percent (for compression and extension springs).

Electrical Conductivity. Percent of copper = 2.8 percent.

Weight. Per cubic inch = 0.306 lb (0.139 kg); per cubic centimetre = 8.47 g.

Heat Treatment. Precipitation harden at 1000°F (538°C) for 8 hours; then drop the temperature slowly to 900°F (480°C), which should take 1 hour,

FIG. 68. Chemical Composition (percent)

Element	Wire MIL-W-4471, comp. A	Strip QQ-N-281a, class B
Nickel	63.0-70.0	63.0-70.0
Copper	Remainder	Remainder
Iron, max.	2.0	2.50
Silicon, max.	1.0	0.50
Manganese, max.	1.5	2.00
Aluminum	2.0-4.0	0.50
Carbon, max.	0.25	0.30
Sulfur, max.	0.10	0.06
Titanium	a	a

[a]May also contain titanium from 0.25 to 1.0 percent with sulfur reduced to 0.01 percent.

then furnace cool or remove from furnace and allow springs to cool in air. Note that this may take slightly longer than the average working day.

Inconel 600: Spring-Quality Wire and Strip

DESCRIPTION

Inconel 600 is a nonmagnetic nickel-chromium-iron alloy, cold drawn to quite high tensile strengths in wire diameters up to ¼ in. (6 mm), although larger sizes and strip are obtainable. It cannot be hardened by heat treatment.

SPECIFICATIONS

Wire: AMS-5687, QQ-W-390; strip: MIL-N-6840, AM-1, AMS-5540.

MANUFACTURE

Rods are cold rolled or cold drawn to desired sizes and shapes by the hard-forming process, just as for Monel 400 alloy.

APPLICATIONS

Although more expensive than stainless steel, this alloy is especially useful in elevated temperatures up to 650°F (345°C) with compression springs stressed as high as 65 000 psi (450 MPa) with a load loss of about 5 percent in 24 hours, but stresses of 50 000 psi (345 MPa) provide a much longer fatigue life. This corrosion-resistant material in the annealed condition is used extensively for heat treating baskets and conveyor belts for furnaces at temperatures up to 2200°F (1200°C). Compression springs made of this material are often used in steam valves, regulating valves, boilers, compressors, turbines, and jet engines.

FIG. 69. Tensile Strength and Elastic Limit (min)

Hardness, Rockwell	Tensile (psi)	Tensile (MPa)	Elastic Limit Tension (psi)	Elastic Limit Tension (MPa)	Elastic Limit Torsion (psi)	Elastic Limit Torsion (MPa)
¾ hard, 36C	165 000	1 140	110 000	760	66 000	455
Full hard, 38C	172 000	1 185	115 000	795	68 000	470
Spring hard, 40C	180 000	1 240	120 000	825	72 000	495

FIG. 70. Chemical Composition

Element	Percent
Nickel	72 min.
Chromium	14-27
Iron	6-10
Manganese	1 max.
Copper	0.50 max.
Silicon	0.50 max.
Carbon	0.15 max.
Sulfur	0.015 max.

FIG. 71. Modulus of Elasticity at Various Temperatures

Temperature (°F)	Temperature (°C)	E (psi)	E (MPa)	G (psi)	G (MPa)
70	21	31 000 000	213 740	11 000 000	75 840
400	200	29 500 000	203 400	10 400 000	71 700
500	260	29 200 000	201 320	10 200 000	70 325
600	315	28 700 000	197 880	9 900 000	68 260
650	345	28 400 000	195 810	9 750 000	67 225
700	370	28 200 000	194 430	9 650 000	66 535

CHEMICAL COMPOSITION

See Fig. 70.

MECHANICAL PROPERTIES OF INCONEL 600
SPRING QUALITY WIRE AND STRIP

Modulus of Elasticity. This material is generally used in elevated temperatures and when the high temperature is known, the reduced value at that temperature (Fig. 71) should be used in design calculations.

Tensile Strength. See Fig. 72.

Elastic Limit. Percent of tensile: tension = 65 to 70 percent (for torsion springs); torsion = 40 to 45 percent (for compression and extension springs).

Electrical Conductivity. Very low; should not be used as a conductor.

Weight. Per cubic inch = 0.304 lb (0.138 kg); per cubic centimetre = 8.415 g.

Heat Treatment. Inconel 600 cannot be hardened by heat treatment, but should be stress equalized after coiling to remove residual stresses:

For average service: 700 to 750°F (370 to 400°C)

For severe service: 800 to 850°F (425 to 450°C)

For high-temperature service: 850 to 900°F (455 to 480°C)

For time at heat, see Fig. 73.

FIG. 72. Tensile Strength, Spring-Quality Wire

Diameter		Tensile (min.)	
(in.)	(mm)	(psi)	(MPa)
Up to 0.010	Up to 0.25	185 000	1 275
Over 0.010 to 0.032	Over 0.25 to 0.80	180 000	1 240
Over 0.032 to 0.25	Over 0.80 to 6.35	170 000	1 170
Over 0.25	Over 6.35	160 000	1 100

FIG. 73. Time at Heat (minutes)

Size		Service		
(in.)	(mm)	Average	Severe	High temperature
Up to 0.025	Up to 0.65	30-45	50-60	1 to 2 h
Over 0.025	Over 0.65	45-60	60-80	2 to 3 h

Inconel X-750: Spring Alloy Bars, Wire, and Strip

DESCRIPTION

Inconel X-750 is a nickel-chromium-iron alloy quite similar to Inconel except that the addition of small amounts of aluminum, columbium, and titanium change it to a precipitation-hardenable alloy. It is often used in larger sizes than are customarily used for Inconel; it can withstand higher temperatures and is nonmagnetic.

SPECIFICATIONS

Wire: AMS 5699A, MIL-S–21997; strip, AMS 5542G, AMS 5598, MIL-N-7786.

MANUFACTURE

Rods are rolled or drawn to required sizes and shapes and supplied in the full-hard or spring-hard condition for general use in temperatures up to 650°F (345°C). For higher temperatures the no. 1 temper condition is more serviceable, but requires a much longer heat treatment.

APPLICATIONS

This corrosion-resistant alloy is nonmagnetic and capable of withstanding 850°F (455°C) temperatures. It is obtainable in a wide variety of sizes and shapes in the form of bars, rods, wire, and strip. It can be coiled or formed hot or cold and then precipitation hardened to higher tensile strengths than Inconel. It can be used at below-freezing temperatures as well as elevated temperatures and is used for rotor blades in gas turbines, steam valves, rocket thrust chambers, and similar high-temperature applications. The hardness, tensile strength, elastic limit, and recommended design stresses vary depending upon temper, heat treatment, and environmental temperature.

CHEMICAL COMPOSITION

See Fig. 74.

MECHANICAL PROPERTIES OF INCONEL X-750
SPRING QUALITY BARS, RODS, WIRE AND STRIP
(After Precipitation Hardening)

Modulus of Elasticity. See Fig. 75.

Tensile Strength. See Fig. 76.

Elastic Limit. Percent of tensile: tension = 65 to 70 percent (for torsion springs); torsion = 40 to 45 percent (for compression and extension springs).

Electrical Conductivity. Very low; should not be used as a conductor.

Weight. Per cubic inch = 0.298 lb (0.135 kg); per cubic centimetre = 8.25 g.

FIG. 74. Chemical Composition

Element	Percent
Nickel (plus cobalt)	70.00 min.
Chromium	14.00-17.00
Iron	5.00-9.00
Titanium	2.25-2.75
Aluminum	0.40-1.00
Columbium (plus tantalum)	0.70-1.20
Manganese	1.00 max.
Silicon	0.50 max.
Sulfur	0.010 max.
Copper	0.50 max.
Carbon	0.08 max.
Cobalt	1.00 max.

FIG. 75. Modulus of Elasticity at Various Temperatures

Temperature		E		G	
(°F)	(°C)	(psi)	(MPa)	(psi)	(MPa)
70	21	31 000 000	213 740	11 200 000	77 220
500	260	29 100 000	200 640	10 200 000	70 330
700	370	28 100 000	193 740	10 000 000	68 950
800	425	27 700 000	190 980	9 800 000	67 570
900	480	27 200 000	187 540	9 400 000	64 810
1 000	540	26 700 000	184 090	9 100 000	62 740
1 100	595	26 100 000	179 950	8 800 000	60 670
1 200	650	25 500 000	175 810	8 100 000	55 850

FIG. 76. Tensile Strength, Spring Temper, Heated at 1200°F (650°C) for 4 Hours

Wire diameter		Tensile (min.)	
(in.)	(mm)	(psi)	(MPa)
Up to 0.25	Up to 6.5	220 000	1 517
Over 0.25 to 0.50	Over 6.5 to 13.00	200 000	1 379
Over 0.50	Over 13.00	190 000	1 310

FIG. 77. Heat Treatment

General service to 650°F (345°C)	Severe service to 800°F (425°C)	High-temperature service[a] to 1000°F (540°C)
1200°F (650°C)	1300°F (700°C)	1350°F (730°C)
for 4 to 6 h	for 6 to 8 h	for 14 to 16 h

Note: The springs may be furnace cooled or cooled in air.

[a]Use lower stresses, not exceeding 55 000 psi (380 MPa) for use in temperatures over 850°F (455°C) and expect a load loss of about 5 percent in 24 hours.

Modulus of Elasticity. See Fig. 75.

Tensile Strength. See Fig. 76.

Heat Treatment. There are several different precipitation-hardening heat treatments for Inconel X-750, depending upon the hardness of the stock before heating, the temperature at which the material is to be used, and the tensile required.

Some spring manufacturers, not wishing to keep their furnaces hot overnight, have used the heat treatment shown in Fig. 77 for all temper conditions and obtained excellent results.

The Society of Automotive Engineers recommends the following heat treatments for elevated-temperature use:

1. For 700 to 850°F (370 to 455°C) at moderate stresses, use no. 1 temper and heat to 1350°F (730°C) for 16 hours and air cool.
2. For higher-stressed springs up to 700°F (370°C), use spring-hard temper and heat to 1200°F (650°C) for 4 hours.
3. For 800 to 1200°F (425 to 650°C), heat 2100°F (1150°C) for 2 hours and air cool, then heat to 1550°F (845°C) for 24 hours and air cool, then heat to 1300°F (700°C) for 20 hours and air cool.

Other Nickel Alloys Occasionally Used for Springs

"A" nickel is a commercially pure, cold-worked nickel, silvery in color, with good spring properties and similar to Monel in corrosion resistance, but magnetic. Its nickel content is 99.0 percent minimum, with small amounts of copper, iron, manganese, and silicon totalling less than 1 percent. The tensile strength of spring-quality wire runs from 125 000 to 150 000 psi (862 to 1034 MPa), with strip running slightly lower. This material is resistant to alkalies, particularly to caustic soda, even at temperatures above 600°F (315°C), and is often used in the food and chemical industries. It cannot be hardened by heat treatment.

E = 30 000 000 psi (206 840 MPa) for torsion springs

G = 11 000 000 psi (75 840 MPa) for compression and extension springs

Recommended design stresses, maximum:

For torsion springs, 65 to 70 percent of tensile

For compression and extension, 38 to 42 percent of tensile

Duranickel is a precipitation-hardened nickel-aluminum-titanium alloy that can be used at 500°F (260°C) temperatures with 50 000 psi (345 MPa) stresses and in larger sizes than "A" nickel. It is only slightly magnetic. Its nickel content is 93 percent minimum, with aluminum up to 4.75 percent maximum and 1 percent titanium. The tensile strength of spring-quality wire after heat treatment runs from 175 000 to 200 000 psi (1200 to 1380 MPa) and higher, with strip running slightly lower. The heat-treatment temperature, time, and method is the same as for Monel K-500.

E = 30 000 000 psi (206 840 MPa) for torsion springs

G = 11 000 000 psi (75 840 MPa) for compression and extension springs

Recommended design stresses, maximum:

For torsion springs, 65 to 70 percent of tensile

For compression and extension, 38 to 42 percent of tensile

Incoloy is a 30 percent nickel, 22 percent chromium, and 45 percent iron alloy. Although rarely used for springs, it has properties almost as high as Inconel, and is also a cold-formed alloy, not hardenable by heat treatment, but it can withstand higher temperatures than Inconel.

Permanickel is a high nickel-titanium-magnesium alloy with properties very similar to Duranickel, but has higher electrical conductivity and thermal conductivity. It is used for thermostat tension springs.

9 Constant-Modulus Alloys

Special nickel alloys having a constant modulus of elasticity over a wide temperature range are highly desirable for springs subjected to temperature changes, especially where the springs must exert uniform loads and deflections. These materials, having a low or zero thermoelastic coefficient, eliminate variations in the stiffness of springs caused by changes in modulus values due to temperature differentials. These corrosion-resistant alloys have uniform and nearly constant elastic characteristics, and low hysteresis and low creep values, making them preferred materials for food-weighing scales, precision instruments, gyroscopes, measuring devices, recording instruments, and computing scales where temperature changes are within the range of -50 to +150°F (-46 to +66°C). These materials are quite expensive; none is regularly stocked in a wide variety of sizes, and they should not be specified without prior discussion with spring manufacturers because some suppliers may not fabricate springs from these alloys because of the special manufacturing processes required. All these alloys are used in small wire diameters and in thin strip only and are covered by U.S. patents.

Elinvar

Elinvar is a nickel-iron-chromium alloy developed in France and was the first of the constant-modulus materials used for hairsprings in watches. It is an austenitic alloy hardened only by cold drawing and cold rolling. Modifications to this basic alloy by the addition of titanium, tungsten, molybdenum, and other alloying elements have brought about improved characteristics and precipitation-hardening abilities, like that exhibited by Ni-Span "C" Alloy 902 and other materials having trade names such as "Elinvar Extra," "Durinval," "Modulvar," "Nivarox," etc. This basic alloy is still used, but other alloys described are more useful.

Iso-Elastic

Iso-Elastic is a popular nickel-iron-chromium-cobalt alloy developed by John Chatillon Sons that is easier to fabricate than Ni-Span C-902. It is used extensively in dynamometers, for computing scales, precision instruments, and food-weighing scales, where its temperature-compensating characteristics meet the requirements of national and state boards of weights and measures.

Some of the properties are as follows:

E = 26 000 000 psi (179 260 MPa) for torsion springs

G = 9 200 000 psi (63 430 MPa) for compression and extension springs

Tensile Strength. 170 000 psi (1 170 MPa) min.

Recommended Design Stresses. For average service are as follows:

For compression springs, 65 000 psi (450 MPa)

For extension springs, 60 000 psi (415 MPa)

For torsion springs, 100 000 psi (690 MPa)

Hardness, Rockwell. C32 to C36.

Elgiloy

Elgiloy, a cobalt-chromium-nickel-iron alloy formerly called 8-J Alloy, Durapower, and Cobenium, was developed by the Elgin National Watch Co. (now Elgiloy Co., Division of American Gage & Machine Co., a Katy Industries, Inc., subsidiary) in cooperation with Battelle Memorial Institute and three steel and wire producing companies. It is a nonmagnetic alloy suitable for below-freezing temperatures and elevated temperatures up to 750°F (400°C), provided that torsional stresses do not exceed 75 000 psi (515 MPa).

Strip is cold rolled 85 percent and then heated at 900°F (480°C) for 5 hours after forming. Wire is cold drawn 45 percent and then heated at 980°F (525°C) for 5 hours after coiling.

It is used for mainsprings in watches, indicating instruments, compasses, and in stepper motors.

Ni-Span C Alloy 902

Ni-Span C Alloy 902, is a nickel-iron-chromium-titanium alloy that is one of the most popular constant-modulus alloys. It was developed by the International Nickel Co. It is usually formed or coiled in the 50 percent cold-worked condition, then precipitation hardened at 900°F (480°C) for 8 hours, although a higher heat of 1250°F (675°C) for 3 hours produces higher hardness. A more useful heat treatment for applications requiring greater precision and uniform performance is to heat at 750°F (400°C) for 2 hours, then bring the oven up to 1200°F (650°C) for another 2 hours, and then air cool. Other properties after heat treatment are as follows:

E = 27 500 000 psi (189 600 MPa) for torsion springs

G = 10 000 000 psi (68 950 MPa) for compression and extension springs

Tensile Strength. Wire and strip = 200 000 psi (1380 MPa) min.

Elastic Limit. Percent of tensile: in tension, 65 to 70 percent (for torsion springs); in torsion, 38 to 42 percent (for compression and extension springs).

Recommended Design Stresses. For average service are as follows:

For compression springs, 60 000 to 70 000 psi (415 to 480 MPa)

For extension springs, 50 000 to 55 000 psi (345 to 380 MPa)

For torsion springs, 120 000 to 130 000 psi (825 to 895 MPa)

Some of the properties after the above-described heat treatments are as follows:

E = 29 500 000 psi (203 400 MPa) for torsion springs

G = 11 200 000 psi (77 220 MPa) for compression and extension springs

Tensile Strength. Strip, 360 000 psi (2 480 MPa); wire, 340 000 psi (2 345 MPa).

Recommended Design Stresses. Same as for stainless steel type 302.

Hardness. Strip, Rockwell C56 to 59; wire, C51 to C55.

Elinvar Extra

Elinvar Extra is a nickel-chromium, titanium-iron alloy with small amounts of cobalt, aluminum, and silicon, developed by Hamilton Technology (Lancaster, Pa.). It is a precipitation-hardened alloy which when cold worked 50 percent and aged 2 hours at 1250°F (675°C) obtains a tensile strength of 200 000 psi (1380 MPa) with a modulus of elasticity E of 28 000 000 psi (193 000 MPa) and a hardness of Rockwell C42. It is used for hairsprings, watch springs, and in instruments.

Other Materials Occasionally Used for Springs

ROCKET WIRE

Rocket wire, made by National Standard Co. of Niles, Michigan, has a chemical composition similar to music wire, but is specially selected for uniformity with carbon from 0.80 to 0.100 percent and manganese from 0.25 to 1.00 percent and then hard drawn to tensile strengths 10 to 20 percent higher than music wire, in sizes from 0.015 in. (0.38 mm) to 0.125 in. (3.2 mm) and 25 to 28 percent higher in lighter sizes. The modulus of elasticity E in tension is 29 500 000 psi (203 400 MPa), and in torsion G is 11 500 000 psi (79 290 MPa). The elastic limit in tension is about 70 percent of the tensile. In torsion, as used for compression and extension springs, the value is about 50 percent. This high-strength wire is especially useful (if the higher cost is justified) for small springs that may be failing due to high operating stresses. Springs should be stress relieved at the same temperatures as for music wire.

HASTELLOY ALLOY B

Hastelloy Alloy B, a nickel-molybdenum-iron alloy developed by Haynes Stellite Co., Division of Union Carbide Corp., is especially suitable for use in hydrochloride acid in all concentratons and temperatures, and has good corrosion resistance to sulfuric acid. Spring-temper wire has a tensile strength from 225 000 to 260 000 psi (1550 to 1800 MPa), which can be increased another 50 000 psi (345 MPa) if necessary by aging at 1450°F (790°C) for 16 hours.

HASTELLOY ALLOY C

Hastelloy Alloy C, like Alloy B, is a nickel-base alloy. It is highly corrosion resistant and can resist attack by wet chlorine gas, strong oxidizing compounds, and sulfuric acid. Its properties are quite similar to Hastelloy B, but slightly lower in all values.

HAVAR

Havar is a cobalt-chromium-nickel-iron alloy with small amounts of tungsten and molybdenum, developed by Hamilton Technology, Lancaster, Pennsylvania. It is nonmagnetic, corrosion resistant, suitable for subzero temperatures and up to 750°F (400°C), with torsional stresses up to 75 000 psi (515 MPa). It can be hardened after forming by heating to 950°F (510°C) for 3 hours to obtain tensile strengths over 300 000 psi (2070 MPa), with Rockwell hardness of 56 to 60C and a modulus of elasticity E of 30 000 000 psi (206 800 MPa). It is used in mainsprings of watches, motor, spiral, and flat springs.

TITANIUM

Several alloy compositions of titanium are available, weighing about one-third that of steel alloys. These alloys are highly corrosion resistant and useful at elevated temperatures. Early compositions were quite brittle, but better control and limitations of the amounts of carbon, oxygen, hydrogen, and nitrogen have reduced the brittle characteristics. Spring alloys can be age hardened at 800°F (425°C) for short periods to obtain high tensile strengths for springs used in equipment for space vehicles.

STAINLESS, TYPE 18-2: ASTM A 313, TYPE XM-28

This low-nickel austenitic alloy was originally developed as a substitute for types 302 and 304 when nickel was difficult to obtain, but it has become an excellent addition to the series of stainless steels. It is nonmagnetic and has good corrosion resistance. In the lighter sizes under 0.026 in. (0.65 mm), it has the same tensile strength as type 302; intermediate sizes up to 0.058 in. (1.50 mm) are slightly lower, but all heavier sizes up to 0.50 in. (13 mm) are higher than 302.

Some of its properties after heating at 700°F (370°C) are as follows:

E = 29 000 000 psi (200 000 MPa) for torsion springs

G = 9 800 000 psi (67 570 MPa) for compression and extension springs

Elastic Limit. Percent of tensile: in tension, 75 percent (approx.) for torsion springs; in torsion, 43 percent (approx.) for compression and extension springs.

WOOD, GLASS, PLASTIC, ALUMINUM, AND FIBERGLASS

A few other materials, including wood, glass, plastic, aluminum, and fiberglass have occasionally been used for springs under special circumstances. Research and development of new materials is a never-ending project; improved alloy compositions and better manufacturing methods will produce other new materials in the near future.

10 Fatigue Failure; Recommended Design Stresses

One of the oldest controversial subjects among engineers is the determination of safe working stresses for springs. One reason is that a high stress which may be satisfactory for one application may be quite unsafe for a different application. Conversely, a low stress needed for a severe-service application may be far too conservative and uneconomical for another application.

Much progress, too, has been made in the steel industry, in the wire-drawing mills, and in spring manufacturing. The tensile strengths and elastic limits for some spring materials have risen. More uniformity and homogeneity in the orientation of molecular structures have occurred. More fatigue testing and a better understanding of fatigue fractures plus an improved method of interpreting the results of fatigue testing have helped produce more useful curves to show recommended design stresses. The ASTM Standards for many spring materials have been revised several times in the past few years. And several ASTM committees are presently examining other materials to see if further revisions are desirable. Such revisions are a continuing process, proving once again that progress and change are inevitable.

Most spring failures are due to high stresses caused by high forces for too many deflections. Such severe service causes a breakdown in the grain structure and eventual failure. Metal fatigue is a phenomenon that develops at a point of maximum stress at some incipient crack, fault, or nick on the surface that progresses at a right angle to the tensile stress due to repeated deflections.

Fatigue Theories

Why should springs and other machine components fail when stressed far below the elastic limit? The answer to this question has intrigued engineers and metallurgists for over a hundred years, but the correct answer has been found only recently. Studies on fatigue have been made in England, Germany, Sweden, and other countries; by many engineering and technical societies including ASTM, ASME, SAE, and others. Many spring companies have fatigue testers, and continuing research is always in progress.

The old theories were incorrect, but usually paved the way to more research. Several of the obsolete theories are the amorphous film theory, the dynamic theory, the attrition theory, the thermodynamic theory, the reversed shear stress theory, and dozens of others.

CRYSTALLIZATION THEORY

Special mention should be made of the erroneous crystallization theory because it keeps cropping up year after year. Mechanics and some untrained technicians observing a broken shaft or a broken spring will often point to that portion of the fracture that appears to have large grains and say, "It broke because it crystallized." Fortunately, such is not the case and never has been. Steel is always crystalline from the time it is solidified until it is remelted. Hardening and tempering will change the appearance of the crystalline structure, and overheating will cause an undesirable growth in the grains, but the part did not break because it crystallized.

The grains near the area where fracture started are small and smooth, usually for about one third or more of the cross-sectional area, because these grains were rubbed together possibly thousands of times before sudden fracture of the central zone resulted in complete breakage. The relative coarseness of the other grains often is mistakenly pointed to as evidence of a crystallized metal, but it can easily be shown that a fresh fracture of the entire section taken elsewhere will have the identical crystalline appearance.

CORRECT THEORY

Wood's theory appears to be the correct theory. From the early theories it is evident that fatigue failures result from the conversion of deformation under cyclic stresses, first into fine slip, then into coarse slip. Fine slip occurs at low stresses and coarse slip at higher stresses. Wood observed that the crystalline structure responds differently to small and large deflections, that work hardening is not present with small deflections, and that coarse slip is an avalanche of fine slip movements. In cold-worked material (used in cold-coiled springs), a number of coarse slip groups can pile up, causing several fatigue failures almost simultaneously. This explains why breakage of several coils in a spring sometimes occurs at about the same time.

Fatigue can be defined as the result of fine slip which gradually develops into a minute crack which propagates until breakage occurs. A formula for predicting exactly when failure will occur may never be developed, but fatigue curves can be derived to present data that can be used for recommended design stresses.

Fatigue Tests

Testing springs in fatigue testing machines is the best method for determining how long a spring will last under repeated deflections. Such testing must be done with care, and a detailed account of all of the spring characteristics should first be made and at least 5 or 10 similar springs should be tested under the same conditions to establish a point on a fatigue curve. Speed of testing is highly important. Fast cycling generates heat, makes the first few coils deflect further than others, and causes early breakage. From 200 to 350 cycles per minute is preferred; under 200 and over 350 are often used, but fast cycling should be avoided except when making environmental tests to duplicate actual operating conditions.

A batch of similar compression springs made from stainless steel type 302, oil-tempered MB spring steel, and music wire were tested at 200 cycles per minute. All were made from 0.063 in. (1.6 mm) diameter wire, 9/16 in (14.3 mm) outside diameter, 2 in. (50 mm) free length, with ends squared and ground. They were all subjected to an initial deflection of ¼ in. (6.35 mm) and for an additional ¾ in. (19 mm). The total corrected stresses in each spring were approximately 30,000 psi (200 MPa), initially to 107,500 psi (740 MPa). The average life number of cycles follows:

Material	*Broke at*
Stainless steel 302	195,700
Oil-tempered MB	202,400
Music wire	1,853,185

Note: springs made from music wire, rotary straightened at the wire mill for use in torsion coilers, lasted only 151,800 cycles. Straightened wire should not be used if a long fatigue life is required. The reason for such early failure of straightened wire is that the tensile strength on cold drawing is longitudinal, whereas rotary straightening is transverse, thereby reducing tensile properties and lowering the elastic limit.

FATIGUE TESTING HINTS

1. Test at least five similar springs simultaneously, and keep accurate data of all spring characteristics including dimensions, materials, range of stress, and other details.

2. Keep the speed of testing between 200 to 350 cycles (300 is best) per minute for most uniform results.
3. Stop the fatigue tester after the first 1000 cycles to check spring set or loss of force and readjust the tester, if necessary, to compensate for the set or loss of force, if any, so that preselected forces and stresses can be maintained.
4. Determine the average life of each batch of five springs as follows: eliminate the lowest and highest figures, then average the two that are closest together. Other batches of five can then be averaged in a similar manner if desired.
5. The shorter the range of stresses, the longer the life.
6. Compression springs with set removed last longer.
7. Compression springs made on the half coil such as 9½ or 10½ active coils last longer than springs with an even number of coils, because of better stress distribution and less tendency to buckle.
8. Springs tested at AC motor speeds of 1800 rpm are cycling at 1/11 of their natural frequency so that they are actually cycling 11 times as fast or 19,800 cycles per minute, which causes early fatigue.
9. True fatigue failures start on the inside of a coil where the stress is highest. Other areas of breakage may be due to tool marks, inclusions or poor quality wire.
10. Keep the tester and the springs well lubricated.

Definitions

A few simplified definitions of terms frequently used are given below.

Fatigue Failure. When a spring is deflected continually, the metal becomes fatigued or tired and failure may occur at a stress far below the elastic limit. This type of failure is called fatigue failure and usually occurs without warning. Failures do not fall within a small range of stresses. Typical failures of five springs may run about as follows: breakage occurred at 150,000; 190,000; 210,000; 240,000; and 300,000 deflections. To determine the average, the lowest and highest are discarded and the two that are closest together are then averaged. In this case the two closest are 190,000 and 210,000, so the average fatigue life is 200,000. Five groups of five springs could be tested to determine a still better average, in which case the five results would be averaged in the same manner. Only four of the five springs in each group of five need to be broken to arrive at the average.

Fatigue Life. This is the number of cycles of deflection until failure occurs at a predetermined amount of stress.

Fatigue Strength. This is the stress at which failure occurs, either by fracture or too much permanent set, after a specified number of deflections. It is always necessary to state the number of deflections when describing fatigue strength. Fatigue strength is also called endurance strength.

Endurance Limit (Fig. 110). This is the highest stress, or range of stress, that can be repeated indefinitely without failure. Usually 10 million deflections is called infinite life and is satisfactory for determining this limit. However, it has been noticed on many occasions that springs sustaining 3 million cycles of deflection usually can withstand the prescribed 10 million cycles, and some researchers are now ending their fatigue tests at this lower figure.

Tensile Strength (Fig. 113). This is the stress required to rupture in tension. It is also called breaking strength, ultimate strength, and ultimate tensile strength.

Yield Point. This is the stress to cause an elongation of 0.50 percent of the original length of the specimen. An overloaded specimen stretches rapidly at this stress, and is easily seen during a tensile test.

Yield Strength. This is the stress, usually taken by the offset method, as 0.20 percent of the original length of the specimen.

Proof Stress. This is the stress that causes no more than 0.001 in. (0.025 mm) elongation per inch (25.4 mm) of the original length, after this stress is removed. Usually called yield strength at 0.01 percent offset.

Proportional Limit. This is the maximum load at which strain or deformation is directly proportional to stress, at zero percent offset.

Elastic Limit. This is the stress to which a spring can be stressed without taking a permanent set. Two values are used. The elastic limit in tension is used for torsion springs and parts subjected to bending such as flat springs, Belleville washers, and spiral springs. The elastic limit in torsion is used for springs subjected to torsional stresses such as compression, extension, and torsion bars. The elastic limit, just below the yield point, is roughly proportional to the tensile strength. Spring steels have a true yield point, easily seen during tensile testing. However, such is not the case with copper-base and nickel-base alloys. For such alloys, the term proportional limit or proof stress is specified, but the term "elastic limit" is used for all materials herein as it has more general usefulness and is better understood by designers.

Modulus of Elasticity (Fig. 112). This term, taken from the Greek, literally means "measure of the elastic ability." It is a measurement of the stiffness or rigidity of any material. The higher the value, the stiffer the material. The lower the modulus, the more deflection takes place. By definition, it is the ratio of the stress intensity on a material to the unit strain produced, within the proportional limit of the material. Two values are used. The modulus in tension E is used for torsion springs and springs subjected to bending such as flat springs, spiral springs, and Belleville washers. The modulus in torsion G is used for compression and extension springs and for torsion bars. The value is expressed in pounds per square inch or in megapascals.

Stress. Stress is the cohesive strength of the grains, and is their reaction or internal force to an applied external force. Whenever a load or deflection is applied, a slight distortion is caused which tends to separate the grains. This internal stress can be calculated and then compared with the recommended design stress curves to see if this internal stress is safe. Stress has been described as a boxer who has received a blow in the stomach–you can't see anything on the outside, but on the inside his muscles are all tightened up, and it is just a matter of time or number of blows he receives before fatigue failure occurs.

Permanent Set. The difference in free length, if any, before and after deflecting a spring to its maximum deflection, is called permanent set. The best designed springs, usable in more than one application, should have no appreciable amount of permanent set.

Recommended Design Stresses*

The three curves for compression and extension springs and the three curves for torsion springs for each material are based upon several conditions, including tensile strength, elastic limit, results from thousands of fatigue tests, comparisons with the recommendations of other researchers, and an examination of thousands of spring designs used in industry.

The curves cover the most popular spring materials in the range of wire diameters commercially available. They do not cover square, rectangular, or specially shaped wire. They should not be interpreted for spiral, clock, flat, volute, or hot-rolled springs, since such items usually require special fatigue testing.

The type of service to which a spring is subjected, the kind of material selected, and the stresses imposed are the three most important factors in determining the life of a spring. Other conditions which affect the spring life are size of material; range of stress; type of loading, such as static, dynamic, or shock; temperature environment, either below freezing or above normal; and the general design, including spring index, sharp bends, sharply bent hooks, and corrosive atmospheres.

The curves are for springs with average stress ranges usually deflecting from 25 to 75 percent of their total deflection, at room temperature, with normal spring indexes, without buckling, and are based on springs properly made and heated to relieve residual stresses caused by coiling.

The curves may be raised 20 to 30 percent for springs that are baked, set relieved, and shotpeened. Although the upper set of three curves are primarily for compression springs, they may be used for extension springs by reducing the values 10 to 15 percent and after checking the bending stresses in the hooks and comparing the hook stresses with the stress curves for torsion springs, as they are based on bending stresses.

The curvature correction factor should be included in calculated stresses

*See Figs. 78 to 107, pp. 144-158.

before comparing those stresses with the curves, particularly for the average and severe stress curves.

The curves have all values expressed in both the inch-pound system and in the metric SI system.

The tensile strength curves (Fig. 113) show only the minimum tensile strengths on which the upper light-service curves are based after considering the elastic limit.

The endurance limit curves (Fig. 110) are based on the range of stress between the first or initial stress and the second or final stress. The narrower this range, the longer the life. An example of use indicated by the dashed lines show that a compression spring made from oil-tempered MB-grade wire, if stressed between 17,000 psi (117 MPa– and 71,500 psi (493 MPa), should have unlimited life.

The recommended design stresses may be used to determine a safe stress before designing a spring or may be used after a spring has been designed to determine its probable life.

EXAMPLE

From the recommended design stress curves for compression springs (Fig. 78) made from music wire, the dashed lines in the lower left section show that a vertical line running upward from 0.085 in. (2.16 mm) intersects the lowest curve at 103,000 psi (710 MPa), which is satisfactory for severe service; the line intersects the middle curve at 125,000 psi (860 MPa), which is satisfactory for average service; and it intersects the top curve at 138,000 psi (950 MPa), which is for light service.

Also, for metric design, the dashed lines in the lower right section show that a vertical line running downward from 3.85 mm intersects the top curve at 860 MPa, which is satisfactory for light service and short life; the line intersects the middle curve at 780 MPa, which is satisfactory for average service; and it intersects the lowest curve at 650 MPa, which is for severe service and long life.

Properly stress-relieved springs made from hard-drawn steels including music wire, hard-drawn steels, and stainless steel 300 series will have an increase in hardness of about 2 to 3 Rockwell C scale with a corresponding increase in their minimum elastic limits, which will add assurance to their meeting the requirements of the curves.

ROUNDING OFF UNWIELDY VALUES

Stresses. The stress values are in pounds per square inch (psi) and in megapascals (MPa), which equals 1 million pascals (Pa). One megapascal equals one meganewton per square metre (1 MPa = 1 MN/m^2), and one pascal = 1 N/m^2, so the values can easily be transposed on a basis of 1 for 1.

Also, MN/m^2, N/mm^2, and MPa are equal numerically.

psi $\times$ 0.006 894 757 = MPa (or rounded off = 0.0069 or roughly 0.007).

The stress values have been rounded off to the nearest logical figure to simplify their usefulness. For example, 100 000 psi actually equals 689.4757 MPa (or MN/m^2), which is an unwieldy figure when analyzing stresses, so it has been rounded off to 690 MPa.

Inches. The conversion of inches and millimetres also causes unwieldy figures, so they have been rounded off to even figures. For example, 0.120 in. actually equals 3.048 mms which has been rounded off to 3 mm. Also, $^3/_8$ in., which equals 9.525 mm, is shown as 9.5 mm and so on.

Temperatures. These, too, have been rounded off for convenience. For example, 480°F actually equals 248.88°C, but is rounded off to 250°C. And 1450°F, which equals 787.77°C, is shown as 790°C.

For more exact conversions refer to Fig. 215.

LIGHT SERVICE

Light service, for 1000 to 10 000 deflections, includes springs subjected to static forces and those having small deflections with low stress ranges and seldom-used springs including those used in bomb fuses, projectiles, shock absorbers, and safety devices. Such springs can be designed with relatively high stresses up to the minimum elastic limit as shown by the top curve for each material.

AVERAGE SERVICE

Average service, for 100 000 to 1 000 000 deflections, includes most springs in general use for machines, brakes, motors, switches, and mechanical products of all types. Normal frequency of deflections under 300 cycles per minute and average use without shock forces permit such springs to operate up to 100,000 deflections with infrequent breakage, and the stress values under the middle curves have often permitted over 1 000 000 deflections where the material and operating conditions were normal. The lower the stress, the longer the life.

SEVERE SERVICE

Severe service, for over 1 000 000 deflections, includes springs subjected to rapid deflections over long periods of time, as in automobile engine valve springs, pneumatic hammers, presses, hydraulic controls, and similar applications. The stresses shown by the lowest curves provide for a minimum of 1 000 000 deflections, but should be lowered at least 10 percent to obtain 10 million deflections, which is generally considered an infinite life for springs. For severe-service use, the endurance curves (Fig. 110) should be considered before determining a safe allowable stress.

Other curves and tables which may be used for both compression and extension springs are included in this section. For recommended design stresses, see Figs. 78-109. For endurance limit curves, see Fig. 110. For causes of spring failure, see Fig. 111. For a summary of modulus of elasticity, see Fig. 112; for a summary of tensile strengths, see Fig. 113. For curvature correction, see Fig. 114. For tables of spring characteristics, see Fig. 115.

FIG. 78. Recommended Design Stresses, Music Wire, ASTM A 228; Compression and Extension Springs

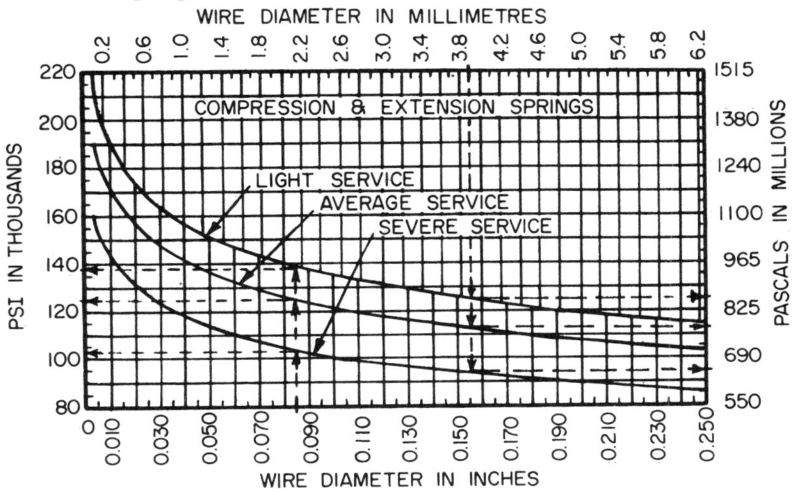

FIG. 79. Recommended Design Stresses, Music Wire, ASTM A 228; Torsion Springs

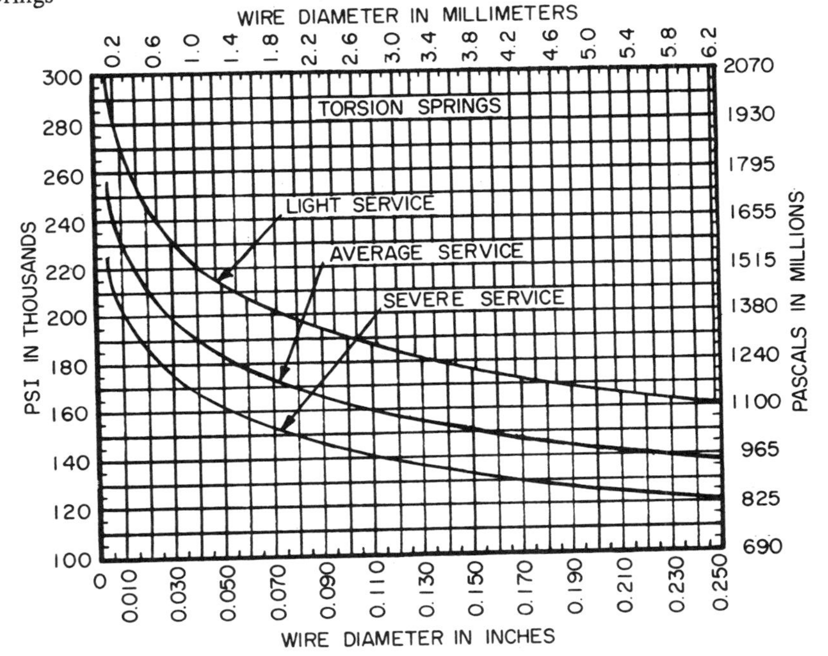

FIG. 80. Recommended Design Stresses, Hard-Drawn Steel Wire, ASTM A 227 (MB Grade); Compression and Extension Springs

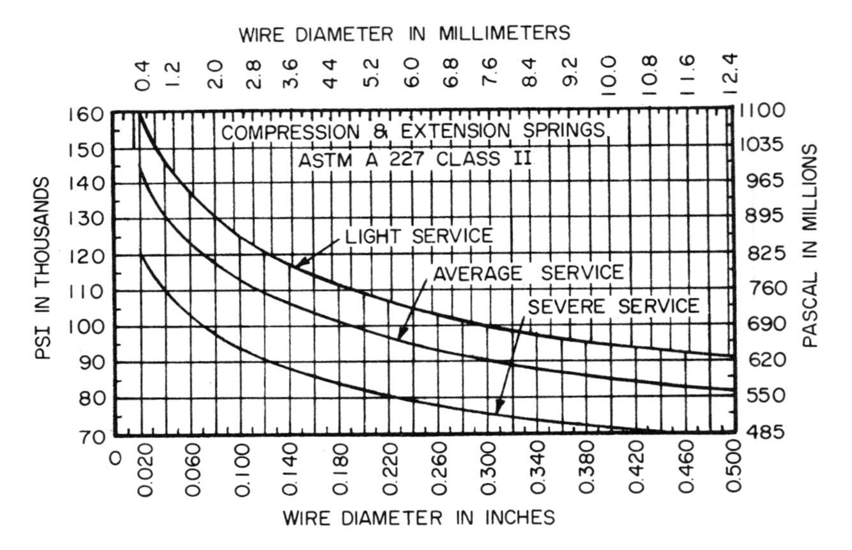

FIG. 81. Recommended Design Stresses, Hard-Drawn Steel Wire, ASTM A 227 (MB Grade); Torsion Springs

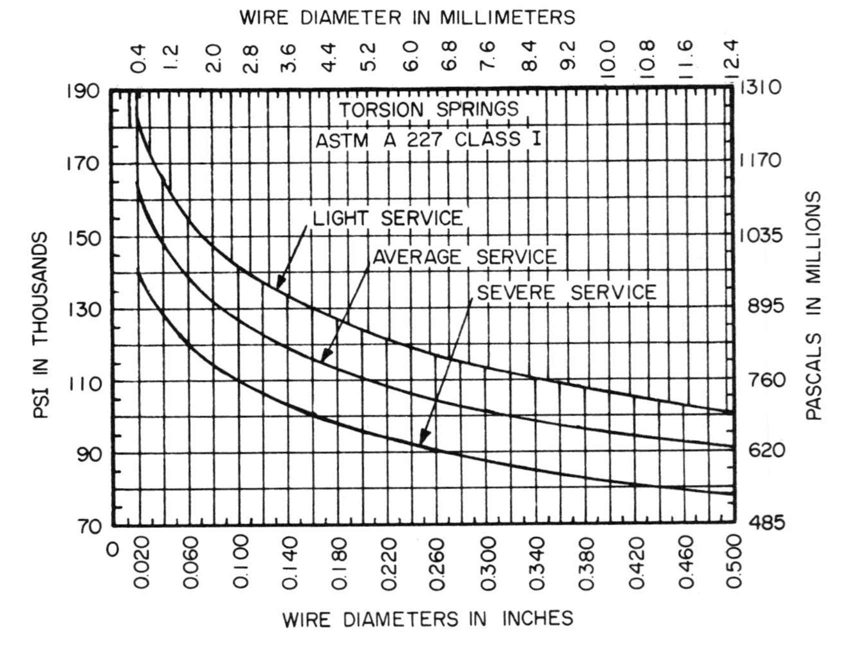

FIG. 82. Recommended Design Stresses, High-Tensile, Hard-Drawn Steel Wire, ASTM A 679 (HB Grade); Compression and Extension Springs

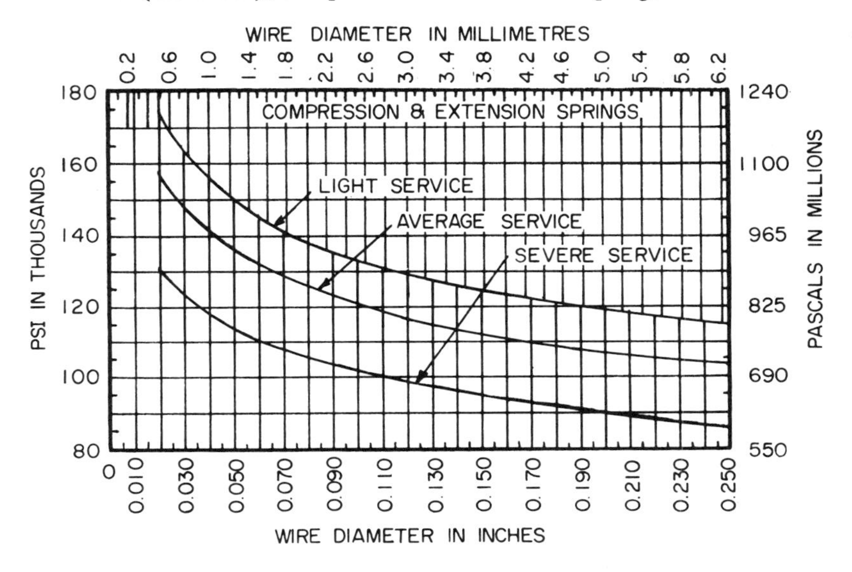

FIG. 83. Recommended Design Stresses, High-Tensile, Hard-Drawn Steel Wire, ASTM A 679 (HB Grade); Torsion Springs

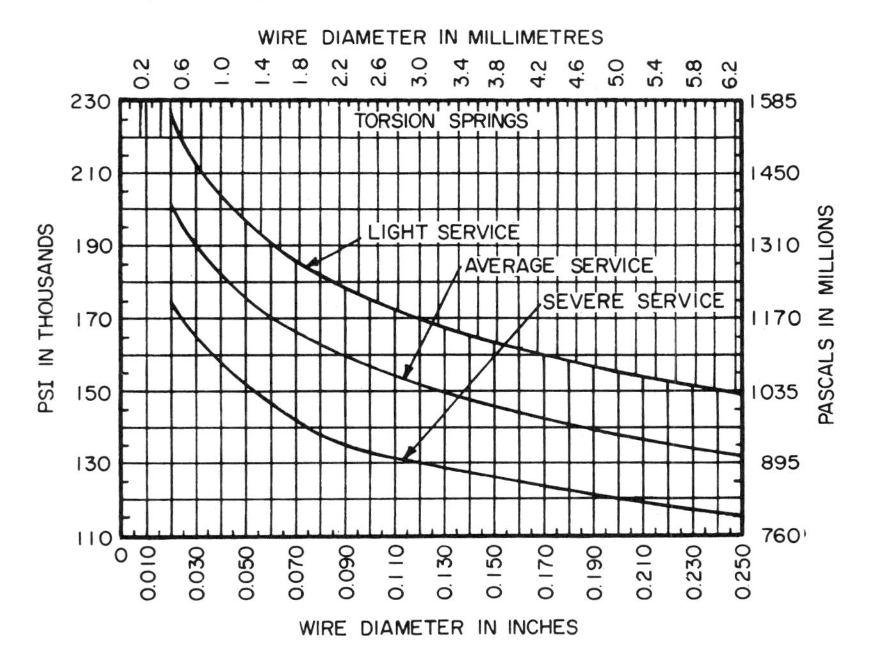

FIG. 84. Recommended Design Stresses, Oil-Tempered Steel Wire, ASTM A 229 (MB Grade); Compression and Extension Springs

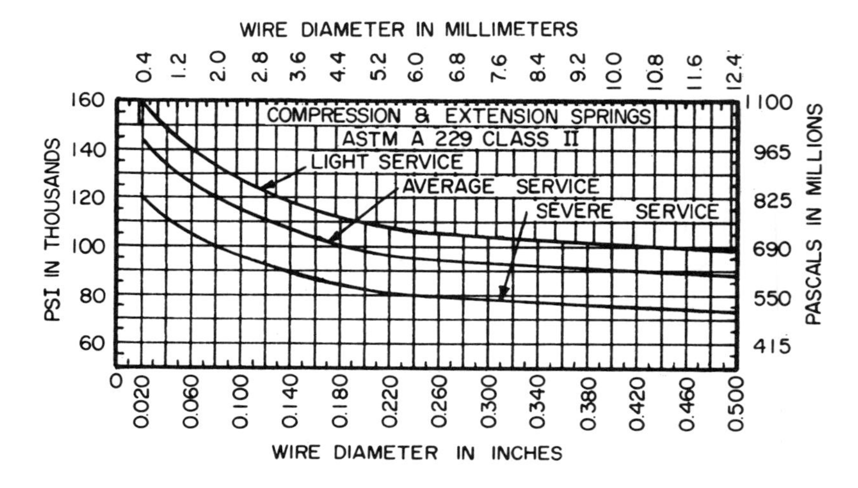

FIG. 85. Recommended Design Stresses, Oil-Tempered Steel Wire, ASTM A 229 (MB Grade); Torsion Springs

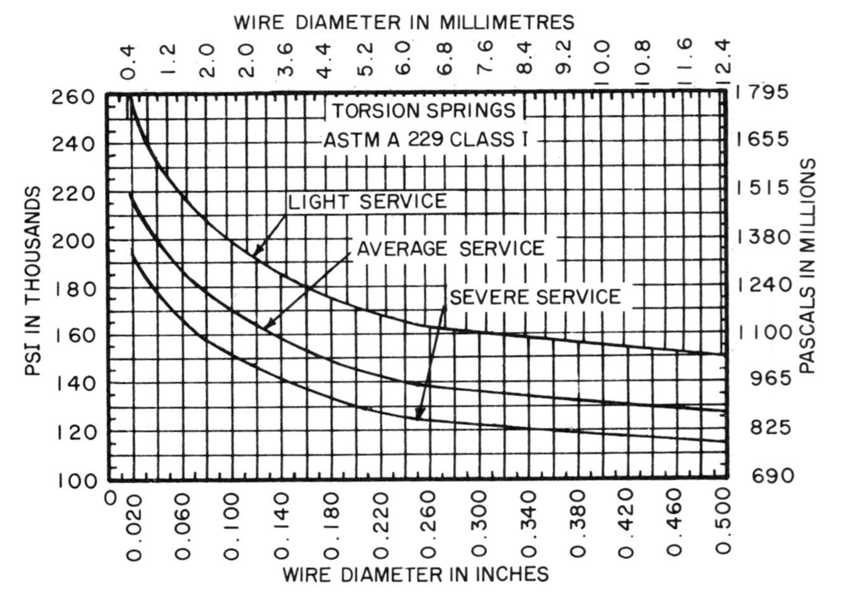

FIG. 86. Recommended Design Stresses, Chromium-Vanadium Alloy Steel Wire, ASTM A 231; Compression and Extension Springs

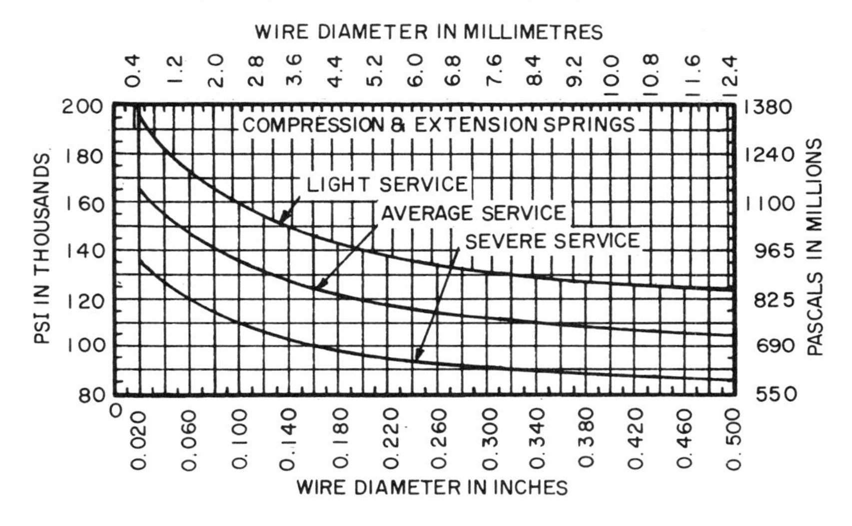

FIG. 87. Recommended Design Stresses, Chromium-Vanadium Alloy Steel Wire, ASTM A 231; Torsion Springs

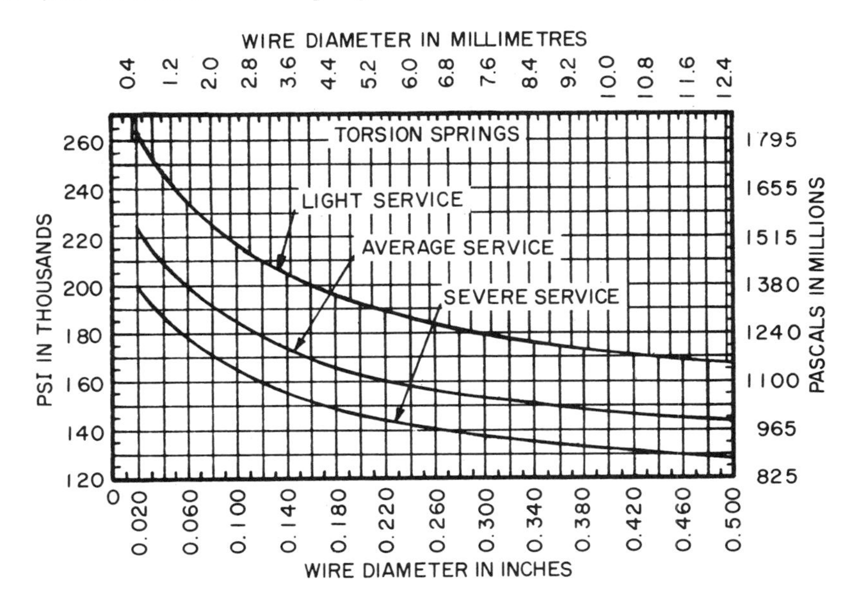

FIG. 88. Recommended Design Stresses, Oil-Tempered Chromium-Silicon Steel Wire,[a] ASTM A 401; Compression and Extension Springs

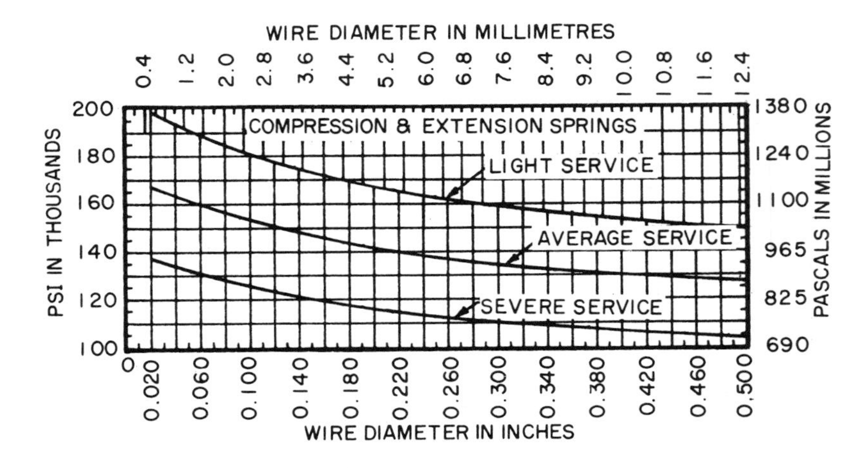

FIG. 89. Recommended Design stresses, Oil-Tempered Chromium-Silicon Steel Wire,[a] ASTM A 401; Torsion Springs

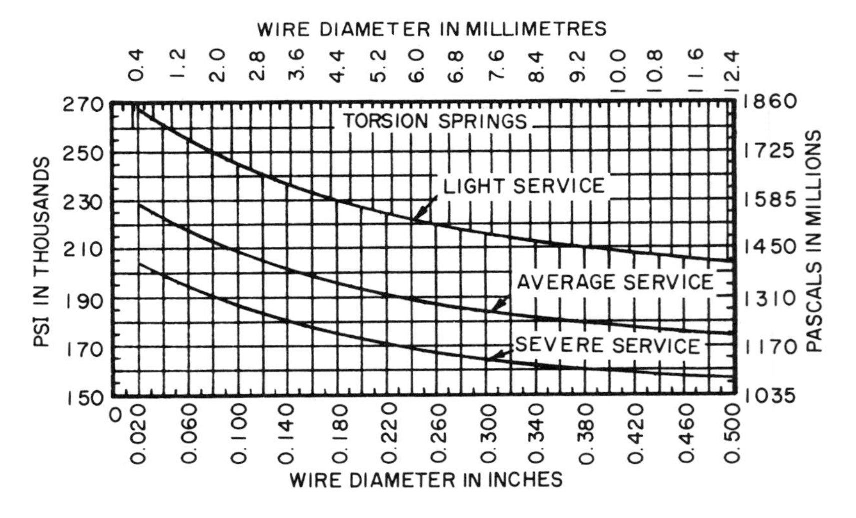

[a]For hard drawn up to 0.100 in. (2.54 mm), reduce values 4 percent; over 0.100 in. (2.54 mm) to 0.150 in. (3.81 mm), 7½ percent; and over 0.150 in. (3.81 mm) to 0.200 in. (5.08 mm), 10 percent. (Larger diameters are not readily available.)

FIG. 90. Recommended Design Stresses, Stainless Steel Wire Types 302 and 304, ASTM A 313; Compression and Extension Springs

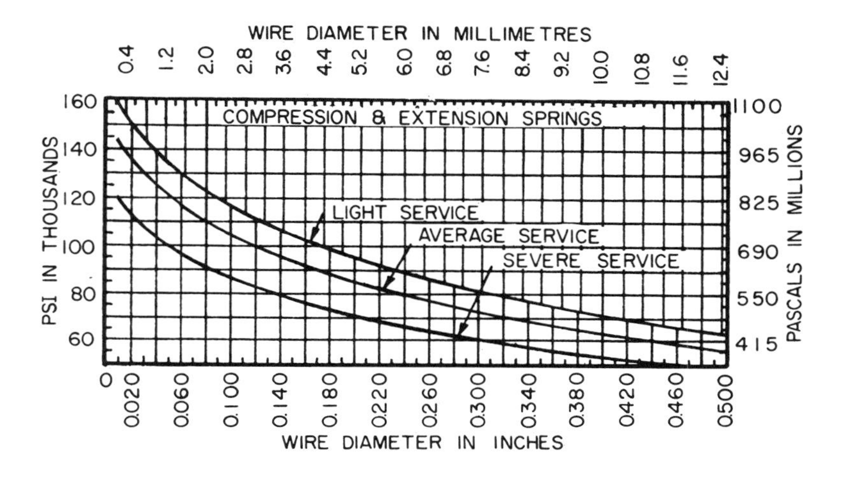

FIG. 91. Recommended Design Stresses, Stainless Steel Wire Types 302 and 304, ASTM A 313; Torsion Springs

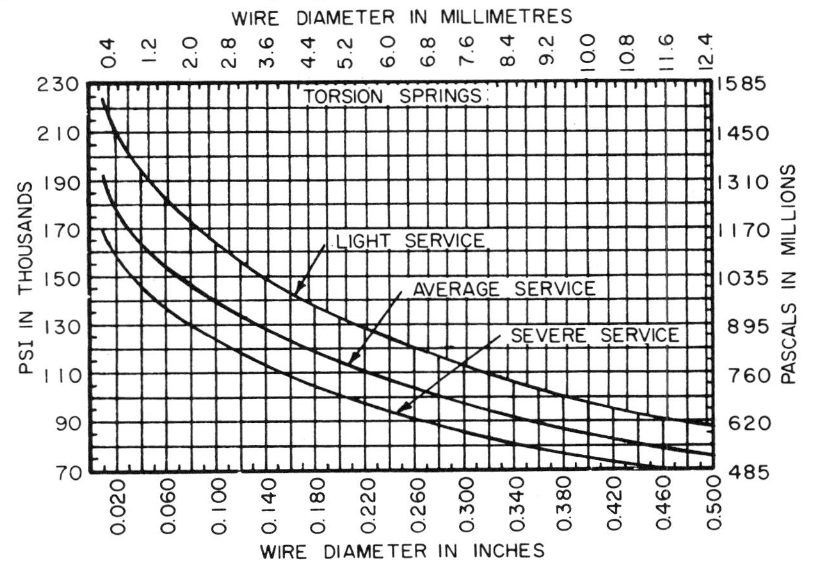

FIG. 92. Recommended Design Stresses, Stainless Steel Wire Type 316, ASTM A 313; Compression and Extension Springs

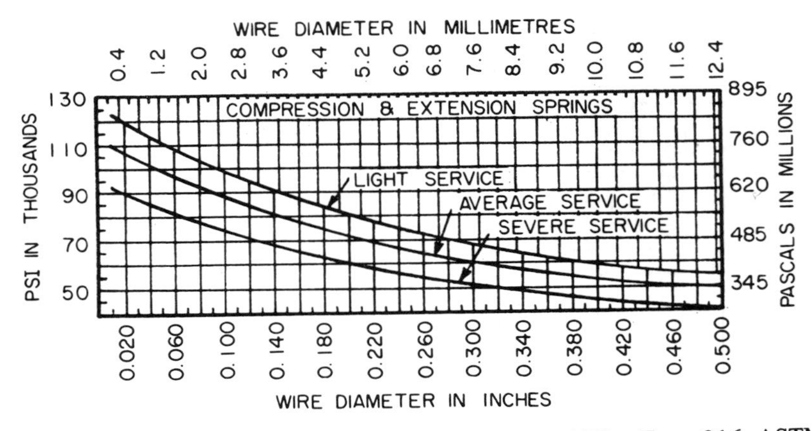

FIG. 93. Recommended Design Stresses, Stainless Steel Wire Type 316, ASTM A 313; Torsion Springs

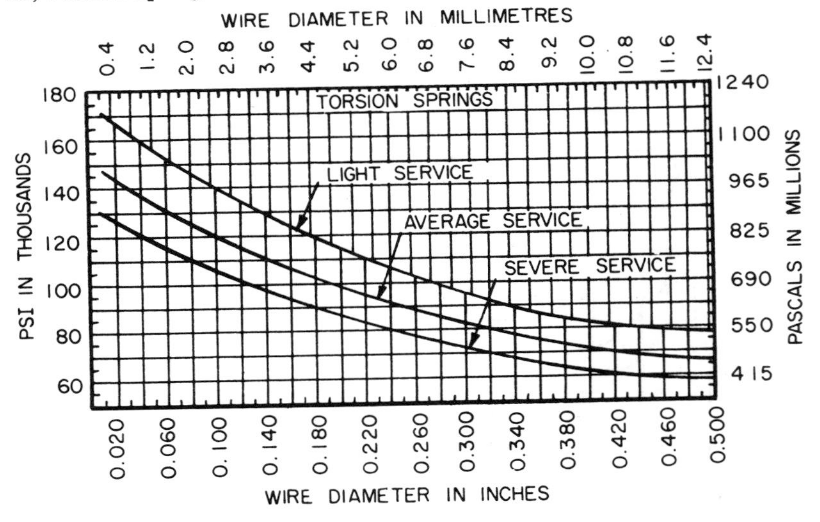

FIG. 94. Recommended Design Stresses, Stainless Steel Wire Type 17-7 PH, Heated to 900°F (482°C) for 1 Hour, ASTM A 313, Type 631; Compression and Extension Springs

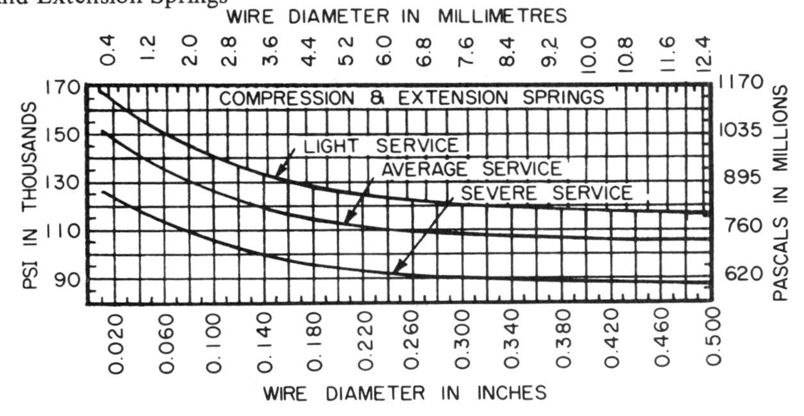

FIG. 95. Recommended Design Stresses, Stainless Steel Wire Type 17-7 PH, Heated to 900°F (482°C) for 1 Hour, ASTM A 313, Type 631; Torsion Springs

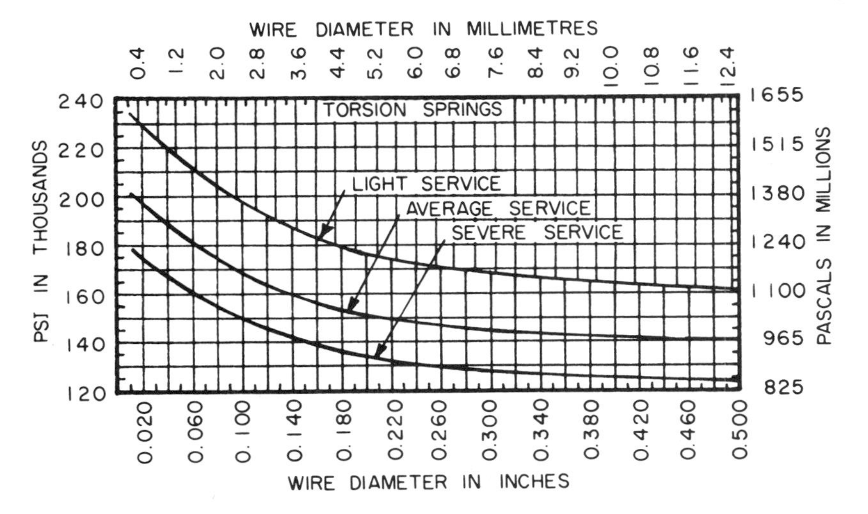

FIG. 96. Recommended Design Stresses, Stainless Steel Wire Type 431 (16-2), Hardened, Tempered, and Cold Drawn; Compression and Extension Springs

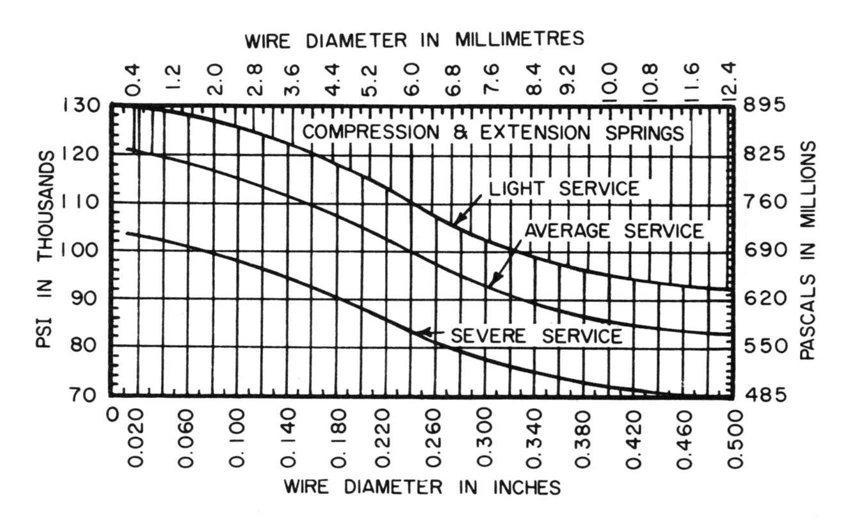

FIG. 97. Recommended Design Stresses, Stainless Steel Wire Type 431 (16-2), Hardened, Tempered, and Cold Drawn; Torsion Springs

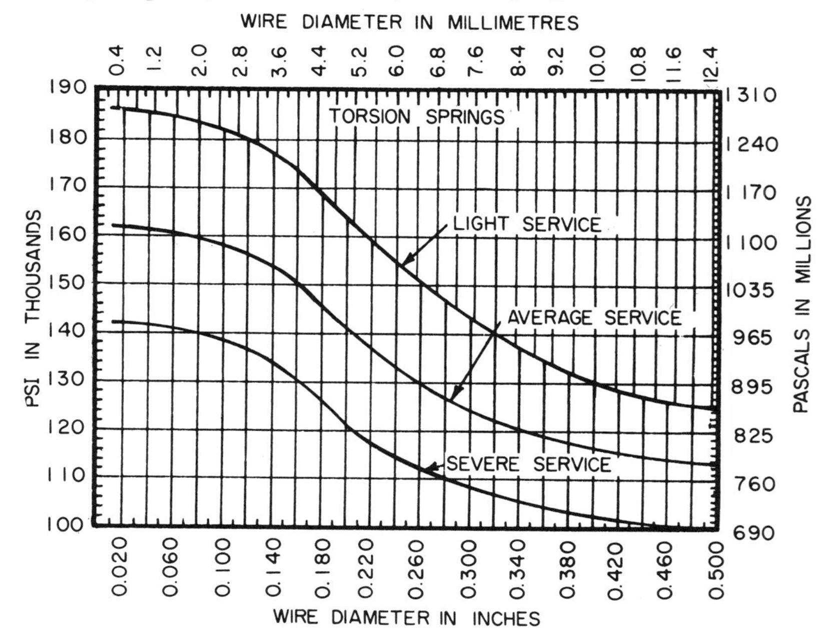

FIG. 98. Recommended Design Stresses, Stainless Steel Wire Type 420, Hardened and Tempered After Coiling; Compression and Extension Springs

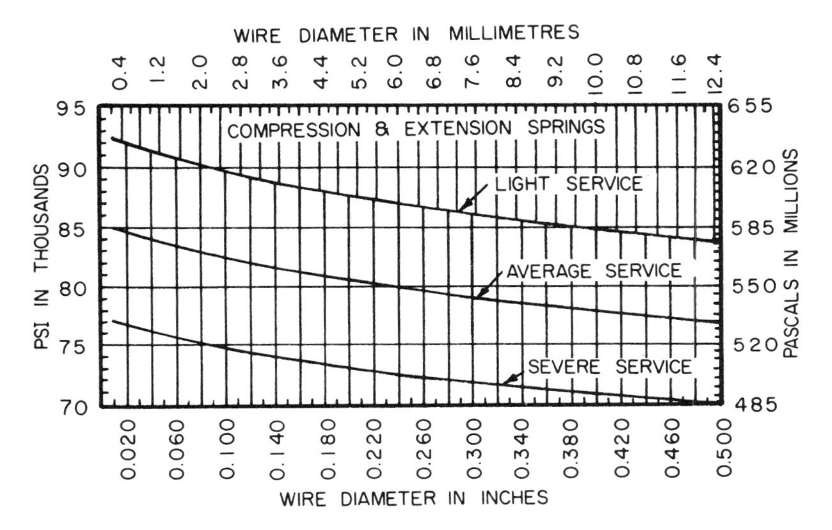

FIG. 99. Recommended Design Stresses, Stainless Steel Wire Type 420, Hardened and Tempered After Coiling; Torsion Springs

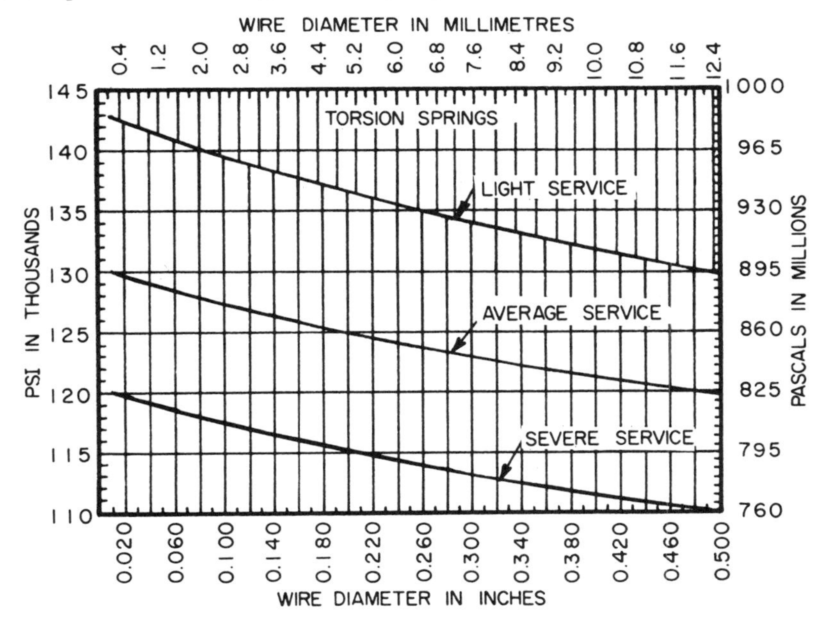

FIG. 100. Recommended Design Stresses, Brass Wire, ASTM B 134, Alloy 260 (70-30); Compression and Extension Springs

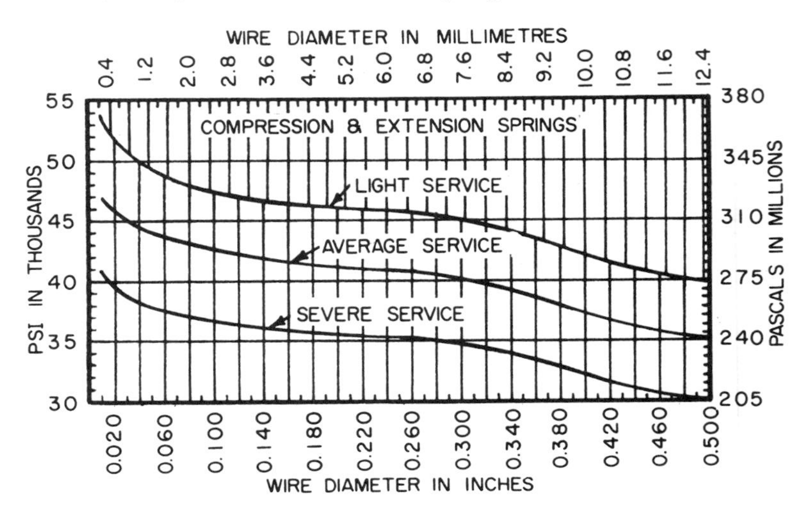

FIG. 101. Recommended Design Stresses, Brass Wire, ASTM B 134, Alloy 260 (70-30); Torsion Springs

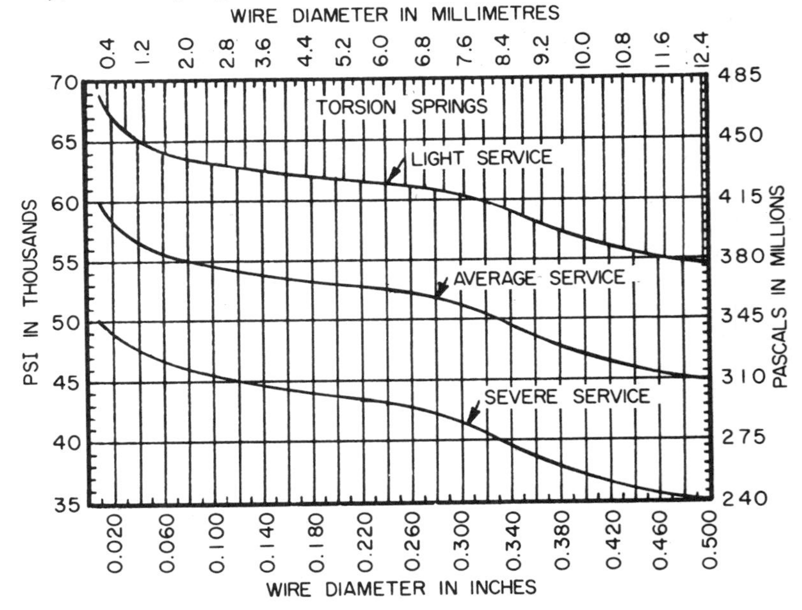

FIG. 102. Recommended Design Stresses, Phosphor-Bronze Wire, ASTM B 159; Compression and Extension Springs

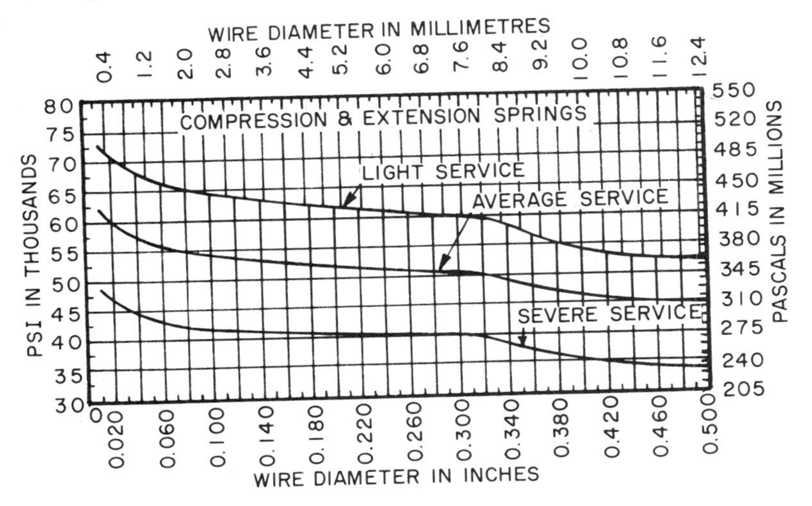

FIG. 103. Recommended Design Stresses, Phosphor-Bronze Wire, ASTM B 159; Torsion Springs

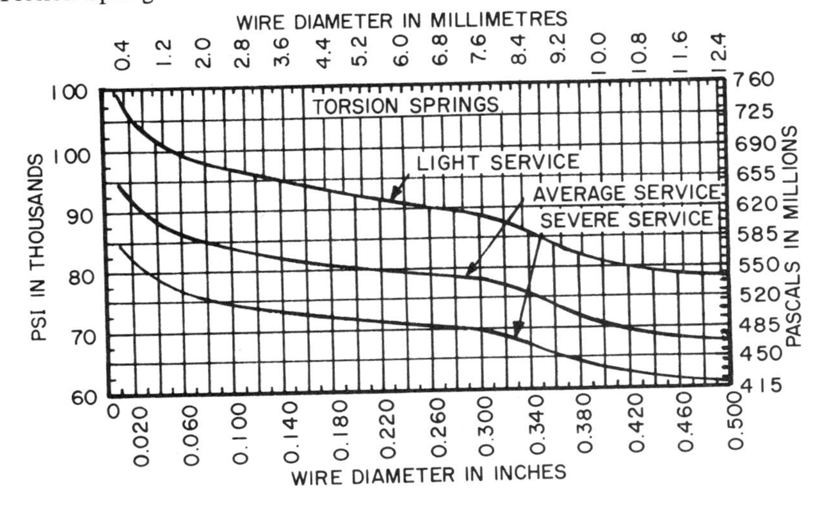

FIG. 104. Recommended Design Stresses, Beryllium-Copper Wire, for ¾ Hard, Heated to 600°F (315°C) for 1 Hour, ASTM B 197; Compression and Extension Springs

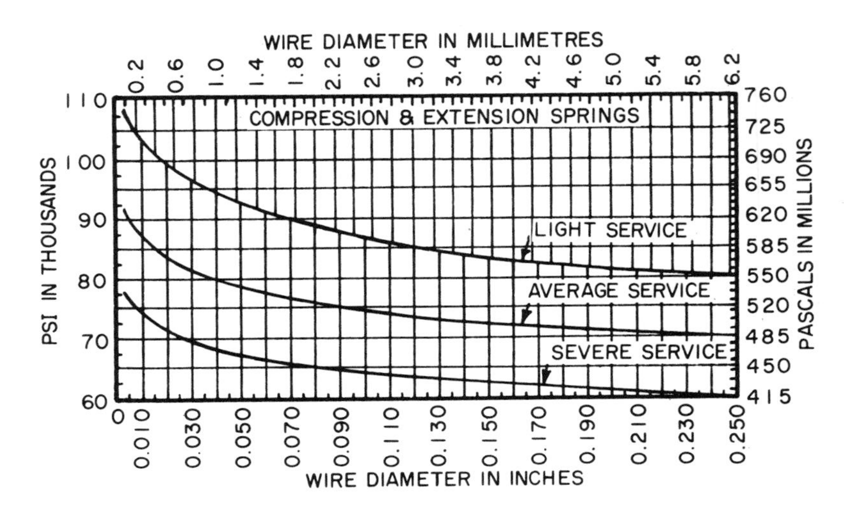

FIG. 105. Recommended Design Stresses, Beryllium-Copper Wire, for ¾ Hard, Heated to 600°F (315°C) for 1 Hour, ASTM B 197; Torsion Springs

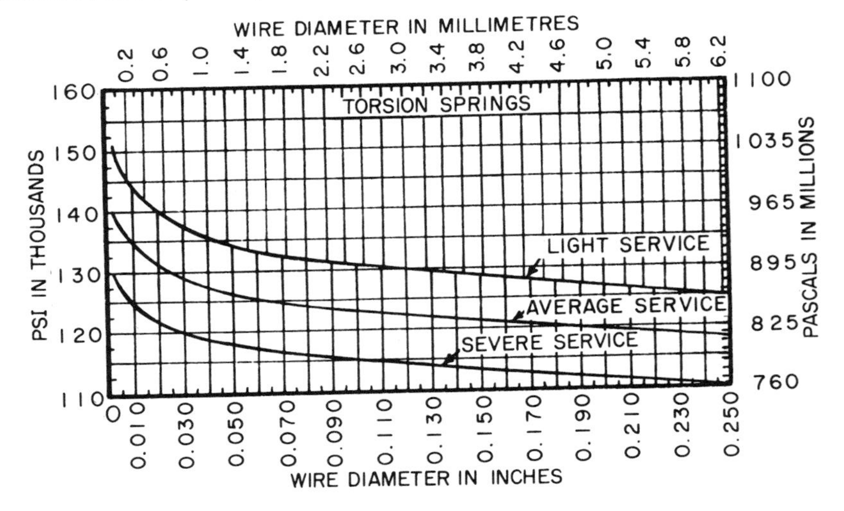

FIG. 106. Recommended Design Stresses, Inconel 600 Spring-Quality Wire, AMS 5687; Compression and Extension Springs

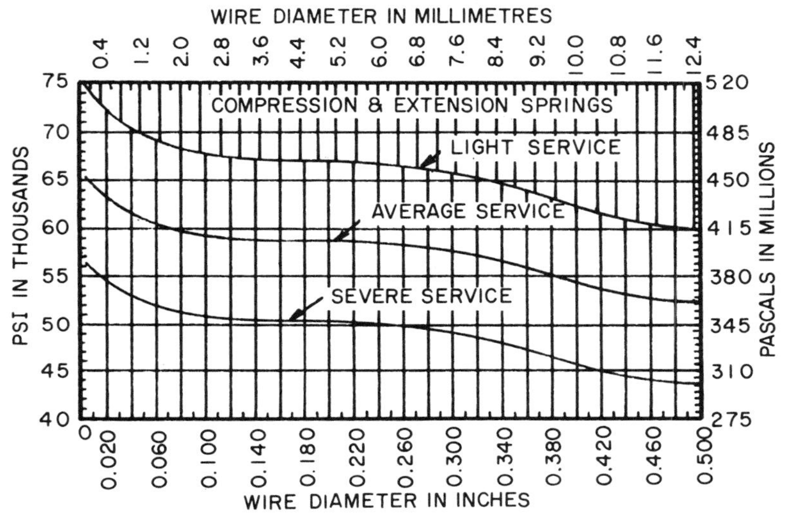

FIG. 107. Recommended Design Stresses, Inconel 600 Spring-Quality Wire, AMS 5687; Torsion Springs

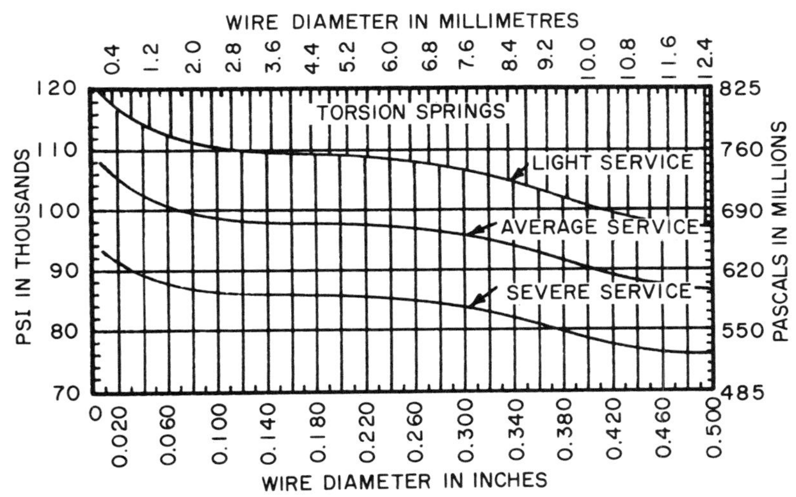

Note: Use these curves for other nickel alloys as follows: for Duranickel use the same curves; for Monel multiply by 0.90; for K-Monel multiply by 0.97; for Inconel-X multiply by 0.95.

FIG. 108. Recommended Design Stresses for Heavy Hot-Rolled Springs; Compression and Extension Springs

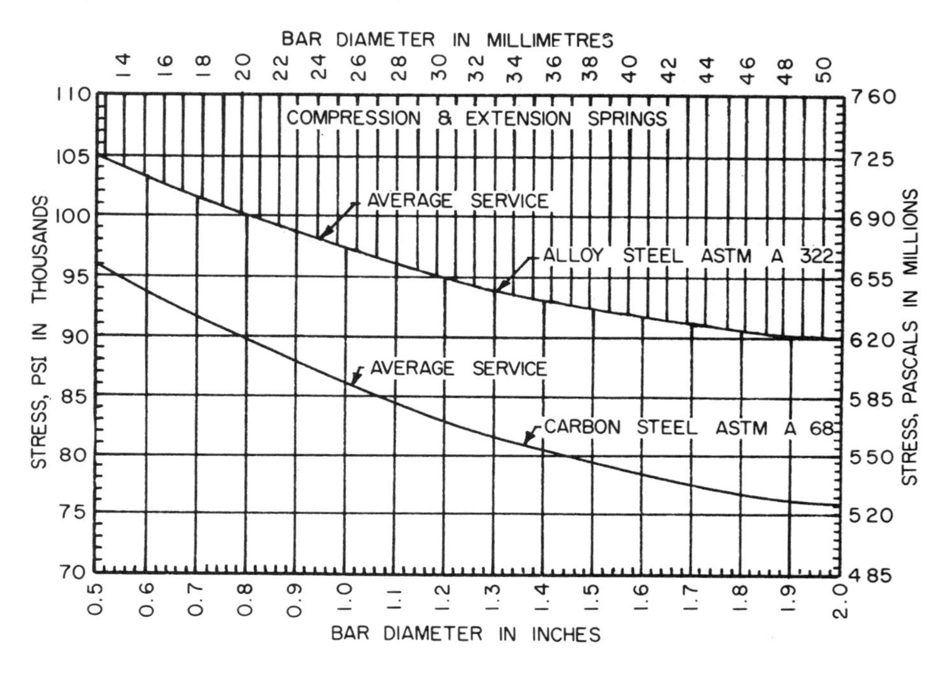

FIG. 109. Recommended Design Stresses for Heavy Hot-Rolled Springs; Torsion Springs

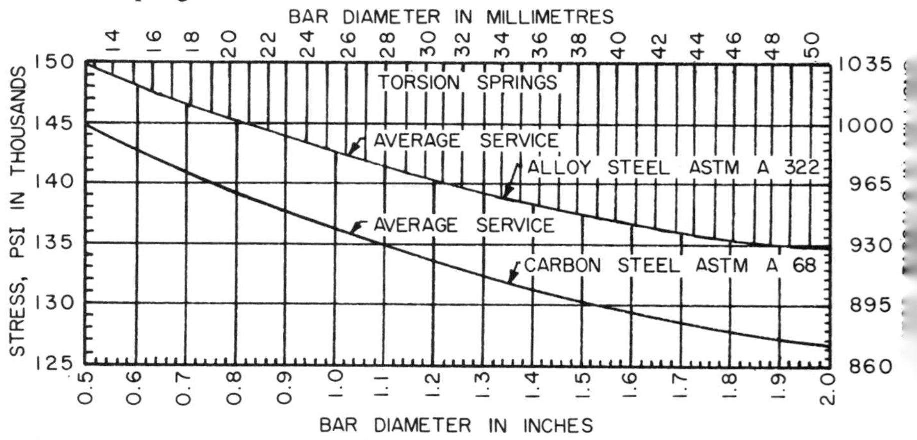

Average service stresses are shown and include curvature correction. For light service, increase 10 percent. For severe service, reduce 15 percent. Alloy steels are recommended in sizes over 1¼ in. (32 mm). Centerless-ground rods, presetting, and shotpeening are highly recommended for a longer fatigue life, especially for severe service. Refer also to ASTM A 125.

FIG. 110. Endurance Limit Curves for Compression Springs

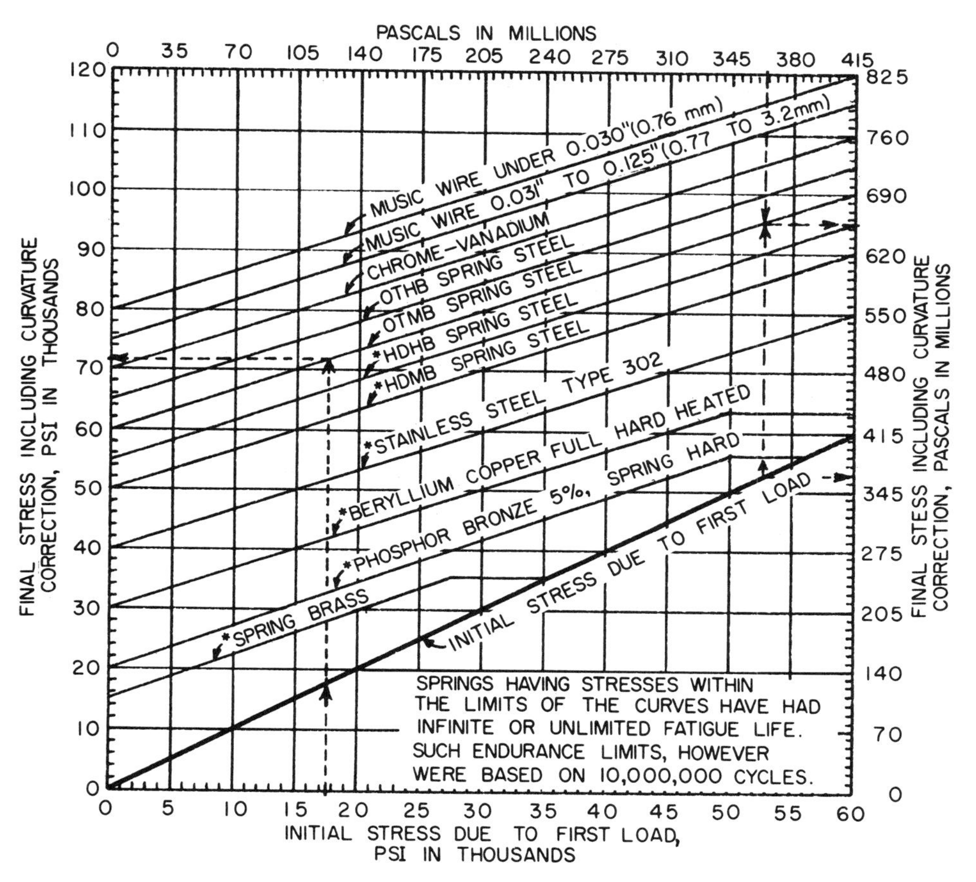

Materials preceded by * are not recommended for severe service. These curves are for commercial spring materials with wire diameters up to ¼ in. (6.35 mm) except as noted. Stress ranges may be increased approximately 30 percent for properly heated, set relieved, shotpeened compression springs.

FIG. 111. Causes of Spring Failure Listed in Sequence According to Frequency of Failure[a]

	CAUSE	COMMENT
GROUP I	High stress	The large majority of spring failures is due to high stresses caused by large deflections and high loads. High stresses should be used only for statically loaded springs. Low stresses lengthen fatigue life.
	Hydrogen embrittlement	Improper electroplating methods and acid cleaning of springs, without proper baking treatment, causes spring steel to become brittle, and is a frequent cause of failure. Nonferrous springs are immune.
	Sharp bends & holes	Sharp bends on extension, torsion, and flat springs, and holes or notches in flat springs, cause high concentration of stress resulting in failure. Bend radii should be as large as possible.
	Fatigue	Repeated deflections of springs, especially above 1,000,000 operations, even with medium stresses, may cause failure. Low stresses should be used for severe operating conditions.
GROUP II	Shock loading	Impact, shock, and rapid loading cause far higher stresses than those computed by the regular spring formulas. High-carbon spring steels do not withstand shock loading as well as the alloy steels.
	Corrosion	Slight rusting or pitting caused by acids, alkali, galvanic corrosion, stress corrosion cracking, or corrosive atmosphere weakens the material and causes higher stresses in the corroded area.
	Faulty heat treatment	Keeping spring materials at the hardening temperature for longer periods than necessary causes an undesirable growth in grain structure, resulting in brittleness even though the hardness may be correct.
	Faulty material	Poor material containing inclusions, seams, slivers, and flat material with rough, slit, or torn edges cause early failure. Overdrawn wire, improper hardness, and poor grain structure result in early failure.
GROUP III	High Temperature	High temperatures reduce spring temper (or hardness), lower the modulus of elasticity thereby causing lower loads, reduce the elastic limit and increase corrosion. Stainless or nickel alloys should be used.
	Low Temperature	Temperatures below −40°F (4°C) lessen the ability of carbon steels to withstand shock loads. Carbon steels become brittle at −70°F (−21°C). Stainless, nickel, or nonferrous alloys should be used.
	Friction	Close fits on rods or in holes result in a wearing away of material and occasional failure. The outside diameters of compression springs expand during deflection but they become smaller on torsion springs.
	Other Causes	Enlarged hooks on extension springs increase the stress at the bend. Electrical overload will cause failure. Welding and soldering frequently destroy the spring temper. Tool marks, nicks, and cuts often become stressraisers. Deflecting torsion springs outward causes high stresses. Winding them tightly causes binding on supporting rods. High speed of deflection, vibration, and surging due to operation near natural periods of vibration or their harmonics causes increased stresses.

[a]Spring failure may be caused by breakage, high permanent set, or loss of force. Group I lists the causes of failures that occur most frequently. Group II contains the less frequent causes of failure, and Group III lists causes of failure that occur occasionally.
Courtesy Spring Manufacturers Institute.

FIG. 112. Modulus of Elasticity

For steel and stainless

Material (in.)	(mm)	Modulus of elasticity G (psi) (MPa)	Modulus of elasticity E (psi) (MPa)
Hard-drawn MB			
up to 0.032	Up to 0.80	11 700 000	28 800 000
		80 700	198 600
0.033-0.063	1.60	11 600 000	28 700 000
		80 000	197 900
0.064-0.125	3.00	11 500 000	28 600 000
		79 300	197 200
0.126-0.625	16.00	11 400 000	28 500 000
		78 600	196 500
Music wire			
up to 0.032	Up to 0.80	12 000 000	29 500 000
		82 700	203 400
0.033-0.063	1.60	11 850 000	29 000 000
		81 700	200 000
0.064-0.125	3.00	11 750 000	28 500 000
		81 000	196 500
0.126-0.250	6.35	11 600 000	28 000 000
		80 000	193 000
Oil-tempered MB		11 200 000	28 500 000
		77 200	196 500
Chrome-vanadium		11 200 000	28 500 000
		77 200	196 500
Chrome-silicon		11 200 000	29 500 000
		77 200	203 400
Silicon-manganese		10 750 000	29 000 000
		74 100	200 000
Stainless steel			
Types 302, 304, 316		10 000 000	28 000 000
		69 000	193 000
Type 17-7 PH		10 500 000	29 500 000
		72 400	203 400
Type 420		11 000 000	29 000 000
		75 800	200 000
Type 431		11 400 000	29 500 000
		78 600	203 400

For nonferrous materials

Material	Modulus of elasticity G (psi) (MPa)	Modulus of elasticity E (psi) (MPa)
Spring brass		
Type 70-30	5 000 000	15 000 000
	34 500	103 400
Phosphor-bronze 5% tin	6 000 000	15 000 000
	41 400	103 400
Beryllium-copper cold-drawn 4 nos.	7 000 000	17 000 000
	48 300	117 200
Pretempered full hd.	7 250 000	19 000 000
	50 000	131 000
Inconel 600	10 500 000	31 000 000 [a]
	72 400	213 700
Inconel X-750	10 500 000	31 000 000
	72 400	213 700
Monel 400	9 500 000	26 000 000
	65 500	179 300
Monel K-500	9 500 000	26 000 000
	65 500	179 300
Duranickel 300	11 000 000	30 000 000
	75 800	206 800
Permanickel	11 000 000	30 000 000
	75 800	206 800
Ni-Span C 902	10 000 000	27 500 000
	69 000	189 600
Elgiloy	12 000 000	29 500 000
	82 700	203 400
Iso-Elastic	9 200 000	26 000 000
	63 500	179 300

[a]May drop 2 000 000 if not full hard.

G is used for compression and extension springs; E is used for torsion, flat, and spiral springs.

Note: The reduced values of G and E for hard-drawn MB and music wire produce more accurate results. A single value may cause errors in forces up to 5 percent.

FIG. 113. Tensile Strength (Minimum) for the Most Popular Spring Materials, Spring-Quality Wire

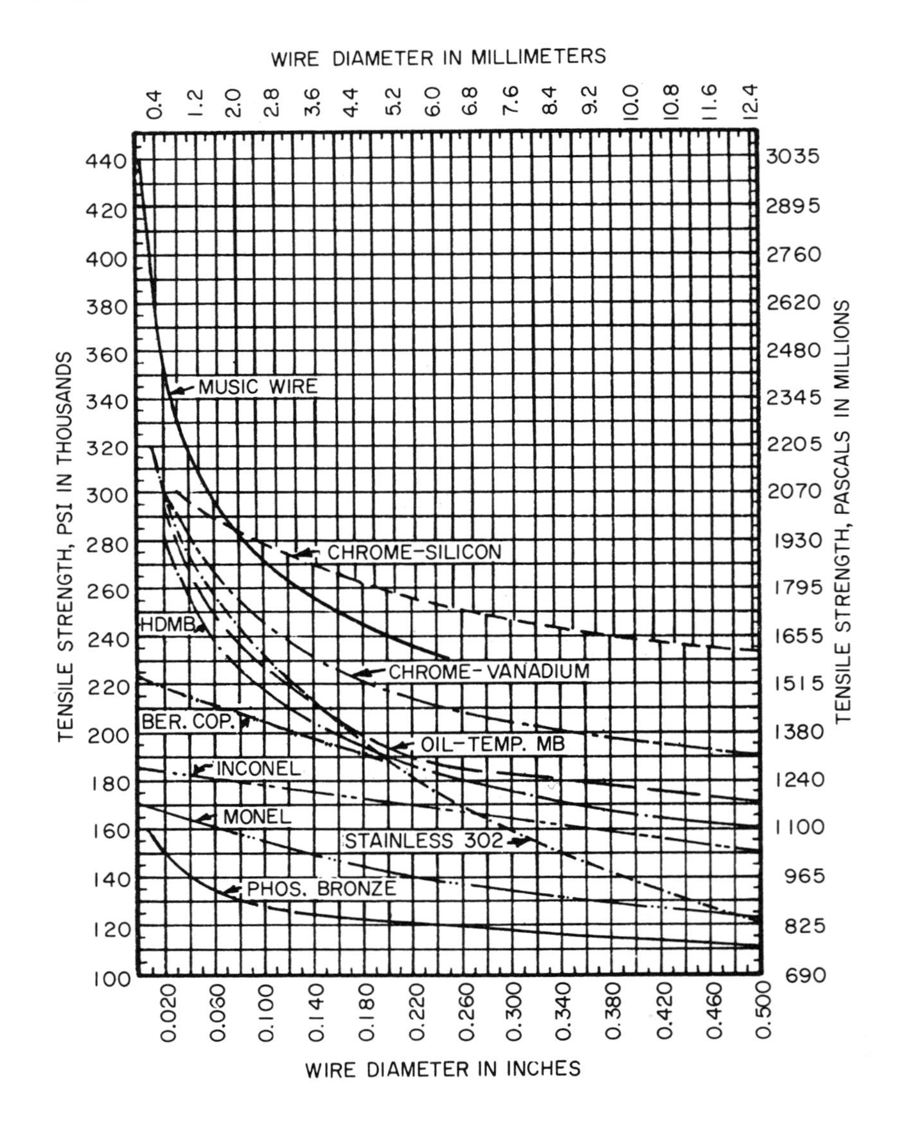

FIG. 114. Curvature Correction. In addition to the usual stress in a spring caused by the force or deflection, there is a direct shearing stress and an increased stress on the inside of the section due to curvature. Therefore, the corrected or total stress in a spring is the stress calculated for the force multiplied by the value of K taken from the proper curvature-correction curve. This then corrects for the uneven distribution of shear along the cross section of a curved wire coiled into a spring. It is best to calculate the total stress after all other calculations have been made and then compare it with the recommended design stress curves to determine if the stress is safe. This total stress should not be used in other formulas for deflection, as it will cause errors.

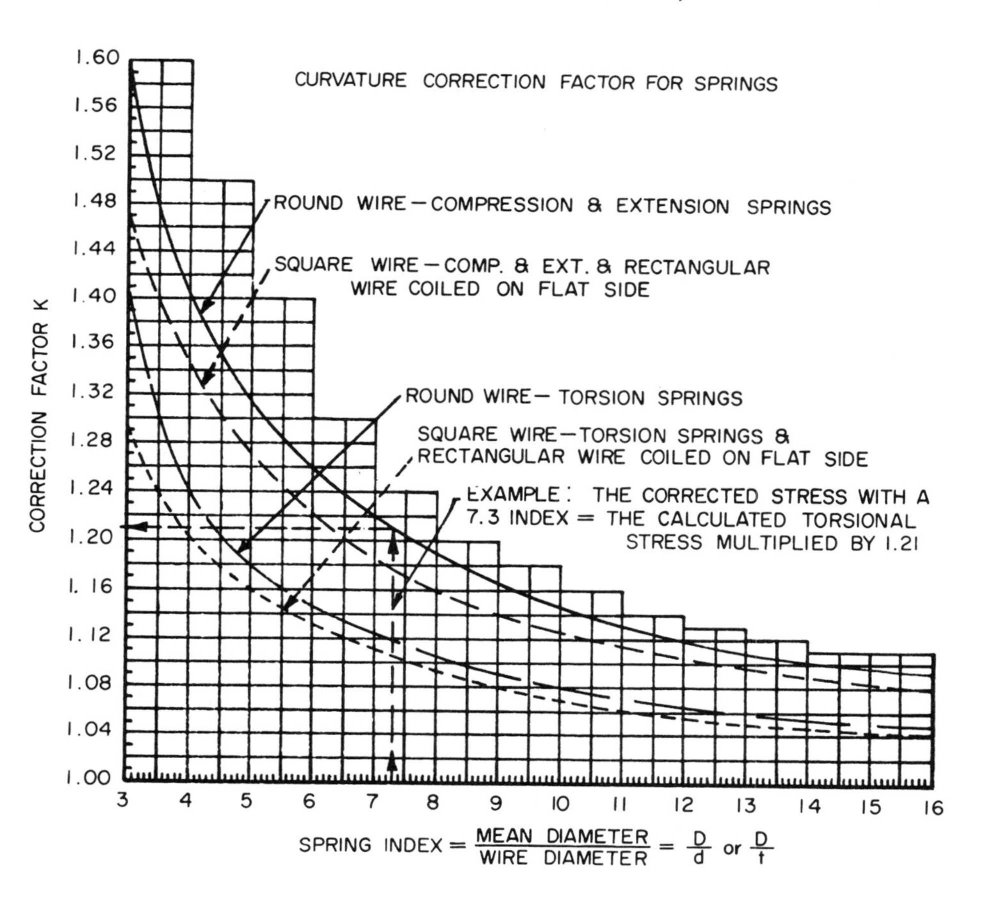

FIG. 115. Tables of Spring Characteristics for Compression and Extension Springs.[a] Oil-Tempered Steel Wire QQ-W-428, Type I, ASTM A229 (Courtesy Dept. of Defense MIL-STD-29A)

WIRE SIZE

OUTSIDE DIAMETER			SIZE W&M									
	DECIMAL		.010	.012	.014	.016	.018	.020	.022	.024	.026	.028
3/32	.0938	f	.01970	.01564	.01276	.01061	.00895	.00764	.00657	.00570	.00496	
		P	.469	.830	1.351	2.07	3.02	4.26	5.83	7.78	10.18	
7/64	.1094	f	.0277	.0222	.01824	.01529	.01302	.01121	.00974	.00853	.00751	.00664
		P	.395	.697	1.130	1.722	2.51	3.52	4.79	6.36	8.28	10.59
1/8	.125	f	.0371	.0299	.0247	.0208	.01784	.01548	.01353	.01192	.01058	.00943
		P	.342	.600	.971	1.475	2.14	2.99	4.06	5.37	6.97	8.89
9/64	.1406	f	.0478	.0387	.0321	.0272	.0234	.0204	.01794	.01590	.01417	.01271
		P	.301	.528	.852	1.291	1.868	2.61	3.53	4.65	6.02	7.66
5/32	.1563	f	.0600	.0487	.0406	.0345	.0298	.0261	.0230	.0205	.01832	.01649
		P	.268	.470	.758	1.146	1.656	2.31	3.11	4.10	5.30	6.72
11/64	.1719	f	.0735	.0598	.0500	.0426	.0369	.0324	.0287	.0256	.0230	.0208
		P	.243	.424	.683	1.031	1.488	2.07	2.79	3.67	4.73	5.99
3/16	.1875	f	.0884	.0720	.0603	.0516	.0448	.0394	.0349	.0313	.0281	.0255
		P	.221	.387	.621	.938	1.351	1.876	2.53	3.32	4.27	5.40
13/64	.2031	f	.1046	.0854	.0717	.0614	.0534	.0470	.0418	.0375	.0338	.0307
		P	.203	.355	.570	.859	1.237	1.716	2.31	3.03	3.90	4.92
7/32	.2188	f		.1000	.0841	.0721	.0628	.0555	.0494	.0444	.0401	.0365
		P		.328	.526	.793	1.140	1.580	2.13	2.79	3.58	4.52
15/64	.2344	f		.1156	.0974	.0836	.0730	.0645	.0575	.0518	.0469	.0427
		P		.305	.489	.736	1.058	1.465	1.969	2.58	3.21	4.18
1/4	.250	f			.1116	.0960	.0839	.0742	.0663	.0597	.0541	.0494
		P			.457	.687	.987	1.366	1.834	2.40	3.08	3.88
17/64	.2656	f			.1272	.1095	.0955	.0846	.0757	.0682	.0620	.0566
		P			.428	.644	.925	1.279	1.717	2.25	2.88	3.63
9/32	.2813	f			.1432	.1234	.1080	.0958	.0857	.0774	.0703	.0643
		P			.403	.606	.870	1.202	1.613	2.11	2.70	3.40
19/64	.2969	f				.1383	.1212	.1076	.0964	.0871	.0792	.0725
		P				.572	.821	1.135	1.521	1.989	2.55	3.21
5/16	.3125	f				.1541	.1351	.1200	.1076	.0973	.0886	.0811
		P				.542	.778	1.074	1.440	1.881	2.41	3.03
21/64	.3281	f				.1708	.1498	.1332	.1195	.1081	.0985	.0902
		P				.515	.739	1.020	1.366	1.785	2.29	2.87
11/32	.3438	f					.1633	.1470	.1321	.1196	.1090	.0999
		P					.703	.970	1.300	1.697	2.17	2.73
23/64	.3594	f					.1817	.1616	.1451	.1315	.1200	.1100
		P					.671	.926	1.239	1.618	2.07	2.60
3/8	.375	f						.1768	.1589	.1440	.1314	.1206
		P						.885	1.185	1.546	1.978	2.48

[a]For the inch-pound method of spring design, these tables are for compression and extension springs made of oil-tempered MB spring steel with G = 11 200 000 with an uncorrected torsional stress (S) = 100 000 psi. The deflection (f) for one coil under a force (P) is shown in the body of the table.

For: Stainless steel, multiply f by 1.067
Spring brass, multiply f by 2.240
Phosphor bronze, multiply f by 1.867
Monel metal, multiply f by 1.244
Beryllium copper, multiply f by 1.725
Inconel (nonmagnetic), multiply f by 1.045

WIRE SIZE

					19	18	17	16	15	14	13	3/32
.030	.032	.034	.036	.038	.041	.0475	.054	.0625	.072	.080	.0915	.0938
.00589 13.35												
.00844 11.16	.00758 13.83	.00683 16.95	.00617 20.6									
.01144 9.58	.01034 11.84	.00937 14.47	.00852 17.51	.00777 21.0								
.01491 8.39	.01354 10.35	.01234 12.62	.01128 15.23	.01033 18.22	.00909 23.5							
.01883 7.47	.01716 9.19	.01569 11.19	.01439 13.48	.01324 16.09	.01172 21.8	.00914 33.8						
.0232 6.73	.0212 8.27	.01944 10.05	.01788 12.09	.01650 14.41	.01468 18.47	.01157 30.07	.00926 46.3					
.0280 6.12	.0257 7.52	.0236 9.13	.0218 10.96	.0201 13.05	.01798 16.69	.01430 27.1	.01155 41.5					
.0333 5.61	.0306 6.88	.0282 8.35	.0260 10.02	.0241 11.92	.0216 15.22	.01733 24.6	.01411 37.5	.01096 61.3				
.0391 5.19	.0359 6.35	.0331 7.70	.0307 9.23	.0285 10.97	.0256 13.99	.0206 22.5	.01690 34.3	.01326 55.8				
.0453 4.82	.0417 5.90	.0385 7.14	.0357 8.56	.0332 10.17	.0299 12.95	.0242 20.8	.01996 31.6	.01578 51.1	.01234 82.4			
.0519 4.50	.0478 5.51	.0443 6.66	.0411 7.98	.0382 9.47	.0345 12.05	.0281 19.30	.0233 29.2	.01851 47.2	.01475 75.3			
.0591 4.22	.0545 5.16	.0505 6.24	.0469 7.47	.0437 8.86	.0395 11.26	.0323 18.01	.0268 27.2	.0215 43.8	.01707 70.0	.01421 99.9		
.0666 3.97	.0615 4.86	.0570 5.87	.0530 7.02	.0495 8.32	.0448 10.57	.0367 16.88	.0306 25.5	.0247 40.9	.01971 65.2	.01650 92.7		
.0746 3.75	.0690 4.58	.0640 5.54	.0596 6.63	.0550 7.85	.0504 9.97	.0415 15.89	.0347 23.9	.0281 38.3	.0225 61.0	.01895 86.5		
.0831 3.56	.0769 4.34	.0714 5.25	.0665 6.27	.0621 7.43	.0564 9.43	.0465 15.00	.0390 22.6	.0318 36.1	.0256 57.2	.0216 81.1	.01718 127.1	
.0921 3.38	.0852 4.12	.0792 4.98	.0733 5.95	.0690 7.05	.0627 8.94	.0518 14.21	.0436 21.3	.0355 34.1	.0288 53.9	.0244 76.2	.01952 119.2	.01869 129.6
.1014 3.22	.0940 3.93	.0874 4.74	.0815 5.67	.0763 6.71	.0694 8.50	.0574 13.50	.0484 20.3	.0396 32.3	.0322 51.0	.0274 72.0	.0220 112.3	.0212 122.0
.1113 3.07	.1031 3.75	.0960 4.53	.0895 5.40	.0839 6.40	.0764 8.10	.0634 12.85	.0535 19.27	.0438 30.7	.0358 48.4	.0305 68.2	.0247 106.1	.0237 115.2

f is the deflection of each active coil caused by a load P, a torsional stress of 100 000 pounds per square inch and a torsional modulus of 11 200 000

FOR SQUARE WIRE
Multiply f by .707
P by 1.2

The method for using this data is explained in the problems section on compression and extension spring design. Tables of spring characteristics are not included for the metric SI system, but spring design examples in that system are included, following those shown in the inch-pound method.

FIG. 115 (continued)

WIRE SIZE

OUTSIDE DIAMETER

SIZE W&M												19	18	17	16
DECIMAL			.022	.024	.026	.028	.030	.032	.034	.036	.038	.041	.0475	.054	.0625
25/64	.3906	f	.1733	.1571	.1434	.1318	.1216	.1127	.1049	.0980	.0918	.0836	.0695	.0589	.0483
		P	1.135	1.480	1.893	2.38	2.94	3.59	4.33	5.17	6.11	7.74	12.27	18.37	29.2
13/32	.4063	f	.1883	.1709	.1560	.1434	.1324	.1228	.1143	.1068	.1001	.0913	.0760	.0645	.0531
		P	1.088	1.420	1.815	2.28	2.82	3.44	4.15	4.95	5.85	7.41	11.73	17.56	27.9
27/64	.4219	f	.2039	.1851	.1691	.1555	.1436	.1332	.1242	.1160	.1088	.0993	.0828	.0703	.0580
		P	1.046	1.364	1.744	2.19	2.71	3.30	3.98	4.75	5.61	7.10	11.24	16.81	26.7
7/16	.4375	f	.220	.1999	.1827	.1680	.1553	.1441	.1343	.1256	.1178	.1075	.0898	.0764	.0631
		P	1.006	1.312	1.678	2.11	2.60	3.17	3.82	4.56	5.39	6.82	10.79	16.13	25.6
29/64	.4531	f	.237	.215	.1968	.1811	.1674	.1554	.1449	.1356	.1272	.1162	.0971	.0827	.0685
		P	.970	1.265	1.616	2.03	2.51	3.05	3.68	4.39	5.19	6.57	10.38	15.50	24.5
15/32	.4688	f		.231	.212	.1947	.1800	.1673	.1560	.1459	.1370	.1252	.1048	.0894	.0741
		P		1.220	1.559	1.956	2.42	2.94	3.55	4.23	5.00	6.33	9.99	14.91	23.6
31/64	.4844	f		.248	.227	.209	.1931	.1794	.1674	.1567	.1471	.1345	.1127	.0962	.0799
		P		1.179	1.506	1.889	2.33	2.84	3.43	4.09	4.83	6.10	9.64	14.37	22.7
1/2	.500	f			.243	.223	.207	.1920	.1792	.1678	.1575	.1441	.1209	.1033	.0859
		P			1.456	1.826	2.26	2.75	3.31	3.95	4.67	5.90	9.30	13.87	21.9
17/32	.5313	f			.276	.254	.235	.219	.204	.1911	.1796	.1645	.1382	.1183	.0987
		P			1.366	1.713	2.12	2.58	3.10	3.70	4.37	5.52	8.70	12.96	20.5
9/16	.5625	f				.286	.265	.247	.230	.216	.203	.1861	.1566	.1343	.1122
		P				1.613	1.991	2.42	2.92	3.48	4.11	5.19	8.18	12.16	19.17
19/32	.5938	f					.297	.277	.259	.242	.228	.209	.1762	.1514	.1267
		P					1.880	2.29	2.76	3.28	3.88	4.90	7.71	11.46	18.04
5/8	.625	f					.331	.308	.288	.270	.254	.233	.1969	.1693	.1420
		P					1.782	2.17	2.61	3.11	3.67	4.63	7.29	10.83	17.04
21/32	.6563	f						.342	.320	.300	.282	.259	.219	.1884	.1582
		P						2.06	2.48	2.95	3.49	4.40	6.92	10.27	16.14
11/16	.6875	f							.352	.331	.311	.286	.242	.208	.1753
		P							2.36	2.81	3.32	4.19	6.58	9.76	15.34
23/32	.7188	f								.363	.342	.314	.266	.230	.193
		P								2.68	3.17	3.99	6.27	9.31	14.6
3/4	.750	f									.374	.344	.291	.252	.21
		P									3.03	3.82	5.99	8.89	13.9
25/32	.7813	f										.375	.318	.275	.23
		P										3.66	5.74	8.50	13.3
13/16	.8125	f										.407	.346	.299	.25
		P										3.51	5.50	8.15	12.7
27/32	.8438	f											.374	.324	.27
		P											5.29	7.83	12.2

FIG. 115 (continued)

WIRE SIZE

15	14	13	3/32	12	11	1/8	10	9	5/32	8	7	3/16	6	5
.072	.080	.0915	.0938	.1055	.1205	.125	.135	.1483	.1563	.162	.177	.1875	.192	.207
.0395 46.0	.0338 64.8	.0275 100.5	.0264 109.2	.0216 161.7										
.0436 43.9	.0373 61.6	.0304 95.6	.0292 103.7	.0241 153.3										
.0477 41.9	.0410 58.8	.0335 91.0	.0322 98.8	.0266 145.7	.0212 228.									
.0521 40.1	.0448 56.3	.0367 86.9	.0353 94.3	.0293 138.9	.0234 217.	.0219 245.								
.0566 38.5	.0488 53.9	.0401 83.2	.0368 90.2	.0321 132.6	.0258 207.	.0242 234.								
.0614 37.0	.0530 51.7	.0437 79.7	.0420 86.4	.0351 126.9	.0282 197.3	.0265 223.	.0232 289.							
.0663 35.6	.0573 49.7	.0473 76.6	.0456 83.0	.0382 121.7	.0308 188.8	.0290 213.	.0259 277.							
.0714 34.3	.0619 47.9	.0512 73.6	.0494 80.0	.0414 116.9	.0335 181.1	.0316 205.	.0277 265.							
.0822 31.9	.0714 44.6	.0593 68.4	.0572 74.1	.0482 108.3	.0393 167.3	.0371 188.8	.0327 244.	.0277 335.						
.0937 29.9	.0816 41.7	.0680 63.9	.0657 69.1	.0555 100.9	.0455 155.5	.0430 175.3	.0380 226.	.0325 309.	.0296 369.	.0278 417.				
.1061 28.1	.0926 39.1	.0774 60.0	.0748 64.8	.0634 94.4	.0522 145.2	.0493 163.6	.0437 211.	.0375 288.	.0344 343.	.0323 386.				
.1191 26.5	.1041 36.9	.0873 56.4	.0844 61.0	.0718 88.7	.0593 136.2	.0561 153.4	.0499 197.1	.0430 269.	.0395 320.	.0371 360.	.0318 486.			
.1330 25.1	.1164 34.9	.0978 53.3	.0946 57.6	.0807 83.7	.0668 128.3	.0634 144.3	.0565 185.3	.0488 252.	.0449 300.	.0423 337.	.0364 454.	.0329 552.		
.1476 23.8	.1294 33.1	.1089 50.5	.1054 54.6	.0901 79.2	.0748 121.2	.0710 136.3	.0634 174.0	.0550 238.	.0507 282.	.0478 317.	.0413 426.	.0374 518.	.0359 561.	
.1630 22.7	.1431 31.5	.1206 48.0	.1168 51.9	.1000 75.2	.0833 114.9	.0791 129.2	.0708 165.9	.0616 225.	.0568 267.	.0537 299.	.0465 402.	.0424 487.	.0406 527.	.0355 681.
.1791 21.6	.1574 30.0	.1329 45.7	.1288 49.4	.1105 71.5	.0923 109.2	.0877 122.7	.0786 157.1	.0685 213.	.0633 252.	.0598 283.	.0520 380.	.0473 460.	.0455 498.	.0400 641.
.1960 20.7	.1724 28.7	.1459 43.6	.1413 47.1	.1214 68.2	.1017 104.0	.0967 116.9	.0868 149.5	.0758 202.	.0701 240.	.0664 269.	.0579 360.	.0528 436.	.0508 472.	.0447 607.
.214 19.80	.1881 27.5	.1594 41.7	.1545 45.1	.1329 65.2	.1115 99.3	.1061 111.5	.0954 142.6	.0834 192.9	.0773 228.	.0733 256.	.0640 343.	.0584 414.	.0563 448.	.0497 575.
.232 18.99	.205 26.3	.1735 40.0	.1682 43.2	.1449 62.4	.1218 95.0	.1159 106.7	.1044 136.3	.0915 184.2	.0849 218.	.0805 244.	.0705 327.	.0644 395.	.0621 426.	.0550 547.

FIG. 115 (continued)

WIRE SIZE

OUTSIDE DIAMETER

SIZE W&M			17	16	15	14	13	3/32	12	11	1/8	10	9	5/32
DECIMAL			.054	.0625	.072	.080	.0915	.0938	.1055	.1205	.125	.135	.1483	.1563
7/8	.875	f	.350	.296	.251	.222	.1882	.1825	.1574	.1325	.1262	.1138	.0999	.0928
		P	7.53	11.80	18.26	25.3	39.4	41.5	59.9	91.1	102.3	130.5	176.3	209.
29/32	.9063	f	.377	.320	.271	.239	.204	.1974	.1705	.1438	.1370	.1236	.1087	.1010
		P	7.26	11.36	17.57	24.3	36.9	39.9	57.6	87.5	98.2	125.2	169.0	199.9
15/16	.9375	f	.405	.344	.292	.258	.219	.213	.1841	.1554	.1479	.1338	.1178	.1096
		P	7.00	10.96	16.94	23.5	35.6	38.4	55.4	84.1	94.4	120.4	162.3	191.9
31/32	.9688	f	.435	.369	.313	.277	.236	.229	.1982	.1675	.1598	.1445	.1273	.1183
		P	6.76	10.58	16.35	22.6	34.3	37.0	53.4	81.0	90.9	115.9	156.1	184.5
1	1.000	f	.465	.395	.336	.297	.253	.246	.213	.1801	.1718	.1555	.1372	.1278
		P	6.54	10.23	15.80	21.9	33.1	35.8	51.5	78.1	87.6	111.7	150.4	177.6
1 1/32	1.031	f	.496	.421	.359	.317	.271	.263	.228	.1931	.1843	.1669	.1474	.1374
		P	6.33	9.90	15.28	21.1	32.0	34.6	49.8	75.5	84.6	107.8	145.1	171.3
1 1/16	1.063	f	.529	.449	.382	.338	.289	.281	.244	.207	.1972	.1788	.1580	.1474
		P	6.13	9.59	14.80	20.5	31.0	33.5	48.2	73.0	81.8	104.2	140.1	165.4
1 3/32	1.094	f		.477	.407	.360	.308	.299	.260	.221	.211	.1910	.1691	.1578
		P		9.30	14.34	19.83	30.0	32.4	46.7	70.6	79.2	100.8	135.5	159.9
1 1/8	1.125	f		.507	.432	.383	.328	.318	.277	.235	.224	.204	.1804	.1685
		P		9.02	13.92	19.24	29.1	31.4	45.2	68.4	76.7	97.6	131.2	154.7
1 5/32	1.156	f		.537	.458	.406	.348	.338	.294	.250	.239	.217	.1921	.1795
		P		8.76	13.52	18.69	28.2	30.5	43.9	66.3	74.4	94.6	127.1	149.9
1 3/16	1.188	f		.568	.485	.431	.368	.358	.311	.265	.254	.231	.204	.1908
		P		8.52	13.14	18.15	27.5	29.6	42.6	64.4	72.1	91.7	123.3	145.4
1 7/32	1.219	f		.600	.513	.455	.389	.379	.330	.281	.269	.244	.217	.203
		P		8.29	12.78	17.66	26.7	28.8	41.4	62.5	70.1	89.1	119.7	141.0
1 1/4	1.250	f		.623	.541	.480	.412	.400	.349	.297	.284	.258	.230	.215
		P		8.07	12.44	17.19	26.0	28.0	40.3	60.8	68.2	86.6	116.2	137.0
1 9/32	1.281	f			.570	.506	.434	.422	.368	.314	.300	.273	.243	.227
		P			12.13	16.74	25.3	27.3	39.2	59.2	66.3	84.3	113.1	133.2
1 5/16	1.313	f			.600	.533	.457	.444	.387	.331	.317	.288	.256	.240
		P			11.81	16.31	24.6	26.6	38.2	57.7	64.6	82.0	110.1	129.7
1 11/32	1.344	f			.630	.560	.481	.467	.408	.348	.334	.304	.270	.253
		P			11.53	15.91	24.0	25.9	37.2	56.2	62.9	79.9	107.2	126.2
1 3/8	1.375	f			.662	.588	.506	.491	.429	.367	.351	.320	.285	.267
		P			11.25	15.53	23.4	25.3	36.3	54.8	61.4	77.9	104.4	123.0
1 13/32	1.406	f			.694	.616	.530	.516	.450	.385	.368	.336	.299	.281
		P			10.99	15.17	22.9	24.7	35.4	53.4	59.9	76.0	101.8	119.9
1 7/16	1.438	f			.727	.647	.556	.540	.472	.404	.387	.353	.314	.295
		P			10.73	14.81	22.3	24.1	34.6	52.2	58.4	74.1	99.4	117.0

FIG. 115 (continued)

WIRE SIZE

8	7	3/16	6	5	7/32	4	3	1/4	2	9/32	0	5/16	2-0	11/32	3-0
.162	.177	.1875	.192	.207	.2188	.2253	.2437	.250	.2625	.2813	.3065	.3125	.331	.3438	.3625
0880 234.	.0772 312.	.0707 377.	.0682 407.	.0605 521.	.0552 626.	.0526 691.	.0459 900.								
0959 224.	.0843 299.	.0772 360.	.0746 389.	.0663 498.	.0606 598.	.0577 660.	.0505 858.								
1041 215.	.0917 286.	.0842 345.	.0812 373.	.0723 477.	.0662 572.	.0632 631.	.0554 819.	.0530 893.							
1127 207.	.0994 275.	.0913 332.	.0882 358.	.0786 457.	.0721 548.	.0688 604.	.0605 784.	.0580 854.	.0533 1006.						
1216 98.8	.1074 264.	.0986 319.	.0954 344.	.0852 439.	.0783 526.	.0747 580.	.0658 751.	.0631 818.	.0581 963.	.0515 1216.					
1308 91.6	.1157 255.	.1065 307.	.1029 331.	.0921 423.	.0845 506.	.0809 557.	.0714 721.	.0685 786.	.0632 924.	.0561 1165.					
1404 85.0	.1243 246.	.1145 296.	.1107 319.	.0993 407.	.0913 487.	.0873 537.	.0772 694.	.0740 755.	.0684 888.	.0609 1119.	.0523 1496.				
.1503 78.8	.1332 238.	.1229 286.	.1188 308.	.1066 393.	.0982 470.	.0939 517.	.0832 668.	.0799 727.	.0739 855.	.0657 1076.	.0567 1437.	.0548 1533.			
.1604 73.0	.1424 230.	.1315 276.	.1272 298.	.1142 379.	.1053 454.	.1008 499.	.0894 645.	.0859 702.	.0795 824.	.0710 1036.	.0613 1382.	.0592 1474.			
.1711 67.6	.1520 222.	.1404 267.	.1359 288.	.1221 367.	.1127 439.	.1079 483.	.0960 622.	.0922 677.	.0854 795.	.0763 1000.	.0661 1331.	.0639 1420.	.0577 1725.		
.1812 162.4	.1620 215.	.1496 259.	.1448 279.	.1303 355.	.1203 424.	.1153 467.	.1025 602.	.0986 655.	.0915 768.	.0819 965.	.0710 1284.	.0687 1369.	.0622 1663.	.0581 1891.	
.1934 157.6	.1721 209.	.1590 251.	.1541 271.	.1388 344.	.1282 411.	.1229 452.	.1094 583.	.1053 634.	.0978 743.	.0876 932.	.0762 1240.	.0737 1322.	.0668 1604.	.0625 1824.	
.205 153.1	.1824 203.	.1690 244.	.1635 263.	.1474 334.	.1363 399.	.1308 438.	.1165 565.	.1122 614.	.1042 719.	.0936 902.	.0815 1199.	.0789 1278.	.0716 1550.	.0670 1761.	
.217 148.9	.1932 197.2	.1791 237.	.1733 255.	.1562 324.	.1449 387.	.1388 425.	.1240 547.	.1193 595.	.1110 697.	.0997 874.	.0870 1160.	.0843 1237.	.0765 1498.	.0717 1702.	.0653 2035.
.229 144.7	.205 191.6	.1894 230.	.1836 248.	.1657 315.	.1535 376.	.1472 413.	.1316 532.	.1268 577.	.1178 677.	.1052 848.	.0926 1124.	.0898 1198.	.0816 1451.	.0766 1647.	.0698 1968.
.242 141.0	.216 186.5	.1999 224.	.1938 241.	.1752 306.	.1623 365.	.1556 402.	.1393 517.	.1343 561.	.1250 657.	.1127 822.	.0984 1091.	.0954 1162.	.0869 1406.	.0816 1596.	.0745 1906.
.255 137.3	.227 181.7	.211 218.	.204 235.	.1848 298.	.1713 356.	.1650 391.	.1472 502.	.1420 546.	.1324 638.	.1194 799.	.1046 1058.	.1014 1127.	.0924 1364.	.0867 1548.	.0794 1846.
.268 133.9	.239 177.1	.222 212.	.215 229.	.1948 291.	.1809 346.	.1737 380.	.1557 489.	.1499 530.	.1399 620.	.1262 770.	.1107 1028.	.1074 1095.	.0980 1325.	.0922 1501.	.0843 1791.
.282 130.6	.252 172.6	.234 207.	.227 223.	.205 283.	.1905 337.	.1829 371.	.1641 476.	.1583 517.	.1476 605.	.1332 756.	.1170 1000.	.1136 1065.	.1038 1286.	.0977 1459.	.0895 1740.

FIG. 115 (continued)

WIRE SIZE

OUTSIDE DIAMETER			3/32	12	11	1/8	10	9	5/32	8	7	3/16	6	5
SIZE W&M			3/32	12	11	1/8	10	9	5/32	8	7	3/16	6	5
DECIMAL			.0938	.1055	.1205	.125	.135	.1483	.1563	.162	.177	.1875	.192	.207
1 15/32	1.469	f	.565	.494	.423	.405	.370	.329	.310	.296	.265	.250	.238	.216
		P	23.6	33.8	51.0	57.1	72.4	97.1	114.2	127.5	168.5	202.	218.	276.
1 1/2	1.500	f	.591	.517	.443	.424	.387	.350	.324	.310	.277	.258	.250	.227
		P	23.1	33.1	49.8	55.8	70.8	94.8	111.5	124.5	164.6	197.1	213.	269.
1 9/16	1.563	f	.645	.565	.484	.464	.424	.378	.355	.340	.305	.283	.275	.249
		P	22.1	31.7	47.7	53.3	67.7	90.6	106.6	118.9	157.1	188.3	203.	257.
1 5/8	1.625	f	.701	.614	.527	.505	.461	.413	.387	.370	.332	.309	.300	.273
		P	21.2	30.3	45.7	51.1	64.8	86.7	102.0	113.9	150.3	180.0	193.9	246.
1 11/16	1.688	f	.760	.666	.572	.548	.501	.448	.421	.403	.362	.337	.327	.297
		P	20.3	29.1	43.9	49.1	62.2	83.2	97.9	109.2	144.1	172.6	185.8	235.
1 3/4	1.750	f	.820	.720	.619	.593	.542	.485	.456	.437	.392	.366	.355	.323
		P	19.58	28.0	42.2	47.2	59.8	80.0	94.0	104.9	138.5	165.6	178.4	226.
1 13/16	1.813	f	.884	.775	.667	.639	.585	.526	.492	.472	.424	.395	.384	.350
		P	18.86	27.0	40.6	45.4	57.6	77.0	90.5	100.8	133.1	159.3	171.4	217.
1 7/8	1.875	f	.948	.833	.717	.687	.629	.564	.530	.508	.457	.426	.414	.377
		P	18.20	26.1	39.2	43.8	55.5	74.2	87.2	97.3	128.2	153.4	165.1	209.
1 15/16	1.938	f		.892	.769	.738	.676	.605	.569	.546	.492	.458	.446	.405
		P		25.2	37.8	42.3	53.6	71.6	84.2	93.8	123.6	147.9	159.2	201.
2	2.000	f		.955	.823	.789	.723	.649	.610	.585	.527	.492	.478	.436
		P		24.3	36.6	40.9	51.8	69.2	81.3	90.6	119.4	142.8	153.7	194.3
2 1/16	2.063	f		1.018	.878	.843	.768	.693	.652	.626	.564	.526	.512	.467
		P		23.6	35.4	39.6	50.1	66.9	78.7	87.6	115.4	138.1	148.5	187.7
2 1/8	2.125	f			.936	.898	.823	.739	.696	.667	.602	.562	.546	.499
		P			34.3	38.3	48.5	64.8	76.1	84.9	111.8	133.6	143.8	181.6
2 3/16	2.188	f			.995	.955	.876	.786	.740	.711	.641	.598	.582	.532
		P			33.3	37.2	47.1	62.8	73.8	82.2	108.3	129.5	139.2	175.8
2 1/4	2.250	f			1.056	1.013	.930	.835	.787	.755	.681	.637	.619	.566
		P			32.3	36.1	45.7	60.9	71.6	79.8	105.7	125.5	135.0	170.5
2 5/16	2.313	f			1.119	1.074	.986	.886	.834	.801	.723	.676	.657	.601
		P			31.4	35.1	44.4	59.2	69.5	77.5	101.9	121.8	131.0	165.4
2 3/8	2.375	f			1.184	1.136	1.043	.938	.884	.848	.763	.716	.696	.637
		P			30.5	34.1	43.1	57.5	67.6	75.3	99.1	118.3	127.3	160.7
2 7/16	2.438	f				1.201	1.102	.991	.934	.897	.810	.757	.737	.674
		P				33.2	42.0	56.0	65.7	73.2	96.3	115.1	123.7	156.1
2 1/2	2.500	f				1.266	1.162	1.046	.986	.946	.855	.800	.778	.713
		P				32.3	40.9	54.5	64.0	71.3	93.7	111.6	120.4	151.9
2 5/8	2.625	f					1.288	1.160	1.094	1.050	.950	.889	.865	.792
		P					38.8	51.7	60.7	67.7	88.6	106.2	114.2	144.0

FIG. 115 (continued)

WIRE SIZE

4	3	¼	2	9/32	0	5/16	2–0	11/32	3–0	3/8	4–0	13/32	7/16	15/32	½
.2253	.2437	.250	.2625	.2813	.3065	.3125	.331	.3438	.3625	.375	.3938	.4063	.4375	.4688	.500
.1924 361.	.1728 464.	.1667 504.	.1554 589.	.1407 736.	.1235 973.	.1199 1036.	.1097 1251.	.1033 1419.	.0964 1691.	.0895 1893.	.0823 2231.	.0780 2478.			
.202 352.	.1815 452.	.1754 499.	.1612 574.	.1482 717.	.1305 947.	.1267 1008.	.1158 1218.	.1090 1381.	.1002 1643.	.0947 1841.	.0871 2168.	.0826 2408.	.0724 3093.		
.223 336.	.200 431.	.1934 468.	.1807 547.	.1636 682.	.1444 901.	.1403 958.	.1286 1156.	.1212 1309.	.1114 1558.	.1055 1743.	.0974 2051.	.0922 2279.	.0812 2923.		
.244 321.	.220 411.	.212 446.	.1986 521.	.1801 650.	.1592 858.	.1547 912.	.1419 1101.	.1339 1246.	.1234 1481.	.1169 1657.	.1079 1948.	.1026 2161.	.0905 2768.	.0799 3499.	
.266 307.	.240 394.	.232 427.	.217 499.	.1971 622.	.1745 819.	.1697 871.	.1560 1049.	.1474 1188.	.1359 1411.	.1290 1577.	.1192 1853.	.1133 2056.	.1002 2631.	.0889 3318.	
.290 295.	.261 377.	.253 409.	.237 477.	.215 595.	.1908 783.	.1856 833.	.1707 1004.	.1613 1135.	.1491 1347.	.1414 1506.	.1310 1768.	.1247 1959.	.1105 2504.	.0982 3158.	.0877 3927.
.314 283.	.283 362.	.274 393.	.257 458.	.234 571.	.208 751.	.202 799.	.1861 961.	.1761 1086.	.1627 1290.	.1547 1440.	.1435 1690.	.1365 1876.	.1212 2391.	.1081 3010.	.0967 3739.
.339 272.	.306 348.	.296 378.	.278 440.	.253 548.	.225 721.	.219 767.	.202 922.	.1912 1042.	.1771 1236.	.1683 1381.	.1562 1619.	.1490 1793.	.1326 2287.	.1183 2877.	.1061 3570.
.365 262.	.331 335.	.320 364.	.300 425.	.273 528.	.243 693.	.237 737.	.219 886.	.207 1001.	.1920 1187.	.1827 1325.	.1698 1553	.1618 .1720.	.1442 2192.	.1291 2754.	.1160 3414.
.392 253.	.355 324.	.344 351.	.323 409.	.295 509.	.263 668.	.256 710.	.236 853.	.224 964.	.208 1142.	.1975 1274.	.1837 1493.	.1754 1653.	.1566 2104.	.1402 2642.	.1262 3273.
.421 245.	.381 312.	.369 339.	.346 395.	.316 491.	.282 644.	.275 685.	.254 822.	.241 928.	.224 1100.	.213 1227.	.1984 1437.	.1893 1591.	.1693 2023.	.1520 2538.	.1371 3141.
.449 236.	.407 302.	.395 327.	.371 381.	.339 474.	.303 622.	.295 661.	.273 794.	.259 .896	.240 1061.	.229 1183.	.213 1385.	.204 1532.	.1826 1948.	.1641 2443.	.1482 3021.
.479 229.	.435 292.	.421 317.	.396 369.	.362 459.	.324 601.	.316 639.	.292 767.	.277 866.	.258 1025.	.246 1142.	.229 1337.	.219 1479.	.1963 1879.	.1768 2353.	.1598 2908.
.511 222.	.463 283.	.449 307.	.423 357.	.387 444.	.346 582.	.337 618.	.312 742.	.296 837.	.276 991.	.263 1105.	.245 1292.	.235 1428.	.211 1814.	.1898 2271.	.1718 2805.
.542 215.	.493 275.	.478 298.	.449 347.	.411 430.	.368 564.	.359 599.	.333 719.	.316 811.	.294 959.	.281 1069.	.262 1250.	.251 1382.	.225 1754.	.203 2194.	.1844 2708.
.576 209.	.523 267.	.507 289.	.477 336.	.437 417.	.392 547.	.382 581.	.354 697.	.337 786.	.314 929.	.299 1036.	.280 1210.	.268 1338.	.241 1697.	.217 2122.	.1972 2618.
.609 203.	.554 259.	.537 281.	.506 327.	.464 405.	.416 531.	.405 564.	.376 676.	.358 762.	.333 901.	.318 1004.	.298 1173.	.285 1297.	.256 1644.	.232 2054.	.211 2533
.644 97.5	.586 252.	.568 273.	.536 317.	.491 394.	.441 516.	.430 548.	.399 657.	.379 741.	.354 875.	.338 975.	.316 1139.	.303 1258.	.273 1594.	.247 1992.	.224 2455.
.717 87.2	.653 239.	.633 259.	.597 301.	.548 373.	.492 488.	.480 518.	.446 621.	.425 700.	.396 826.	.379 920.	.355 1075.	.340 1187.	.307 1503.	.278 1876.	.253 2310.

FIG. 115 (continued)

WIRE SIZE

OUTSIDE DIAMETER			9	5/32	8	7	3/16	6	5	7/32	4	3	1/4
SIZE W&M			9	5/32	8	7	3/16	6	5	7/32	4	3	1/4
DECIMAL			.1483	.1563	.162	.177	.1875	.192	.207	.2188	.2253	.2437	.250
2¾	2.750	f	1.280	1.208	1.159	1.049	.983	.956	.876	.821	.794	.723	.701
		P	49.2	57.8	64.4	84.6	101.0	108.6	137.0	162.4	177.9	227.	246.
2⅞	2.875	f	1.406	1.327	1.274	1.154	1.081	1.052	.965	.904	.874	.797	.773
		P	47.0	55.1	61.7	80.7	96.3	103.6	130.5	154.8	169.5	216.	234.
3	3.000	f		1.452	1.394	1.263	1.184	1.152	1.057	.992	.959	.874	.849
		P		52.7	58.7	77.1	92.0	99.0	124.7	147.8	161.9	206.	223.
3⅛	3.125	f			1.520	1.378	1.291	1.257	1.154	1.083	1.047	.955	.927
		P			56.3	73.9	88.1	94.8	119.4	141.5	154.9	197.2	214.
3¼	3.250	f				1.497	1.404	1.366	1.255	1.178	1.139	1.040	1.010
		P				70.9	84.5	90.9	114.5	135.6	148.5	189.0	205.
3⅜	3.375	f				1.621	1.520	1.480	1.360	1.277	1.235	1.128	1.096
		P				68.1	81.2	87.3	109.9	130.3	142.6	181.5	196.4
3½	3.500	f					1.643	1.589	1.469	1.379	1.336	1.220	1.185
		P					78.1	84.0	105.8	125.3	137.2	174.5	188.9
3⅝	3.625	f					1.768	1.723	1.583	1.487	1.439	1.316	1.278
		P					75.3	81.0	101.9	120.7	132.1	168.1	181.9
3¾	3.750	f						1.850	1.701	1.599	1.548	1.415	1.374
		P						78.1	98.3	116.4	127.4	162.1	175.4
3⅞	3.875	f							1.822	1.714	1.658	1.517	1.474
		P							95.0	112.4	123.1	156.5	169.3
4	4.000	f							1.950	1.833	1.774	1.624	1.578
		P							91.8	108.7	119.0	151.3	163.7
4¼	4.250	f								2.08	2.02	1.847	1.795
		P								102.0	111.6	141.8	153.5
4½	4.500	f									2.28	2.08	2.03
		P									105.1	133.5	144.4
4¾	4.750	f										2.34	2.27
		P										126.1	136.4
5	5.000	f											2.53
		P											129.2
5¼	5.250	f											
		P											
5½	5.500	f											
		P											
5¾	5.750	f											
		P											
6	6.000	f											
		P											

FIG. 115 (continued)

WIRE SIZE

2	9/32	0	5/16	2-0	11/32	3-0	3/8	4-0	13/32	7/16	15/32	1/2
.2625	.2813	.3065	.3125	.331	.3438	.3625	.375	.3938	.4063	.4375	.4688	.500
.662 286.	.608 354.	.547 463.	.534 491.	.496 589.	.472 663.	.441 783.	.422 872.	.395 1018.	.379 1124.	.343 1422.	.311 1773.	.284 2182.
.730 272.	.671 337.	.604 440.	.590 467.	.548 560.	.523 631.	.489 744.	.468 828.	.438 967.	.421 1067.	.381 1349.	.346 1681.	.317 2067.
.801 260.	.737 322.	.664 420.	.649 446.	.604 534.	.576 601.	.539 709.	.515 789.	.484 920.	.465 1015.	.421 1283.	.383 1598.	.351 1964.
.876 248.	.807 307.	.727 401.	.710 426.	.662 510.	.631 574.	.591 677.	.566 753.	.531 878.	.510 969.	.463 1223.	.422 1523.	.387 1870.
.954 238.	.879 294.	.793 384.	.775 408.	.722 488.	.689 549.	.645 648.	.618 720.	.581 840.	.558 926.	.507 1169.	.463 1455.	.424 1785.
1.036 228.	.955 283.	.862 369.	.842 391.	.785 468.	.750 527.	.703 621.	.673 690.	.633 804.	.609 887.	.553 1119.	.505 1392.	.464 1708.
1.120 219.	1.033 272.	.933 354.	.912 376.	.851 449.	.813 506.	.762 596.	.731 .663	.687 .772	.661 .851.	.602 1073.	.550 1335.	.505 1636.
1.209 211.	1.115 261.	1.008 341.	.986 362.	.919 432.	.878 486.	.824 573.	.790 637.	.744 742.	.715 818.	.651 1031.	.596 1282.	.548 1571.
1.301 204.	1.200 252.	1.085 328.	1.061 349.	.991 417.	.946 469.	.888 552.	.852 614.	.802 715.	.772 788.	.704 993.	.644 1233.	.592 1511.
1.395 196.7	1.288 243.	1.166 317.	1.139 336.	1.064 402.	1.017 452.	.955 532.	.916 592.	.863 689.	.830 759.	.758 956.	.694 1188.	.639 1455.
1.493 190.0	1.379 235.	1.249 306.	1.221 325.	1.141 388.	1.091 437.	1.025 514.	.983 571.	.926 665.	.892 733.	.814 923.	.746 1146.	.687 1403.
1.700 178.1	1.570 220.	1.424 287.	1.392 304.	1.302 363.	1.245 409.	1.170 481.	1.123 535.	1.059 622.	1.020 685.	.932 862.	.856 1070.	.789 1309.
1.920 167.6	1.775 207.	1.610 270.	1.574 286.	1.473 342.	1.409 384.	1.325 452.	1.273 502.	1.201 584.	1.157 643.	1.058 809.	.972 1003.	.898 1227.
2.15 158.3	1.991 195.6	1.807 255.	1.768 270.	1.655 322.	1.584 362.	1.490 426.	1.432 474.	1.351 551.	1.303 606.	1.192 762.	1.097 945.	1.013 1155.
2.40 149.9	2.22 185.2	2.02 241.	1.973 256.	1.847 305.	1.769 343.	1.664 403.	1.600 448.	1.512 521.	1.457 573.	1.335 721.	1.228 893.	1.136 1091.
2.66 142.4	2.47 175.6	2.24 229.	2.19 243.	2.05 290.	1.964 325.	1.849 383.	1.778 425.	1.680 494.	1.620 544.	1.485 683.	1.368 846.	1.266 1034.
	2.72 167.5	2.47 218.	2.42 231.	2.26 276.	2.17 310.	2.04 364.	1.965 404.	1.857 470.	1.792 517.	1.640 649.	1.504 804.	1.403 982.
		2.71 208.	2.65 220.	2.49 263.	2.38 295.	2.25 347.	2.16 385.	2.04 448.	1.972 493.	1.810 619.	1.669 766.	1.546 935.
			2.90 211.	2.72 251.	2.61 282.	2.46 332.	2.35 368.	2.24 428.	2.16 471.	1.984 591.	1.830 731.	1.697 893.

SPRING DESIGN

11 Compression Springs

Introduction

The design of springs, once considered an art requiring "know-how" and often done empirically or by guesswork, is actually a science that can be performed by a strict engineering discipline. The use of slide rules, charts, tables, and electronic calculators considerably reduces the laborious process of repetitive design procedures which often required several trial-and-error sets of calculations before a satisfactory design could be evolved. The principal characteristics important in spring design are the force, deflection, and stress relationship. Other important factors which must be determined are the solid height, types of materials, types of ends, and similar properties. The design of a spring may be arrived at by several routes—the methods described here should aid those in search of an optimum design by the easiest, shortest, and least painful means. However, competence in spring design, regardless of the method used, still requires a designer to have a working acquaintance with the types of materials available, recommended design stresses, manufacturing methods, and be able to use a slide rule, electronic calculator, or have access to a computer.

MATERIALS

Prior data covering the generally used spring materials are essential for a designer to have at his fingertips. Music wire, hard-drawn MB, oil-tempered MB, and stainless steels are the most popular. Alloy steels such as chromium-silicon and chromium-vanadium are often used. Copper-base alloys such as beryllium-copper, phosphor-bronze, and spring brass are occasionally required, as are the proprietary nickel-base alloys such as Inconel, Monel, Ni-Span C, and so on—as described in Part I.

DESIGN METHODS

There are no secrets in spring design. The methods now available for designing springs have evolved from early trial-and-error observations all the way up to preprogrammed computers. However, all methods descend from the theoretical engineering relationships between strength of materials and stress and strain. From these relationships, basic spring design formulas are derived. Solving these formulas for different spring dimensions provides data for tables of spring characteristics. It is then a relatively easy matter to transfer the tabular data into nomographs, charts, or curves and onto special spring-design slide rules—and, finally, to have it stored in the memory system of a computer so that a series of complete spring design specifications can be printed quickly.

Compression Springs

Compression springs (Fig. 116) are open-wound helical springs that exert a load or force when compressed. They may be cylindrical, conical, barrel shaped, or concave in form, with ends left plain or ground. The least expensive and most

FIG. 116. Compression Springs, Types of Ends

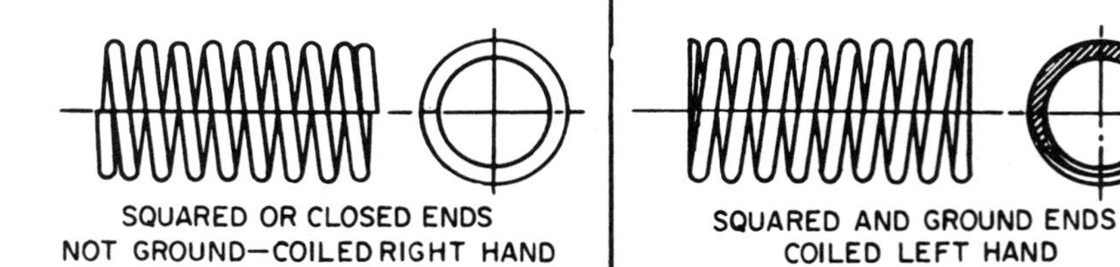

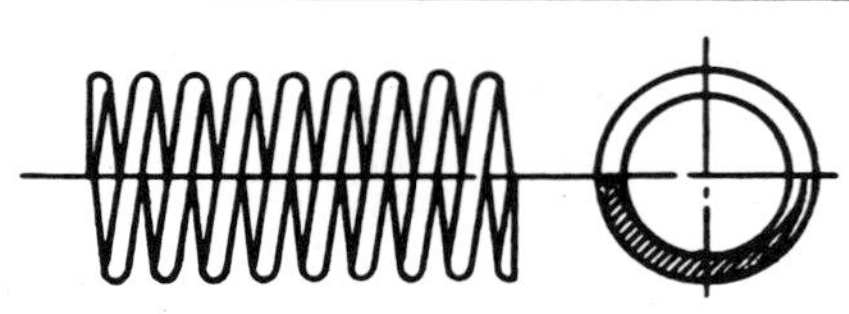

Courtesy U.S. Dept. of Defense, MIL-STD-29A.

FIG. 117. Compression Spring Formulae for Dimensional Characteristics

Dimensional Characteristics	TYPE OF ENDS			
	Open or Plain (not ground)	Open or Plain with ends ground	Squared or Closed (not ground)	Closed and Ground
Pitch (p)	$\frac{FL-d}{N}$	$\frac{FL}{TC}$	$\frac{FL-3d}{N}$	$\frac{FL-2d}{N}$
Solid Height (SH)	(TC + 1)d	TCxd	(TC + 1)d	TCxd
Active Coils (N)	N = TC or $\frac{FL\ d}{p}$	N = TC − 1 or $\frac{FL}{p} - 1$	N = TC-2x or $\frac{FL-3d}{p}$	N = TC-2x or $\frac{FL-2d}{p}$
Total Coils (TC)	$\frac{FL-d}{p}$	$\frac{FL}{p}$	$\frac{FL-3d}{p} + 2$	$\frac{FL-2d}{p} + 2$
Free Length (FL)	(p × TC) + d	pxTC	(p × N) + 3d	(p × N) + 2d

Courtesy Department of Defense, MIL-STD-29A.

used type of end, where the index D/d is above 10, is with closed ends, not ground; this type is especially satisfactory for light wire sizes under 1/32 in. (0.80 mm) in diameter. The second most useful type, especially where small load tolerances or minimum buckling is required, is the "squared and ground"; this type also provides more uniform loading and will stand upright without tipping. The plain ends and the plain ends ground are seldom used, as they often tangle together during shipping; they cannot stand upright and do not exert uniform loads. For compression spring formulas, see Figs. 117 and 118.

A study of the methods for designing springs, using the same spring in each method, was made to determine which methods were the easiest and quickest to use. Compression springs were selected first, as they represent 80 to 90 percent of all springs made by many spring companies. Spring manufacturers often receive blueprints with forces and deflections, and most of the limiting dimensions, and the spring manufacturer is expected to select a suitable spring material, wire size, calculate the stress, and determine the number of coils. A typical example is shown in Fig. 119.

PROBLEM 1.

A compression spring (Fig. 119) has the following specifications: OD = 1 in., free length = 3¼ in., Force = 60 lb at 1¾ in. compressed height (1½ in. deflection from free length), solid height = 1 5/8 in. max., ends closed and ground. Determine the wire diameter d, stress S, number of active coils N, etc., using oil-tempered MB spring steel, ASTM A229.

FIG. 118. Design Formulas for Compression and Extension Springs

Property	Round wire	Square wire
d = wire diameter, t = side of square	$\sqrt[3]{\frac{2.55\ PD}{S}}$ $\frac{\pi SND^2}{GF}$	$\sqrt[3]{\frac{2.40\ PD}{S}}$ $\frac{2.32\ SND^2}{GF}$
S = stress torsional	$\frac{PD}{0.393d^3}$ $\frac{GdF}{\pi ND^2}$	$\frac{PD}{0.416t^3}$ $\frac{GtF}{2.32ND^2}$
N = active coils	$\frac{GdF}{\pi SD^2}$ $\frac{Gd^4F}{8PD^3}$	$\frac{GtF}{2.32SD^2}$ $\frac{Gt^4F}{5.58PD^3}$
F = deflection	$\frac{\pi SND^2}{Gd}$ $\frac{8PND^3}{Gd^4}$	$\frac{2.32SND^2}{Gt}$ $\frac{5.58PND^3}{Gd^4}$
P = force applied	$\frac{0.393Sd^3}{D}$ $\frac{Gd^4F}{8ND^3}$	$\frac{0.416St^3}{D}$ $\frac{Gt^4F}{5.58ND^3}$
S_{IT} = stress due to initial tension	$\frac{S}{P} \times IT$	$\frac{S}{P} \times IT$

D = mean diameter; G = modulus of elasticity in torsion (Fig. 112); 0.393 = $\pi/8$, t = side of square. All dimensional values are in inches or in millimetres, with stresses in pounds per square inch (psi) or in megapascals (MPa).

FIG. 119. Compression Spring

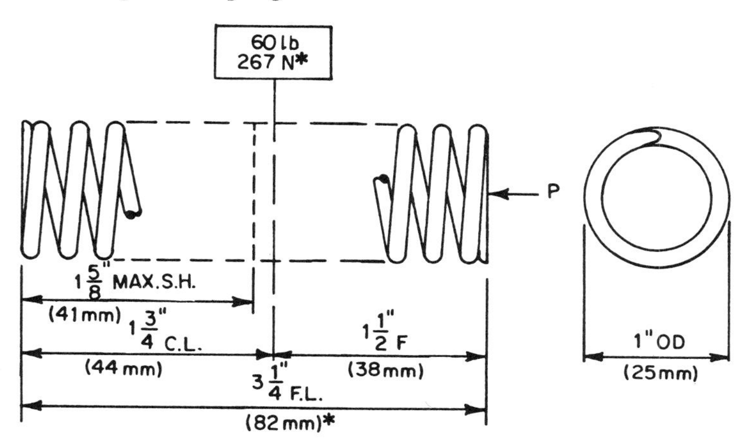

Method 1: By Tables

One of the easiest and quickest methods is by using the tables of spring characteristics as follows: locate 1 in. OD in the left column (Fig. 115), and to the right in a box about one third of the way across the table a force (P) of 78.1 lb appears (which is a little more than required, therefore safe) under a deflection (f) of 0.1801 in. (which is the deflection of one coil under a force of 78.1 lb with an uncorrected stress of 100 000 psi). At the top of that column a wire diameter of no. 11 (0.1205 in.) is found; knowing these values, the spring can be easily and quickly designed with the help of a slide rule or calculator. This is called the "percent method."

1. The stress at 60 lb is obtained as follows:

The 60-lb force divided by 78.1 × 100 = 76.8 percent (rounded off). Therefore the stress S at 60 lb = 76.8 percent of 100 000 = 76 800 psi.

2. The 76.8 percent figure is also used to determine the deflection per coil f: 76.8 percent of 0.1801 in. = 0.1383 in.

3. The number of active coils N = F/f = 1½ ÷ 0.1383 in. = 10.85 (say 11).

4. Total coils = AC + 2 = 11 + 2 = 13. Therefore a quick answer is 13 total coils of 0.1205 in. diameter wire. However, the design should be carried out further as follows:

5. Solid height = TC × d = 13 × 0.1205 in. = 1.57 in. (Fig. 117) (total deflection = FL – SH = 3.25 – 1.57 in. = 1.68 in.).

6. Stress solid = (76 800/1.5) × 1.68 = 86 000 psi.

7. Spring index = OD/d − 1 = 1/0.1205 − 1 = 7.3.
8. Curvature correction factor K = 1.21 (Fig. 114).
9. Total stress at 60 lb = 76 800 × 1.21 = 92 900 psi. [This is a safe design stress, as it is below the 110 000 psi recommended for 0.1205 in. oil-tempered MB shown on the middle curve of the recommended design stresses (Fig. 84)].
10. Total stress at solid = 86 000 × 1.21 = 104 000 psi. (This too is a safe stress, as it is below the 123 000 psi shown on the top curve (Fig. 84), therefore the spring will not take a permanent set.) The entire design has been completed by this quick and easy method without using complicated formulas; the only faster way is by using a programmable calculator or a computer.

Method 2: By Formulas

The same spring can be designed by formulas without using any tables, but this method requires assumptions for stress and mean diameter as follows:

1. Assume a safe stress well below the middle (or lowest curve) (Fig. 84) for oil-tempered MB steel, say 80 000 psi. (One could have selected 85 000 psi.)
2. Assume a mean diameter D slightly below the 1 in. OD, say 7/8 in. The value for G from the table for the modulus of elasticity (Fig. 112) is 11 200 000 psi.
3. A trial wire diameter d is then found from the formula provided in the table of formulas for compression and extension springs (Fig. 118) as follows:

$$d = \sqrt[3]{\frac{2.55PD}{S}} = \sqrt[3]{\frac{2.55 \times 60 \times 0.875}{80\,000}}$$

$$= \sqrt[3]{0.001\,673} = 0.1\,188 \text{ in. (say 0.120 in.)}$$

The nearest wire size is the same as the one used in the previous method, no. 11 (0.1205 in.).

Note: A short-cut method, to avoid solving for the cube root, is to use the table of cubes (Fig. 213), in which the closest number to 0.001 673 is 0.001 64, which is the cube of 0.118 in.; but 0.118 in. is not a standard wire gauge size, so use the nearest, which is no. 11 (0.1205 in.), the same as that determined by Method 1.

Then, calculate the stress S using the correct mean diameter D, which is now 1 in. − 0.1205 in. = 0.8795 in.

4. The stress S is

$$S = \frac{PD}{0.393d^3} \qquad \frac{(60 \times 0.8\,795)}{0.393 \times 0.1\,205^3} = 76\,740 \text{ psi},$$

which is about the same as in Method 1.

The number of active coils N (Fig. 118) is $N = \dfrac{GdF}{\pi SD^2}$

$$N = \frac{11\,200\,000 \times 0.1\,205 \times 1.5}{3.1\,416 \times 76\,740 \times .8\,795^2} = 10.85 \text{ (say 11)}$$

The answer is the same as found in Method 1. The total coils, solid height, total stresses, and other calculations are made exactly as in Method 1.

Method 3: Metric

In the metric system using SI units, the force in pounds is expressed in newtons:

1 lb = 4.448 22 newtons (N)

The dimensions in inches are supposed to be expressed in metres, but this becomes unwieldy, and it is much simpler to express them in millimetres:

1 in. = 25.4 millimetres (mm)

The stress should be expressed in pascals, but this too becomes unwieldy, and it is much easier to use megapascals (millions of pascals):

1 psi = 0.006 894 757 megapascals (MPa)

100 000 psi = 689.4757 MPa or roughly 689 MPa.

The modulus of elasticity can also be expressed in megapascals (Fig. 112).

G of 11 200 000 psi = 77 221.27 (say 77 200 MPa)

If the force and dimensions in Problem 1 are converted to exact values metrically, they become unwieldy. It is far better to "think metric" and use even figures wherever possible as follows:

d = wire diameter = 0.1205 in. × 25.4 = 3.0607 mm (say 3 mm)

OD = outside diameter = 1 in. × 25.4 = 25.4 mm (say 25 mm)

FL = free length = 3¼ in. × 25.4 = 82.55 mm (say 82 mm)

HUF = height under force = 1¾ in. × 25.4 = 44.45 mm (say 44 mm)

F = deflection = 1½ in. × 25.4 = 38.1 mm (say 38 mm)

SH = solid height, max. = 1⅝ in. × 25.4 = 41.25 mm (say 41 mm)

P = force = 60 lb × 4.448 22 = 266.89 N (say 267 N)

Problem 1 may be restated metrically as follows: A compression spring has the following specifications: OD = 25 mm, free length = 82 mm, force = 267 N at 44 mm compressed height (38 mm deflection from free length, solid height = 41 mm max., ends closed and ground. Determine the wire diameter d, stress S, number of active coils, etc.

Method 4: Metrics

If tables of spring characteristics are not available, it is necessary to use the formulas as in Method 2 using metric (SI) units, and making similar assumptions for stress and mean diameter as follows:

1. Assume a safe stress well below the middle (or lowest curve) for oil-tempered MB steel (Fig. 84) as before, say 550 MPa (600 MPa could have been selected).

2. Assume a mean diameter D slightly below the 25-mm OD, say 22 mm. The value for G from the table for the modulus of elasticity (Fig. 112) is 77 200 MPa.

A trial wire diameter d (Fig. 118) is then found from the same formula as before:

$$d = \sqrt[3]{\frac{2.55PD}{S}} = \sqrt[3]{\frac{2.55 \times 267 \times 22}{550}}$$

$$= \sqrt[3]{27.234} = 3.0087 \text{ mm}$$

The nearest even size is 3 mm, and should be used.

The stress S is

$$S = \frac{PD}{0.393d^3} \qquad \frac{267 \times 22}{0.393 \times 3^3} = 553.57 \text{ MPa (say 554 MPa)}$$

The number of active coils N (Fig. 118) is

$$N = \frac{GdF}{\pi SD^2} = \frac{77\,200 \times 3 \times 38}{3.1\,416 \times 554 \times 22^2} = 10.447 \text{ (say } 10\tfrac{1}{2}\text{)}$$

The total number of coils is AC + 2 = 10½ + 2 = 12½.

Therefore a quick answer is 12½ total coils of 3-mm-diam. wire. However, the design should be carried out further as follows:

$$\text{Solid height} = TC \times d = 12\tfrac{1}{2} \times 3 = 37.5 \text{ mm (Fig. 117)}$$

$$\text{(total deflection} = FL - SH = 82 - 37.5 = 44.5 \text{ mm)}$$

$$\text{Stress solid} = \frac{554}{38} \times 44.5 = 648.76 \text{ MPa (say 649)}$$

$$\text{Spring index} = \frac{OD}{d} - 1 = \frac{25}{3} - 1 = 7.3$$

$$\text{Curvature correction factor } K = 1.21 \text{ (Fig. 114)}$$

$$\text{Total stress at 267 N} = 554 \times 1.21 = 670 \text{ MPa)}$$

[This is a safe design stress, as it is below the 760 MPa recommended for 3-mm oil-tempered MB shown on the middle curve of the recommended design stresses (Fig. 84).]

$$\text{Total stress at solid} = 649 \times 1.21 = 785 \text{ MPa}$$

[This too is a safe stress, as it is below the 825 MPa shown on the top curve (Fig. 84), therefore the spring will not take a permanent set.]

Note: The spring designed metrically is quite similar to the spring designed in the inch-pound system. The slight differences are due to using even millimetres on all dimensions and rounding off unwieldy figures.

Conical Compression Springs

Conical compression springs tapering from top to bottom in the form of a cone are useful in several applications as follows:

1. Small Solid Height: A conical spring can be designed so that each active coil fits within the next coil, so the solid height can be equal to one or two thicknesses of wire. This is useful where the solid height is limited.

2. Variable Rate: Conical springs with a constant or uniform pitch, have an increasing force rate instead of a constant force rate as do regular compression springs, because the larger coils gradually begin to bottom as a force is applied. A variable pitch can be designed to give a uniform rate if necessary.

3. Stability: Conical springs have more lateral stability and less tendency to buckle than regular compression springs.

4. Vibration: Conical springs with a uniform pitch have an increasing natural period of vibration (instead of a constant) as each coil bottoms, thereby reducing vibration and resonance.

The good features, however, should not overshadow the facts that conical springs use material inefficiently, require more set-up time, need special grinding fixtures, and occasionally may cost two or three times as much as regular compression springs.

Although conical compression springs can be coiled with a variable pitch, it is often easier to coil them with a uniform pitch. It is best to leave the type of pitch to the judgment of the spring manufacturer, as he may be required to alter the pitch or use a combination of pitches to meet the force requirements.

DESIGN

There are several methods of analyzing the complex stresses in conical springs depending upon the type of pitch, angular relationship of the coils, constant slope, curving contours, and other factors. Analyzing all of the variables and solving for stresses, deflections, and forces can be a mathematician's delight and a playground for higher mathematics, but spring manufacturers can ill afford to devote valuable time to such matters and many use a simplified solution that quickly and easily provides a suitable wire size and number of active coils, as follows:

APPROXIMATE METHOD

First, calculate the average outside diameter. This can be done in several ways depending upon the diameters specified. The easiest way is to add the outside diameter of the top end to the outside diameter of the base and divide by 2. Adding the OD of the base to the ID at the top and dividing by 2 will provide the average D. Second, calculate the approximate force at the solid height. Third, locate the average OD in the tables of spring characteristics (Fig. 115) and find a force, deflection per coil, and a wire size, and proceed with calculations in the same manner as for regular compression springs. This method should be used with care if large deflections cause coils to bottom early.

EXACT METHOD

A more exact method is first to determine the wire size and number of coils by the approximate method and then analyze the force, deflection, and stress of each individual coil and make adjustments to the design to meet the requirements. This may require a large number of individual calculations for each spring, but should be done for large, important springs and where a long fatigue life may be needed.

PROBLEM 2

A conical compression spring (Fig. 120) is to be made of oil-tempered MB spring steel, ASTM A 229. Determine a suitable wire size, stress, and number of coils to meet the force and deflection requirements for a fatigue life of over 1 million cycles.

FIG. 120. Conical Spring

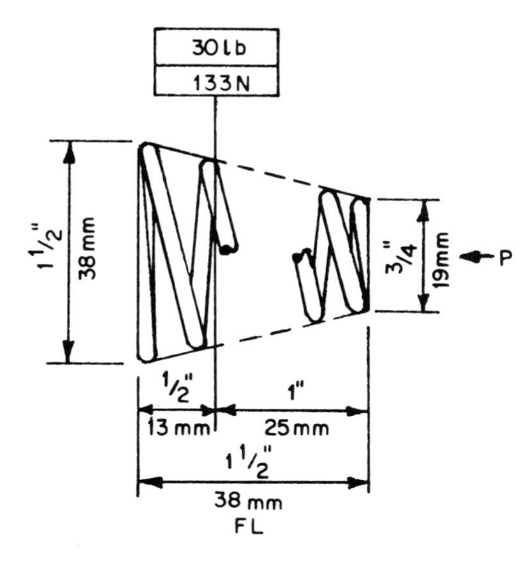

Solution in the inch-pound system:

1. Average OD = 1½ + ¾ ÷ 2 = 1 ⅛ in.
2. Approx. solid force = 30 ÷ 1 × 1½ = 45 lb

Method 1: By Tables

(Fig. 115): Locate 1 ⅛ in. OD in the left column and to the right about one-fourth of the way across, a force of 45.2 lb is found, with a deflection per coil of 0.277 in. with a stress of 100 000 psi, and at the top of that column a wire size no. 12 (0.1055 in. diam.) will be found.

1. Percent Method: The percentage of 30 lb to 45.2 lb = 32 ÷ 45.2 × 100 = 70.796 percent (say 70.8 percent).
2. The stress S = 70.8 percent of 100 000 = 70 800 psi.
3. The deflection per coil f = 70.8 percent of 0.277 = 0.196 in.
4. The number of active coils N = F ÷ f = 1 ÷ 0.196 = 5.1 (say 5 active, 7 total).

Therefore a quick answer is 7 total coils of 0.1055 in. diam. wire.

However, the design should be carried out further, as follows: Assuming that the spring can go flat, i.e., to the thickness of one coil, then:

5. The solid stress is

$$\frac{S}{F} \times (FL - d) = \frac{70\,800}{1} \times (1.5 - 0.1055) = 98\,730 \text{ psi}$$

6. The average spring index = OD/d − 1 = 1.125/0.1055 − 1 = 9.66.
7. The curvature correction factor K = 1.147 (Fig. 114).
8. Total stress at 30 lb = 70 800 × 1.147 = 81 200 psi. This is a safe stress, as it is below the 95 000 psi shown on the bottom curve of the recommended design stresses (Fig. 84) and therefore should have a fatigue life over 1 million cycles.
9. Total stress at solid = 98 730 × 1.147 = 113 240 psi. This too is a safe stress, as it is below the 126 000 psi shown on the top curve (Fig. 84); therefore the spring will not take a permanent set. This spring, if made with a variable pitch so that all coils go flat at the same time, meets the specifications.

Method 2: By Formulas

The same spring can be designed with formulas by assuming a safe stress and an average mean diameter, as follows:

1. Assume a safe stress below the lowest curve (Fig. 84), say 70 000 psi (75 000 psi could have been selected).
2. Assume a mean diameter D slightly below the average OD, say 1 in. The value for G (Fig. 112) is 11 200 000 psi.
3. A trial wire diameter d from the formula (Fig. 118) is

$$d = \sqrt[3]{\frac{2.55PD}{S}} = \sqrt[3]{\frac{2.55 \times 30 \times 1}{70\,000}}$$

$$= \sqrt[3]{0.001093} = 0.103 \text{ in.}$$

The nearest standard wire gauge size is 0.1055 in., the same as in the previous method.

4. The correct mean diameter D = average OD − d is

$$\frac{1.5 + 0.75}{2} - 0.1055 = 1.0195 \text{ (say 1.02)}$$

5. The stress is

$$S = \frac{PD}{0.393d^3} = \frac{30 \times 1.02}{0.393 \times 0.1055^3} = 66\,300 \text{ psi}$$

6. The number of active coils is

$$N = \frac{GdF}{\pi SD^2} = \frac{11\,200\,000 \times 0.1055 \times 1}{3.1416 \times 66\,300 \times 1.02^2} = 5.45$$

The answer is quite close to that of Method 1, the slight difference being due to rounding off some of the values. The total coils, total stresses, and other calculations are made as in Method 1.

Method 3: Metrics

Make assumptions as before:

1. Assume a safe stress below the lowest curve (Fig. 84), say 500 MPa (475 or 525 MPa could have been selected).
2. Determine the average OD = (38 + 19)/2 = 28.5 mm.
3. Assume a mean diameter D slightly below the average OD, say 26 mm.
4. A trial wire diameter d from the formula (Fig. 118) is

$$d = \sqrt[3]{\frac{2.55PD}{S}} = \sqrt[3]{\frac{2.55 \times 133 \times 26}{500}}$$

$$= \sqrt[3]{17.6358} = 2.603 \text{ (say 2.6 mm)}$$

5. The correct mean diameter D = average OD − d = 28.5 − 2.6 = 25.9.
6. The stress is

$$S = \frac{PD}{0.393d^3} = \frac{133 \times 25.9}{0.393 \times 2.6^3} = 499 \text{ MPa}$$

7. The number of active coils is

$$N = \frac{GdF}{\pi SD^2} = \frac{77\,200 \times 2.6 \times 25}{3.1\,416 \times 499 \times 25.9^2} = 4.77 \text{ (say 5)}$$

This is slightly less than the preceding method, because all dimensions were rounded off.

Carrying out the design further:

8. Maximum deflection, assuming that the spring can go flat to one coil thickness = FL − d = 38 − 2.6 = 35.4 mm.
9. Stress solid = (S/F) × 35.4 = (499/25) × 35.4 = 707 MPa.
10. By referring to the recommended design stress curves (Fig. 84) it will be found that the stress of 499 MPa is below the lowest curve and the 707 MPa is below the top curve, so this spring is also safely designed.

Rectangular Wire Compression Springs

Rectangular wire wound on the thinner edge over an arbor is occasionally used for general-purpose compression springs, but not often. However, it has found a place for itself in die springs where high loads with moderate deflections and short solid heights are required. Such springs are usually quite highly stressed and are not always expected to have a long fatigue life. Winding the wire on the larger flat surface against the arbor is seldom done and so will not be described.

Rectangular wire with sharp edges should not be used because the sharp edges cause early breakage. Round wire rolled flat has rounded edges, lasts much longer, and is generally used.

KEYSTONE EFFECT

When square or rectangular wire is coiled into springs, some of the material in the extreme outside edge is drawn into the inner edge and this inner edge upsets, so the wire becomes keystone or trapezoidal in shape. This increase in the upset portion also increases the solid height and should be allowed for when checking the solid height of a spring. This change depends upon the spring index and the thickness, and should be determined by the following formula, which can be used for both square and rectangular sections:

$$t' = 0.48t \left(\frac{OD}{D}\right) + 1$$

t' = the enlarged thickness of the inner edge after coiling

t = thickness before coiling

DESIGN

Theoretical mathematicians have published technical papers describing different methods for stress analysis and for determining the forces and deflections of compression springs made of rectangular wire. The reasons for the differences are due to differing methods of determining the amount of upset during coiling, variations in the amount of roundness, changes in section due to flattening round wire, and so on. A perfect method suitable for all variations in cross sections may never be available, but the simplified formulas using constants based on the ratio of width to thickness work quite well for rectangular wire with rounded edges as follows:

$$S = \frac{AGtF}{ND^2} \qquad N = \frac{AGtF}{SD^2}$$

$$S = \frac{PD}{Bbt^2} \qquad P = \frac{BSbt^2}{D}$$

In all cases,

b = breadth or width

t = thickness of thin section

A and B are taken from curves (Fig. 121).

FIG. 121. Constants for Rectangular Wire with Round Edges

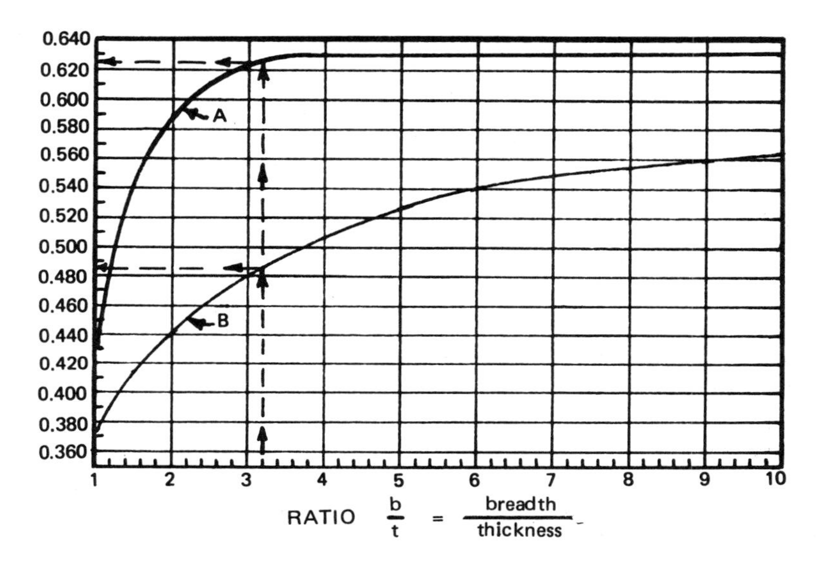

Diameter Changes

The outside diameter of a compression spring increases and the outside diameter of an extension spring decreases when deflection occurs. This change is because the slope or angle of the coils relative to the axis of the spring changes when deflection takes place. Allowance for this change should be made when a compression spring is housed in a tube to prevent friction or binding which can cause increased stresses and early failure. The increased diameter OD_1 for a compression spring deflected to the solid position can be calculated from the following formula:

$$OD_1 = \sqrt{D^2 + \frac{p^2 - d^2}{\pi^2}} + d$$

Where p = pitch, D = mean diameter, d = wire diameter, and the dimensions can be in inches or millimetres.

EXAMPLE

The compression spring (Fig. 119) has an OD = 1 in., wire diameter d = 0.1205 in., D = 0.8795 in., the pitch p (Fig. 117) = FL − 2d ÷ N = 3.25 − 0.241 ÷ 11 = 0.274 in.

The increased OD after deflection to solid from the above formula is

$$OD = \sqrt{0.8795^2 + \frac{0.274^2 - 0.1205^2}{3.1416^2}} + 0.1205$$

$$= \sqrt{0.7735 + \frac{0.0751 - 0.0145}{9.8696}} + 0.1205$$

$$= \sqrt{0.7796} + 0.1205 = 0.8830 + 0.1205 = 1.0035 \text{ in.}$$

The increase in this case is only 0.0035 in., but it could have been much more if the index and pitch had been higher.

Dynamic Forces

Springs are generally subjected to one of three types of forces: (1) slowly applied force, usually found on lever arms used in machinery; (2) suddenly applied force, applied quickly without restraint, also found quite often in industrial machinery; and (3) impact force such as a weight dropped from a height and containing kinetic energy. Simplified formulas for both compression and extension springs to determine the deflection F_1 under such forces are shown in Fig. 122.

The equation for impact considers the kinetic energy of a falling body and becomes a quadratic equation in the form $ax^2 + bx + c = 0$ and may be solved by the formula:

$$x = \frac{-b \pm \sqrt{b^2 - 4ac}}{2a} \quad \text{(see Problem 9)}$$

Other interesting applications of the formula for impact force include the speed with which a spring will return, or the speed it will return a weight or propel

FIG. 122. Formulas for Dynamic Forces

Slowly applied force	Suddenly applied force	Impact force
$F_1 = \frac{P}{r}$	$F_1 = \frac{2P}{r}$	$F_1{}^2 = \frac{2P(h+F)}{r}$

Where P = applied force, r = rate, and h = height load is dropped. All dimensions including h are in inches or millimetres, with the force in pounds or newtons.

a mass. The energy of a spring equals one half the load times the distance moved. Energy also equals ½ mv^2, where m is the mass and v is the velocity.

Vibration, Natural Frequency, and Surge

Compression springs subjected to the dynamic effects of vibration, surge, and rapid force oscillations may have stresses increased as much as 40 percent or more. This is due to the inertia effect of the coils because suddenly applied forces make the first, second, and third coils deflect more than other coils, and this explains why so many springs often break on one of these three coils. Most fractures occur on the third coil due to coil clashing.

Springs subjected to suddenly applied forces should first be checked for static force in the usual manner, then checked again with the formula for suddenly applied forces and the stresses compared with the curves for recommended stresses. Then, if satisfactory, the natural period of vibration should be calculated and compared with the induced frequency.

Suddenly applied forces often cause a surge wave to occur and if constantly repeated, as in automobile engine valve springs, can excite the natural period of vibration to such an extent that coils can clash sufficiently to cause a spring to jump out of its constraining holder.

When the natural period of vibration of a spring and its harmonics are low, a spring will surge causing coil clash and early fatigue. The natural period of vibration should be at least 13 times the frequency of the applied force to avoid surging. Some authorities claim that harmonics as high as 20 may have to be considered in some instances. The higher the natural frequency, the better!

Elimination of surge and resonance with harmonics in valve springs is usually accomplished by coiling one or two active coils near one end at a reduced pitch. These coils then close up and come together, thereby changing the natural frequency of the spring. A similar elimination of surging occurs in conical springs coiled with a uniform pitch. Such springs under deflection have the largest active coil bottoming, thereby inducing a change in the natural frequency. Other methods to reduce surging include using two springs, one within the other, using stranded wire springs, or stiffening the spring to increase the force, or by using square or rectangular wire. Rubber bumpers touching a coil will also stop surging.

Springs subjected to deflections or vibrating at standard AC motor speeds of 1800 rpm may be operating at only $^1/_{11}$ of their natural frequency; this causes resonance so they are actually operating 11 times as fast, or 19 800 cycles per minute, thus causing early failure.

Formulas for determining the natural period of vibration (n′) per second for steel springs follow:

1. For an unloaded spring,

$$n'' = \frac{3510 \times d}{(D/2)^2 \times N}$$

2. For an initially loaded spring, (neglecting weight of spring)

$$n' = 187.6\sqrt{\frac{1}{F}}$$

The compression spring in Problem 1, (Fig. 119) has two natural periods of vibration as follows:

1. In the unloaded condition,

$$n'' = \frac{3510 \times d}{(D/2)^2 \times N} = \frac{3510 \times 0.1205}{(0.8795/2)^2 \times 11} = 198.8 \text{ per second}$$

2. In the loaded condition,

$$n' = 187.6\sqrt{\frac{1}{F}} = 187.6\sqrt{\frac{1}{1.5}} = 187.6 \times .8166 = 153 \text{ per second}$$

(Multiply by 60 if required in vibrations per minute)

These natural periods of vibration should exceed 13 times the frequencies to which they are subjected.

Compression Spring Buckling

Buckling of helical compression springs often occurs when the free length is four or more times the mean diameter. The amount of deflection and the slenderness ratio are the dominating factors.

The type of ends is important, and for most applications where buckling occurs, it is highly desirable to have the end coils closed and ground. Such springs should be guided over a rod or in a tube to retard buckling, and the rods or tubes should be well lubricated to reduce friction.

Figure 123 is a buckling curve for compression springs with end coils closed and ground. The example shown by the dashed line indicates that a compression spring having a slenderness ratio of 6.5 will buckle if the deflection is equal to 33 percent of the free length. It is assumed that the ends press against a flat plate and are not restrained.

When a long spring, say 10 to 20 times its diameter, is required, it may be better to use several shorter springs over a rod with washers spaced between the springs, as this will reduce both buckling and friction.

FIG. 123. Buckling curve for compression springs, ends closed and ground. (Courtesy Department of Defense, MIL-STD-29A.)

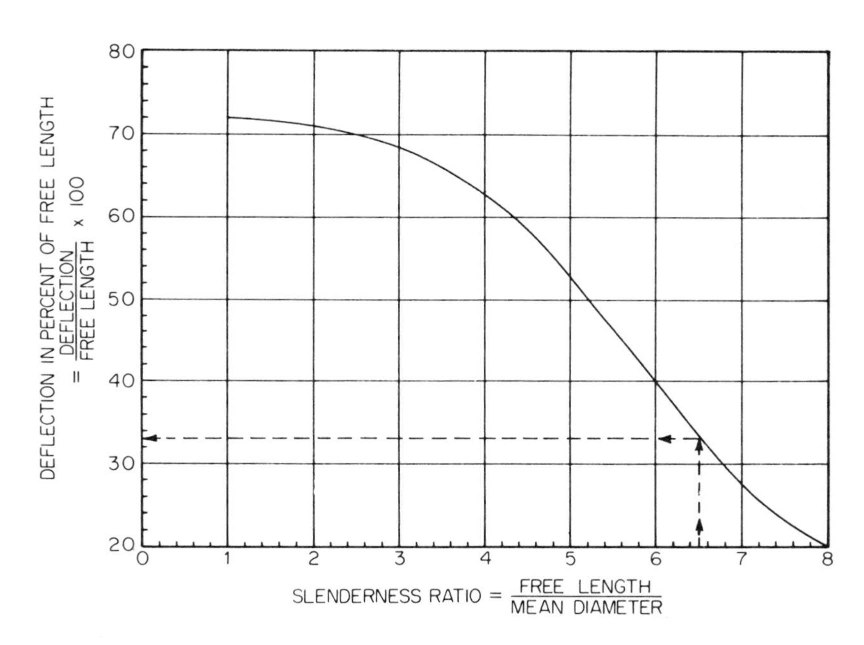

Compression Springs with Drawbars

Compression springs can be used as extension springs when they are fitted with drawbars as per Fig. 124. This is often done when a regular extension spring might be deflected excessively and result in failure.

The compression spring should be designed with a safe stress at the solid height so that it will not take a permanent set. The drawbars can be made from any grade of hard-drawn wire, as they are quite strong if made with the same diameter of wire as used in the spring. The ends should be closed, but not ground. Many sizes of springs have been made in this manner using wire diameters from $^1/_{16}$ to $^1/_2$ in. (1.5 to 13 mm) with outside diameters of $^1/_2$ to 5 in. (13 to 127 mm). One of the features is that such springs have a fixed stop when the spring goes solid, thereby eliminating overstretching. Another good feature is the elimination of buckling. The drawbars should be lubricated during operation to reduce friction.

FIG. 124. Compression spring fitted with drawbars.

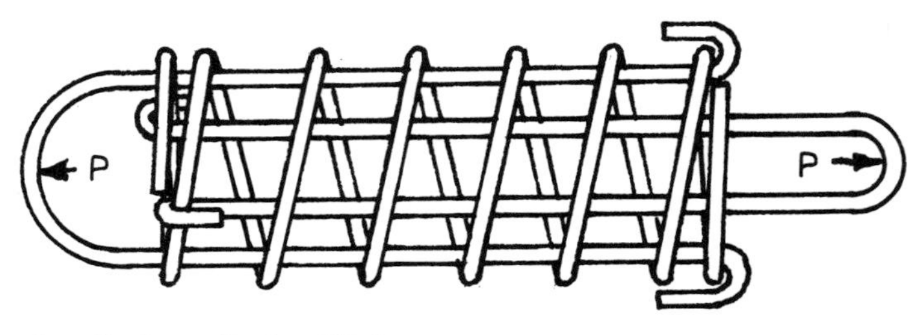

Compression Springs—Design Hints

1. Compression springs are the easiest type to make, can be made more rapidly and accurately than other types, and should be used in preference to other types wherever possible.
2. Avoid using open ends or open ends ground. Such springs usually tangle when shipped. They also buckle when deflected.
3. Use closed ends not ground whenever possible, especially on light wire sizes under 1/32 in. (0.80 mm) or where a large spring index D/d prevails such as 13 or larger. Grinding is a slow, expensive operation.
4. Use closed ends ground and state degree of squareness on important springs and on those used under diaphragms or in regulator valves, or where buckling might occur.
5. Use conical springs when a short, solid height is needed and to reduce buckling and surging.
6. Avoid using conical, barrel shaped, or other special shapes, including tangs on the ends, when a standard helical spring could be used.
7. Design springs with a reasonably safe stress when compressed solid so that they may be adaptable to other installations.
8. Hand—do not specify coiling right or left hand if it is not important. Specify right hand if it must be threaded onto a bolt. If a spring is used inside another, one spring should be wound left hand and the other right hand to avoid meshing of the coils.
9. Specify a force with a tolerance at an exact compressed height rather than a definite deflection. Testing instruments are designed to test at specified heights. Specifying two or more forces automatically determines a rate and should be avoided if possible, but is often necessary on important springs. The rate equals the difference in the two forces divided by the amount of deflection between those forces.
10. Three springs standing side by side will have a rate and a solid force three times that of one spring. Three springs placed one on top of another will

have a rate only one third of one spring and the solid force will be the same as for one spring.

11. Springs are frequently one of the least expensive parts of a machine, but often cause the most trouble. Specify the mandatory requirements, leaving the exact wire size and number of coils to the discretion of the spring manufacturer (if in doubt) so that he can use material in stock, as this will often lower the cost and provide faster delivery.
12. Specify unusual conditions such as high or low temperatures, corrosive surroundings, impact force and fatigue life, or number of cycles of deflection desired on drawings of springs or on purchase orders.
13. Spring indexes D/d between 7 and 13 are best proportioned for design and manufacture. Over 16 may require especially large tolerances, and under 4 cannot always be coiled on automatic coilers.
14. Standard springs are not stocked like screws and bolts, but a few spring manufacturers do stock some small sizes in one or two materials. These may be especially useful when small quantities or prototypes are needed.

12 Extension Springs

Extension springs are the direct opposite of compression springs, in that they are close-coiled helical springs that extend under a pulling force. However, the same formulas for design are used (Fig. 118) plus additional formulas for stresses in hooks and initial tension.

Extension springs represent about 10 percent of all springs made by some companies. They take longer to design because consideration must be given to stress due to initial tension, stress and deflection of hooks, special coiling methods, secondary operations, and allowance for overstretching at assembly. It is often necessary to make two, three, or more trial sets of calculations before arriving at a suitable design.

Figure 125 shows standard hooks and loops. A "hook" is open to fit over a projection; a "loop" is a closed hook. A regular machine hook or loop is made on an automatic looper and is the least expensive and often is the most satisfactory type.

Figure 126 shows special types of hooks. These special end formations are quite expensive, but are often used to suit design conditions. The coned ends with swivel hooks are used to reduce hook stresses and to eliminate hook breakage. Reducing the hook diameter reduces the hook stresses.

Crossover loops are easier to make and less expensive than full loops over center, and if made without sharp radius bends they often last longer than the full loops over center.

FIG. 125. Standard extension spring ends. (Courtesy U.S. Department of Defense, MIL-STD-29A.)

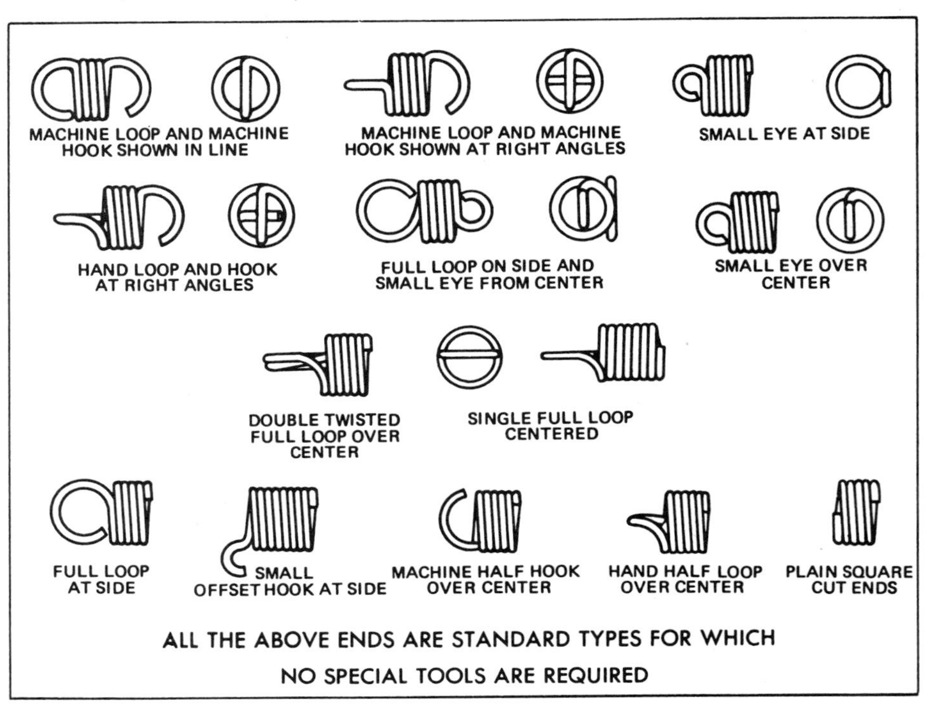

FIG. 126. Special extension spring ends. (Courtesy U.S. Department of Defense, MIL-STD-29A.)

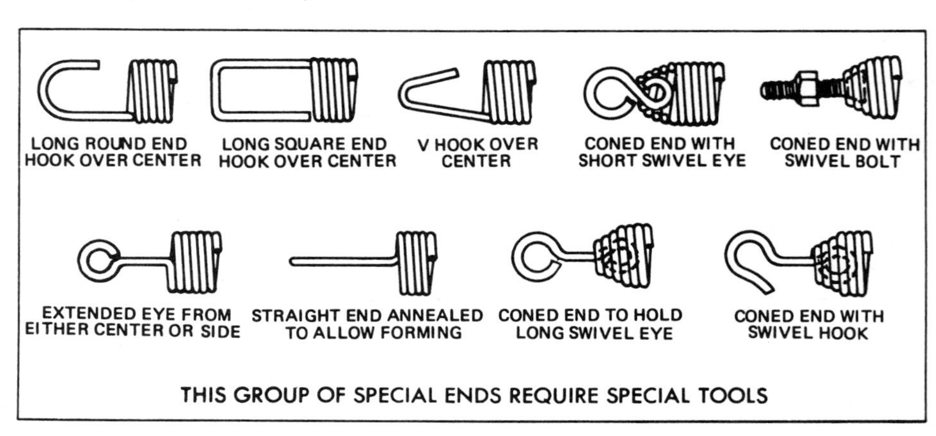

Initial Tension in Extension Springs

Initial tension is a force which presses the coils of a close-wound extension spring against each other and keeps all the coils together. This force must be overcome before the coils of a spring begin to open up. This force is wound into an extension spring by bending each coil as it is wound, away from its normal coiling direction, thereby causing a slight twist in the wire which makes the coils spring back tightly against the adjacent coil. This force should be wound only into extension springs—if wound into torsion springs, this force will cause errors in loads and deflections.

Initial tension can be wound into cold-coiled extension springs made from hard-drawn or oil-tempered wires only, such as music wire, stainless steels of the 300 series, phosphor-bronze, Monel, Inconel, and other pre-hardened materials. It cannot be wound into extension springs made from annealed materials or those that require hardening after coiling, including hot-rolled springs.

Some initial tension should be coiled into extension springs to hold the spring together. Otherwise, a long spring with many coils will have a different length in the horizontal position than when hung vertically. The amount of initial tension that can be wound into an extension spring depends principally upon the spring index, as shown in Fig. 127.

FIG. 127. Permissible stress due to initial tension.

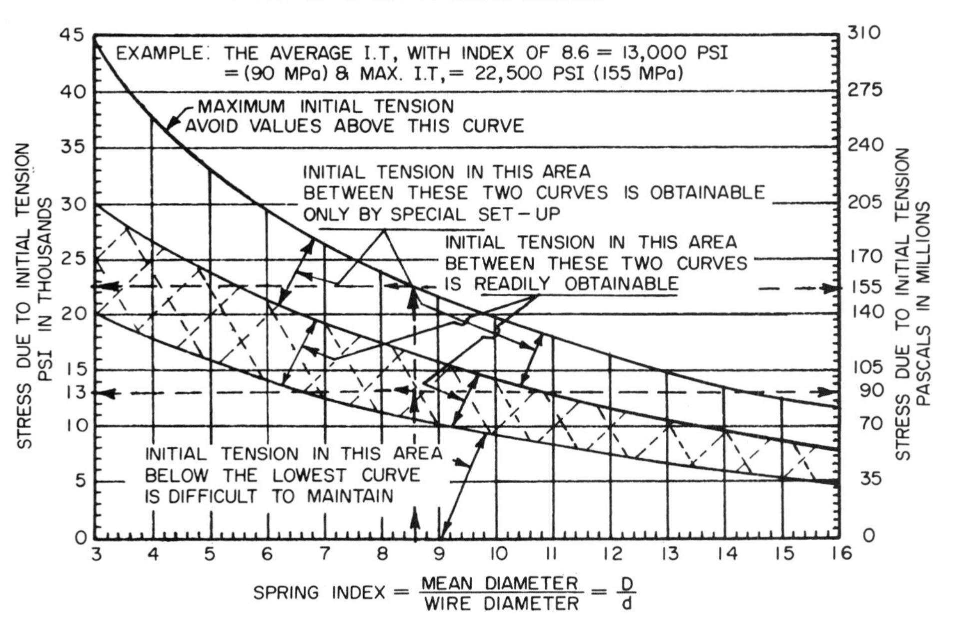

FIG. 128. Extension spring.

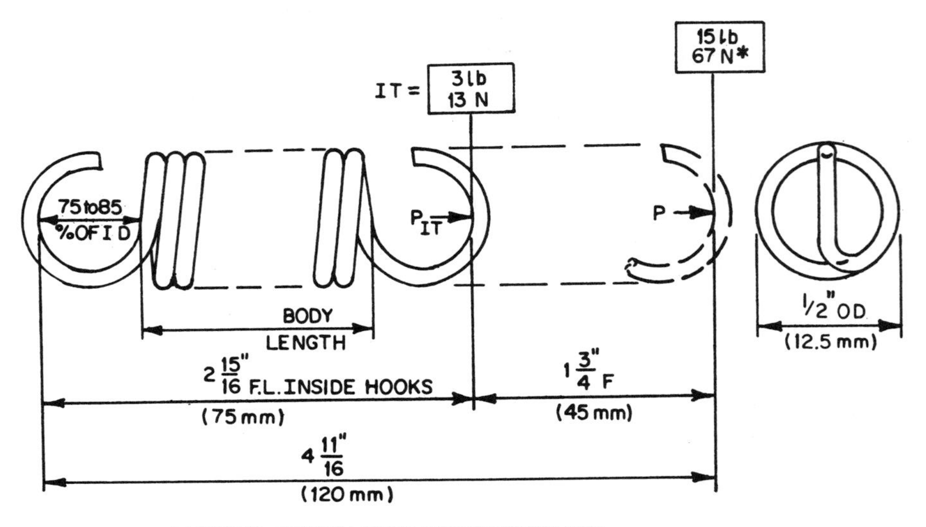

PROBLEM 3

An extension spring (Fig. 128) has the following requirements: OD = ½ in., free length inside hooks = 2 $^{15}/_{16}$ in., force (including initial tension) = 15 lb at 4 $^{11}/_{16}$ in. extended length inside hooks (1$^{3}/_{4}$ in. deflection) with regular full hooks over center and using oil-tempered MB spring steel, ASTM A 229.

The wire size, stress, number of coils, and other data need to be determined. Also, allow about 20 to 25 percent of the 15-lb force for initial tension, say 3 lb, and then design for a 12-lb force (not 15 lb) at 1$^{3}/_{4}$ in. deflection.

Lower stresses should be used for extension springs than for compression springs to allow for overstretching at assembly and to obtain a safe stress on the hooks.

The tables of spring characteristics (Fig. 115) are the same for both compression and extension springs.

Method 1: By Tables

Locate ½ in. OD in the left column (Fig. 115), and to the right about half way across the table a force (P) of 21.9 lb appears under a deflection (f) of 0.0859 in. At the top of that column a wire diameter of no. 16 (0.0625 in.) is found.

Using the percent method:

1. The percent of 12 lb to 21.9 lb equals the 12-lb force divided by 21.9 × 100 = 54.79 percent (say 54.8 percent).

2. The stress therefore is 54.8 percent of 100 000 = 54 800 psi.
3. The deflection per coil is 54.8 percent of 0.0859 = 0.047 in.
4. The number of active coils = F ÷ f = 1.75 ÷ .047 = 37.23 (say 37).

This should be reduced by 1 to allow for the deflection of two hooks (see Hints at the end of this chapter); use 36 active coils.

Therefore a quick answer is 36 active coils in the body of the spring using 1/16 in. diam. oil-tempered MB wire and 3 lb initial tension. However, the design should be carried out further as follows:

5. Body length = TC + 1 X d = 36 + 1 X 0.0625 = 2 5/16 in.
6. Length, body to inside hook, is

$$\frac{\text{FL} - \text{body}}{2} = \frac{2\,^{15}/_{16} - 2\,^{5}/_{16}}{2} = {}^{5}/_{16} \text{ in.}$$

$$\text{Percent of ID} = \frac{0.3125}{\text{ID}} = \frac{0.3125}{0.375} = 83 \text{ percent}$$

This percentage is satisfactory (see Hints).

7. Stress due to initial tension is

$$S_{IT} = \frac{S \times IT}{P} = \frac{54\,800 \times 3}{12} = 13\,700 \text{ psi}$$

(Note: Item 7 is a short-cut method using simple proportions). The stress due to IT is obtainable as shown by Fig. 127.

8. Spring index = OD/d − 1 = 0.500/0.0625 − 1 = 7.
9. Curvature correction factor K = 1.20 (Fig. 114).
10. Total stress at 15-lb force = 54 800 + 13 700 X 1.20 = 82 200 psi.

[This is less than the 105 000 psi for 0.0625-in. oil-tempered MB wire as shown by the lowest curve (Fig. 84) for recommended design stresses, and therefore is a safe stress that permits additional deflection usually necessary for assembly purposes.

11. The large majority of failures of extension springs is due to a high bending stress in the hooks. This stress should be compared with the curves for torsion spring stresses (Fig. 84) and is calculated from the formula (Fig. 129) as follows:

$$S_b = \frac{5PD^2}{IDd^3} = \frac{5 \times 15 \times 0.4375^2}{0.375 \times 0.0625^3} = 156\,800 \text{ psi}$$

This is less than the 165 000 psi shown by the lowest curve for torsion springs (Fig. 85) made from 0.0625-in. oil-tempered MB wire, and therefore should permit a long fatigue life.

This completes the design quickly and easily, using short-cut methods.

Method 2: By Formulas

The same problem can be solved using the formulas for compression springs. This will require an assumption for stress and mean diameter as follows:

1. Assume a safe stress well below the lowest curve for oil-tempered MB steel wire (Fig. 84), say 80 000 psi.

2. Assume a mean diameter D slightly less than the ½-in. OD, say 0.40 in. The value for G from the table for the modulus of elasticity (Fig. 112) is 11 200 000 psi.

3. A trial wire diameter d (Fig. 118) is then determined as follows:

$$d = \sqrt[3]{\frac{2.55PD}{S}} = \sqrt[3]{\frac{2.55 \times 15 \times 0.40}{80\,000}}$$

$$= \sqrt[3]{0.00019} = 0.0575 \text{ in.}$$

The nearest standard wire size is the same one used in the previous example: no. 16 (0.0625 in.).

Using the tables of cubes, the value nearest to 0.000 19 is 0.000 205, which shows 0.059 in., which is again quite close to the standard gauge no. 16 (0.0625 in.) which should be used, and is the same wire as was selected by the previous method.

4. The correct mean diameter is now OD – d = 0.500 – 0.0625 = 0.4375 in.

5. The stress at the 12-lb force is

$$S = \frac{PD}{0.393\,d^3} = \frac{12 \times 0.4375}{0.393 \times 0.0625^3} = 54\,718 \text{ psi}$$

which is quite close to the result by Method 1.

6. The number of active coils is

$$N = \frac{GdF}{\pi SD^2} = \frac{11\,200\,000 \times 0.0625 \times 1.75}{3.1416 \times 54\,718 \times 0.4375^2} = 37.23 \text{ (say 37)}$$

This should be reduced by 1 to allow for the deflection of the 2 hooks, so the answer is exactly the same as in Method 1. The stress due to initial tension, stress on hooks, body dimensions, etc., is done exactly as described in Method 1.

Method 3: Metrics

A somewhat similar extension spring using quite even values for force and dimensions in SI units follows:

An extension spring has the following requirements: OD = 12.5 mm, free length inside hooks = 75 mm, force (including initial tension) = 67 N at

120 mm inside hooks (deflection = 45 mm), with regular full hooks over center and using oil-tempered MB spring steel. This spring is quite close in size to the one in Fig. 127.

Determine the wire size, stress, number of coils, etc. Also, allow about 20 to 25 percent of the 67-N force for initial tension, say 13 N, and then design for a 54-N force (not 67 N) at 45 mm deflection, using safe stresses for long life.

1. Assume a safe stress well below the lowest curve for oil-tempered MB steel wire, say 550 MPa (Fig. 84).

2. Assume a mean diameter D slightly less than the 12.5-mm OD, say 11 mm. The value for G from the table for the modulus of elasticity (Fig. 112) is 77 200 MPa.

3. A trial wire diameter d is then found from the formula (Fig. 118) as before:

$$d = \sqrt[3]{\frac{2.55PD}{S}} = \sqrt[3]{\frac{2.55 \times 67 \times 11}{550}}$$

$$= \sqrt[3]{3.417} = 1.5062 \text{ mm}$$

(The nearest even size is 1.5 mm, and should be used.)

4. The stress at the 54-N force is

$$S = \frac{PD}{0.393d^3} = \frac{54 \times 11}{0.393 \times 1.5^3} = 447.84 \text{ MPa (say 448 MPa)}$$

The number of active coils is

$$N = \frac{GdF}{\pi SD^2} = \frac{77\,200 \times 1.5 \times 45}{3.1416 \times 448 \times 11^2} = 30.59 \text{ (say } 30\tfrac{1}{2}\text{)}$$

5. This should be reduced by 1 to allow for the deflection of the two hooks, so the answer is to use 29½ active coils of 1.5-mm wire. This differs slightly from the results obtained by the other methods because the 1.5-mm wire diameter is slightly smaller than the 0.0625-in. wire previously used, therefore a smaller number of coils is to be expected. To come closer to the 36 coils previously used, it is merely necessary to increase the wire size to the equivalent of 0.0625 in., which is 1.58 mm, but such sizes are not commonly used.

The stress due to initial tension, stress on hooks, body dimensions, etc., is done exactly as described in Method 1, but using metric SI units.

Approximate Stresses in Hooks

The calculations to determine the exact stresses in hooks are quite complicated and lengthy. Also, the radius of the bends forming the hooks are difficult to determine and frequently vary between specifications on a drawing and actual samples. However, regular hooks are higher stressed than the coils in the body of the spring and are most often the cause of failure.

The principal stress S_b is in bending at point A in Fig. 129. The hook is also subjected to torsional stresses lower in value at a point about half way between point A and the body of the spring. This holds true for regular full hooks or loops and crossover center hooks, provided that the radius of the sharp bend is equal to at least 75 percent of the wire diameter. When hooks break due to torsion it usually is due to sharply made bends. If nicely rounded bends are made, the failure is practically always at point A, and for this reason only the simplified formula for the approximate stress in bending is provided here. This formula can be used for all types of hooks including regular hooks over center, crossover center hooks, half hooks, extended hooks, enlarged or reduced hooks, machine hooks, double hooks, or any other kind. However, half hooks and side hooks may require further analysis, as they often are subjected to additional stresses.

Crossover center hooks, when made without sharp bends, often last longer than regular hooks. The bending stresses are the same and the life should be the same. However, if a large bend radius is made on a regular hook, the bending stresses often coincide with some torsional stresses, thereby explaining the reason for earlier breakage. If sharper bends were made on the regular hooks, the life should be the same.

Many types and sizes of hooks with different radii at the bends were fatigue tested and it was found that the hooks on springs made from $^1/_{16}$-in. (1.6-mm) oil-tempered MB steel lasted about 15 percent longer than hooks on springs made of stainless steel type 302, both having identical stresses.

FIG. 129. Formula for bending stress S_b in hook.

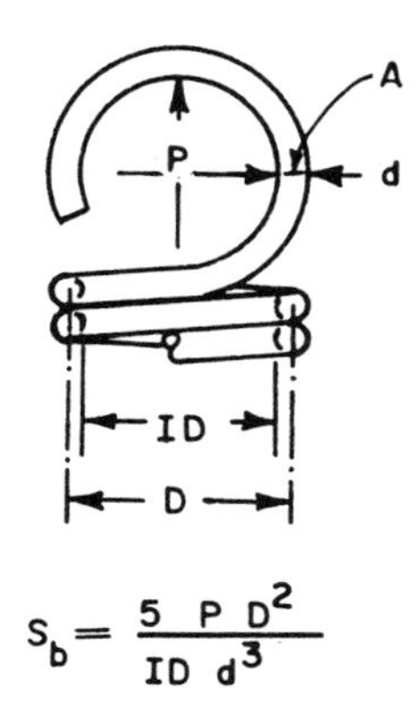

$$S_b = \frac{5\ P\ D^2}{ID\ d^3}$$

However, similar hooks on springs made from music wire lasted over 30 times as long! A comparison of the recommended design stress curves (Fig. 85 and 91) shows higher allowable stresses for oil-tempered MB steel, thus explaining the longer life. Music wire, with its absence of the pits or roughness seen on oil-tempered wire, proves the value of smooth, clean surfaces where long life is desired.

Extension Springs—Design Hints

1. Extension springs are more expensive than compression springs and should be stressed about 10 to 15 percent lower to allow for overstretching at assembly and to reduce the stresses in the hooks.
2. They should be designed with some initial tension to hold the coils together. Springs without initial tension are difficult to coil, and a long heavy spring with no initial tension will have a different length horizontally than when tested vertically. At least 10 percent of the maximum force should be in initial tension.
3. Each full hook deflects under a load equivalent to about half a coil, therefore deduct one coil from the calculated number of coils determined by formula to allow for deflection of two hooks. If not allowed for, a 10-coil spring would be 10 percent low on load! Each half hook deflects approximately equal to one tenth of a coil.
4. All coils are active in an extension spring, including the coil used to form a hook, therefore AC = TC, but make allowance for hook deflection.
5. For regular hooks, the distance from inside the hook to the body of the spring is about 75 to 85 percent of the inside diameter. This reduction is caused by bending a full diameter to an angular position over the center and should be allowed for in design.
6. Hooks are stressed in bending. This stress should be calculated and compared with the three curves for torsion springs, as they also are stressed in bending, for the material used. Refer to the recommended design stresses for this comparison.
7. Do not specify the relative position of the hooks with regard to each other unless it is important. The spring manufacturer may have to vary the position a little to meet the more important force and deflection requirements. Avoid using large, extended, or special hooks where possible, as they may double to triple the cost of the spring.
8. Keep the outside diameter of a hook the same as the OD of the spring so that the hook can be made by bending up a regular coil. Do not specify the hook opening with small tolerances. Swivel hooks and coned ends to reduce hook breakage are quite expensive—a hook with a reduced OD may often suffice and is far less costly.

9. The body length or closed portion of an extension spring equals the number of coils in the body plus 1, multiplied by the wire diameter.
10. Electroplating does not deposit a good coating between the coils nor on the inside diameter, but such springs should not be extended during plating as this causes a higher amount of hydrogen embrittlement.
11. Specify forces at extended lengths between hooks, not at amounts of deflection, as this is the way the testing instruments are constructed for production testing.
12. Heating extension springs to remove residual stresses caused by coiling also may reduce the amount of initial tension by as much as 50 percent. Allowance for this reduction should be made in manufacturing.
13. Specify unusual operating conditions such as elevated temperatures, shock or impact force, cycles per minute, corrosive atmospheres, and expected fatigue life on prints.
14. If stresses are high and danger of overstretching exists, consider using compression springs fitted with drawbars.
15. When two or more forces are specified, they automatically determine the rate and the deflection. Testing for two forces is often expensive, time consuming, and frequently can be avoided by proper design considering normal spring tolerances.

13 Torsion Springs

Torsion springs are helical coil springs that exert a torque or rotary force and are subjected to bending stresses. Such springs are wound right or left hand so that deflection causes the spring to wind up from the free position—never to unwind from the free position. Torsion springs reduce in diameter as they wind up; their length becomes longer and these changes should be calculated to see if binding on a supporting rod occurs. Such springs should be supported over a rod whenever possible. A slight open space should be provided between coils to eliminate coil friction, which would alter the torque and deflection.

Figure 130 shows the most popular types of ends, but dozens of other types are used and many can be formed on automatic torsion coilers. The one often used and the most economical is the straight torsion end, as it can be made quite easily on regular automatic coilers ordinarily used for coiling compression springs, provided that the coiler is fitted with a torsion attachment. The hook ends are second in popularity, with the hinged ends and straight offset ends following in that order.

Figure 131 illustrates torque. Note that $5 \times 2 = 10$ in.-lb and so does 10×1 and $20 \times \frac{1}{2}$, so it makes no difference where the force is applied to the arm of a torsion spring; the torque remains exactly the same.

Torque

By definition, torque is a force that produces rotation. Torsion springs exert a force (torque) in a circular arc, and the arms rotate about the central axis.

FIG. 130. Types of ends for torsion springs.

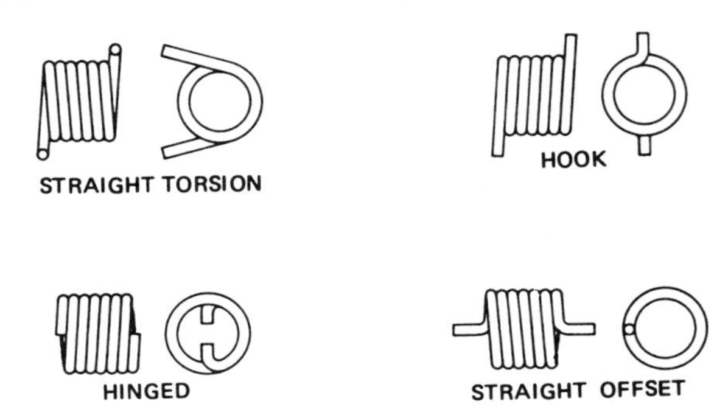

FIG. 131. Right-hand torsion spring. Torque = 10 lb-in. = 1130 N·mm.

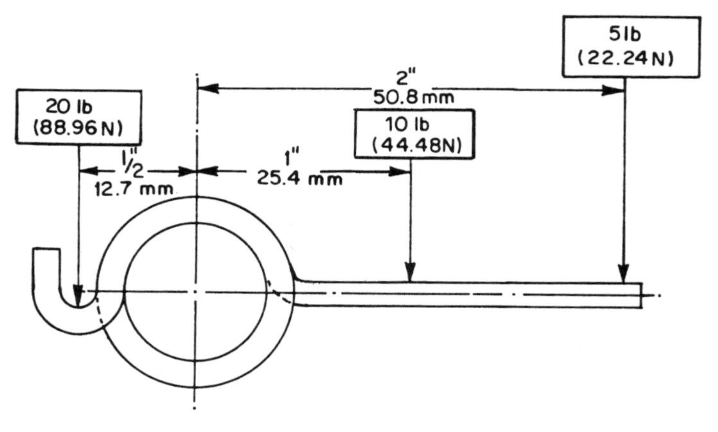

The stress is in bending, not in torsion as might be suspected. In the spring industry it is customary to specify torque with deflection or with the arms at a definite position. Formulas for torque are in pound-inches, (often reversed and called inch-pounds). If ounce-inches are specified, it is necessary to divide this value by 16 in order to use the formulas in the inch-pound system.

When a force is specified at a distance from a centerline, the torque, often called moment, is equal to the force multiplied by the distance. The force can be in pounds or ounces with the distance in inches or the force can be in newton metres with the distance in millimetres.

Formulas for torque are based on the tangent to the arc of rotation with a rod to support the spring. The stress in bending caused by the moment $P \times R$ is identical in magnitude to the torque T, provided that a rod is used.

To further simplify the understanding of torque, observe in both Fig. 131 and Fig. 132 that although the turning force is in a circular arc, the torque is not equal to P times the radius. The torque in both cases equals $P \times R$ because the spring rests against the support rod at point A.

FIG. 132. Left-hand torsion springs. Torque T = P × R, not P × radius.

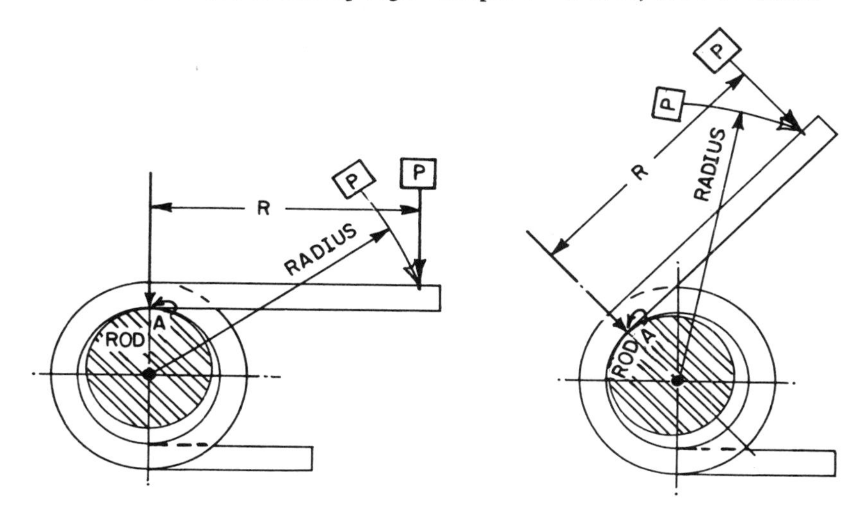

Design Methods

Several methods to design a torsion spring can be used. The easiest and quickest method is by using programmed computers or programmable electronic desk calculators having memory systems. Unfortunately, not all designers have access to preprogrammed equipment. Special spring design slide rules are useful, but they require assumptions; perform the main portion of the job and the remainder such as body length must be done by using a standard slide rule, an electronic calculator, or by hand.

Design Considerations

After determining the force or torque required, a designer usually estimates a suitable space or size limitations. However, the space should be considered approximate until the wire size and number of coils have been determined. The wire size is dependent principally upon the torque. When design tables showing torque, deflection, and wire sizes are not available, it is necessary to solve an equation to determine the proper wire size using the torque established and assume a safe design stress from the recommended design stress curves.

A survey of over 1000 torsion spring designs showed that nearly all required deflections ranging from 30 to 300° instead of in revolutions. Practically all had inside diameters ranging from 1/16 to 1 1/2 in. All required round wire with torques ranging from 0.022 to 700 in.-lb. As a result of this survey, the design tables (Fig. 134) were based on those ranges. However, other designs can be computed by the simplified formulas provided (Fig. 135).

With these design tables the selection of a suitable wire size and number of coils to provide a specified deflection is convenient. Many other factors affecting the design and operation of torsion springs are included in the paragraphs under design hints. Refer also to the section on spring manufacture.

Note: The design stresses used for the torque and deflection (Fig. 134) are taken from the average service curves and are uncorrected for curvature. With correction, the stresses would be between the average and light service curves. An expected fatigue life should be between 10 000 and 50 000 cycles. If spring indexes ÷ d exceed 8, up to 100 000 cycles can be expected. If a longer fatigue life is required, a slightly heavier wire size with more coils is advisable.

PROBLEM 4

Design a torsion spring (Fig. 133) using music wire, having an inside diameter of ¾ in., to exert a torque (T) of 4 in.-lb at a deflection (F) of 200°, suitable for average service. The spring is to be supported over a $^{21}/_{32}$-in.-diam. rod.

The wire size, stress, number of coils, and other dimensions must be determined.

Method 1: By Tables

From the table of torsion spring characteristics (Fig. 134), locate the ¾-in. ID in the left column. Above this column, the words "torque in. lb." are found and about one third of the way across the table a torque of 4.322 in.-lb is found. Above this torque, a stress of 176 000 psi is shown and at the top of this column

FIG. 133. Torsion spring.

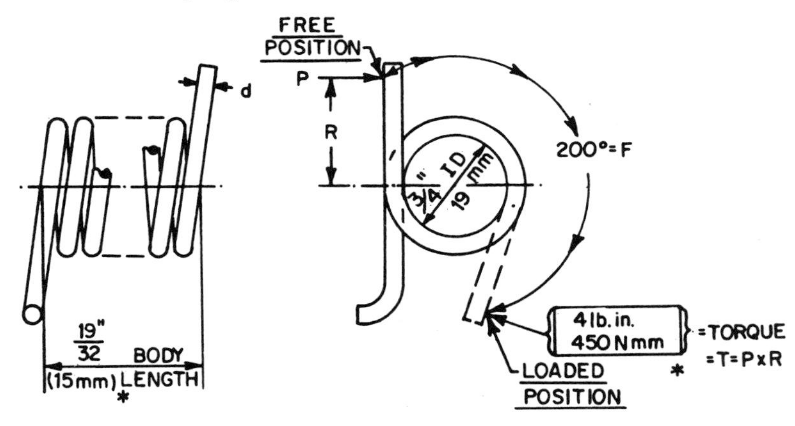

a wire size no. 26 (0.063 in.) is found. Lower down in the same column a deflection of 30.70° for one coil is shown.

The interpretation of these data show that one coil ¾ in. in ID made of 0.063-in. music wire, with a torque of 4.322 in.-lb and a stress of 176 000 psi, will deflect 30.70°. These data can then be used for design as follows:

Using the percent method:

1. The percent of 4 in.-lb to the 4.322 value equals 4 in.-lb divided by 4.322 × 100 = 92.549 (say 92.5 percent).
2. The stress therefore = 92.5 percent of 176 000 = 162 800 psi.
3. The deflection per coil = 92.5 percent of 30.70 = 28.3975° (say 28.4°).
4. The number of active coils = F ÷ f = 200 ÷ 28.4 = 7.04 (say 7) (active coils = total coils in torsion springs).

Therefore, a quick answer is to use 7 total coils of 0.063-in.-diam. music wire. The stress will be 162 800 psi, and this is a safe stress as it is below the middle curve of the recommended design stresses for torsion springs (Fig. 79).

However, the design should be carried out further as follows:

5. Body length = TC + 1 × d = 7 + 1 × 0.063 = 0.504 in. A slight clearance should be allowed between each coil to prevent friction of 20 to 25 percent of the wire diameter, say 0.015 in. There are six spaces, so the total clearance added to the space required for the coils is 6 × 0.015 + 0.504 = 0.594 in. (say $^{19}/_{32}$ in.).
6. Spring index = ID/d + 1 = 0.750/0.063 + 1 = 12.90.
7. Curvature correction factor K from curve = 1.06 (Fig. 114).
8. Total stress at 4 in.-lb = 162 800 × 1.06 = 172 568 psi. This is a safe design stress, as it is below the 175 000 psi recommended for 0.063-in. music wire shown on the middle curve of the recommended design stresses for torsion springs (Fig. 79).
9. The inside diameter (Fig. 135) after deflection is

$$ID_1 = \frac{N(\text{ID free})}{N + F/360} = \frac{7 \times 0.750}{7 + 200/360} = 0.6948 \text{ in.}$$

This provides satisfactory clearance (including allowance for ID tolerances) when used over a $^{21}/_{32}$-in. rod.

Method 2: By Formulas

Designing torsion springs by formulas is a little more difficult, because assumptions must be made for stress and mean diameter as follows:

1. Assume a trial safe stress well below the middle curve for music wire; knowing that a torque of 4 in.-lb requires light wire, say 160 000 psi (Fig. 79).
2. Assume a mean diameter D, slightly above the ¾-in. ID, say 0.8 in. The value for E from the table for the modulus of elasticity (Fig. 112) is 28 700 000 psi.
3. A trial wire diameter (d) is then found from the formula provided in the table of formulas (Fig. 135):

FIG. 134. Helical torsion springs design table.[a]

P.S.I. IN THOUSANDS

INSIDE DIAMETER INCHES

MUSIC WIRE ← E=29,500,000 →

A.M.W. GAUGE		1	2	3	4	5	6	7
DECIMAL		.010	.011	.012	.013	.014	.016	.018
DES. STRESS		232,	229,	226,	224,	221,	217,	214,
TORQUE in.lb.		.0228	.0299	.0383	.0483	.0596	.0873	.1226
1/16	.0625	22.35	20.33	18.64	17.29	16.05	14.15	18.72
5/64	.078125	27.17	24.66	22.55	20.86	19.32	16.96	15.19
3/32	.09375	31.98	28.98	26.47	24.44	22.60	19.78	17.65
7/64	.109375	36.80	33.30	30.38	28.02	25.88	22.60	20.12
1/8	.125	41.62	37.62	34.29	31.60	29.16	25.41	22.59
9/64	.140625	46.44	41.94	38.20	35.17	32.43	28.23	25.06
5/32	.15625	51.25	46.27	42.11	38.75	35.71	31.04	27.53
3/16	.1875	60.89	54.91	49.93	45.91	42.27	36.67	32.47
7/32	.21875	70.52	63.56	57.75	53.06	48.82	42.31	37.40
1/4	.250	80.15	72.20	65.57	60.22	55.38	47.94	42.34

A.M.W. GAUGE		8	9	10	11	12	13	14
DECIMAL		.020	.022	.024	.026	.029	.031	.033
DES. STRESS		210,	207,	205,	202,	199,	197,	196,
TORQUE in.lb.		.1650	.2164	.2783	.3486	.4766	.5763	.6917
1/16	.0625	11.51	10.56	9.818	9.137	8.343	7.896	
5/64	.078125	13.69	12.52	11.59	10.75	9.768	9.215	
3/32	.09375	15.87	14.47	13.36	12.36	11.19	10.53	10.18
7/64	.109375	18.05	16.43	15.14	13.98	12.62	11.85	11.43
1/8	.125	20.23	18.38	16.91	15.59	14.04	13.17	12.68
9/64	.140625	22.41	20.33	18.69	17.20	15.47	14.49	13.94
5/32	.15625	24.59	22.29	20.46	18.82	16.89	15.81	15.19
3/16	.1875	28.95	26.19	24.01	22.04	19.74	18.45	17.70
7/32	.21875	33.31	30.10	27.55	25.27	22.59	21.09	20.21
1/4	.250	37.67	34.01	31.10	28.49	25.44	23.73	22.72

M.W. ← E=29,500,000 → ← E=29,000,000 →

A.M.W. GAUGE		8	9	10	11	12	13	14	15	16	17	18	19	20	21
DECIMAL		.020	.022	.024	.026	.029	.031	.033	.035	.037	.039	.041	.043	.045	.047
DES. STRESS		210,	207,	205,	202,	199,	197,	196,	194,	192,	190,	188,	187,	185,	184,
TORQUE in.lb.		.1650	.2164	.2783	.3486	.4766	.5763	.6917	.8168	.9550	1.107	1.272	1.460	1.655	1.876
9/32	.28125	42.03	37.92	34.65	31.72	28.29	26.37	25.23	23.69	22.32	21.09	19.97	19.06	18.13	17.37
5/16	.3125	46.39	41.82	38.19	34.95	31.-14	29.01	27.74	26.04	24.51	23.15	21.91	20.90	19.87	19.02
11/32	.34375	50.75	45.73	41.74	38.17	33.99	31.65	30.25	28.38	26.71	25.21	23.85	22.73	21.60	20.68
3/8	.375	55.11	49.64	45.29	41.40	36.84	34.28	32.76	30.72	28.90	27.26	25.78	24.57	23.34	22.33
13/32	.40625	59.47	53.54	48.83	44.63	39.69	36.92	35.26	33.06	31.09	29.32	27.72	26.41	25.08	23.99
7/16	.4375	63.83	57.45	52.38	47.85	42.54	39.56	37.77	35.40	33.28	31.38	29.66	28.25	26.81	25.64
15/32	.46875	68.19	61.36	55.93	51.08	45.39	42.20	40.28	37.74	35.47	33.44	31.59	30.08	28.55	27.29
1/2	.500	72.55	65.27	59.48	54.30	48.24	44.84	42.79	40.08	37.67	35.49	33.53	31.92	30.29	28.95

MUSIC WIRE ← E=29,000,000 →

GAUGE		16	17	18	19	20	21	22	23	24	25	26	27	28	29
DECIMAL		.037	.039	.041	.043	.045	.047	.049	.051	.055	.059	.063	.067	.071	.075
DES. STRESS		192,	190,	188,	187,	185,	184,	183,	182,	180,	178,	176,	174,	173,	171,
TORQUE in lb.		.9550	1.107	1.272	1.460	1.655	1.876	2.114	2.371	2.941	3.590	4.322	5.139	6.080	7.084
17/32	.53125	39.86	37.55	35.47	33.76	32.02	30.60	29.29	28.09	25.93	24.07	22.44	21.37	20.18	19.01
9/16	.5625	42.05	39.61	37.40	35.59	33.76	32.25	30.87	29.59	27.32	25.35	23.62	22.49	21.23	19.99
19/32	.59375	44.24	41.67	39.34	37.43	35.50	33.91	32.45	31.10	28.70	26.62	24.80	23.60	22.28	20.97
5/8	.625	46.43	43.73	41.28	39.27	37.23	35.56	34.02	32.61	30.08	27.89	25.98	24.72	23.33	21.95
21/32	.65625	48.63	45.78	43.22	41.10	38.97	37.22	35.60	34.12	31.46	29.17	27.16	25.83	24.37	22.93
11/16	.6875	50.82	47.84	45.15	42.94	40.71	38.87	37.18	35.62	32.85	30.44	28.34	26.95	25.42	23.91
23/32	.71875	53.01	49.90	47.09	44.78	42.44	40.52	38.76	37.13	34.23	31.72	29.52	28.07	26.47	24.89
3/4	.750	55.20	51.96	49.03	46.62	44.18	42.18	40.33	38.64	35.61	32.99	30.70	29.18	27.52	25.87

M.W. E=29,000,000 → ← E=28,500,000 → MUSIC WIRE—ASTM A228

GAUGE		24	25	26	27	28	29	30	31	32	33	34	35	36	37
DECIMAL		.055	.059	.063	.067	.071	.075	.080	.085	.090	.095	.100	.106	.112	.118
DES. STRESS		180,	178,	176,	174,	173,	171,	169,	167,	166,	164,	163,	161,	160,	158,
TORQUE in.lb.		2.941	3.590	4.322	5.139	6.080	7.084	8.497	10.07	11.88	13.81	16.00	18.83	22.07	25.49
13/16	.8125	38.38	35.54	33.06	31.42	29.61	27.83	25.93	24.25	22.90	21.55	20.46	19.19	18.17	17.14
7/8	.875	41.14	38.09	35.42	33.65	31.70	29.79	27.75	25.94	24.48	23.03	21.86	20.49	19.39	18.29
15/16	.9375	43.91	40.64	37.78	35.88	33.80	31.75	29.56	27.63	26.07	24.52	23.26	21.80	20.62	19.44
1	1.000	46.67	43.19	40.14	38.11	35.89	33.71	31.38	29.32	27.65	26.00	24.66	23.11	21.85	20.59
1 1/16	1.0625	49.44	45.74	42.50	40.35	37.99	35.67	33.20	31.01	29.24	27.48	26.06	24.41	23.08	21.74
1 1/8	1.125	52.20	48.28	44.86	42.58	40.08	37.63	35.01	32.70	30.82	28.97	27.46	25.72	24.31	22.89
1 3/16	1.1875	54.97	50.83	47.22	44.81	42.18	39.59	36.83	34.39	32.41	30.45	28.86	27.02	25.53	24.04
1 1/4	1.250	57.73	53.38	49.58	47.04	44.27	41.55	38.64	36.08	33.99	31.94	30.27	28.33	26.76	25.19
1 5/16	1.3125	60.50	55.93	51.94	49.28	46.37	43.51	40.46	37.76	35.58	33.42	31.67	29.63	27.99	26.35
1 3/8	1.375	63.26	58.48	54.30	51.51	48.46	45.47	42.28	39.45	37.17	34.90	33.07	30.94	29.22	27.50
1 7/16	1.4375	66.03	61.03	56.66	53.74	50.56	47.43	44.09	31.14	38.75	36.39	34.47	32.25	30.45	28.65
1 1/2	1.500	68.79	63.58	59.02	55.97	52.65	49.39	45.41	42.83	40.34	37.87	35.87	33.55	31.67	29.80

[a]The *design table* contains the design characteristics of the most generally used torsion springs made from standard wire gauges. The deflection for one coil at a specified torque and stress is shown in the body of the table.

FIG. 134 (continued)

MUSIC WIRE — ←E=29,000,000→ (gauges 15–26) | ← E=28,500,000 → (gauges 27–31)

15 .035	16 .037	17 .039	18 .041	19 .043	20 .045	21 .047	22 .049	23 .051	24 .054	25 .059	26 .063	27 .067	28 .071	29 .075	30 .080	31 .085
194,	192,	190,	188,	187,	185,	184,	183,	182,	180,	178,	176,	174,	173,	171,	169,	167,
.8168	.9550	1.107	1.272	1.460	1.655	1.876	2.114	2.371	2.941	3.590	4.322	5.139	6.080	7.084	8.497	10.07
9.646	9.171															
10.82	10.27	9.771	9.320	8.957												
11.99	11.36	10.80	10.29	9.876	9.447	9.102	8.784									
13.16	12.46	11.83	11.26	10.79	10.32	9.929	9.572	9.244	8.654	8.141						
14.33	13.56	12.86	12.23	11.71	11.18	10.76	10.36	9.997	9.345	8.778	8.279	7.975				
16.67	15.75	14.92	14.16	13.55	12.92	12.41	11.94	11.50	10.73	10.05	9.459	9.091	8.663	8.232	7.772	7.364
19.01	17.94	16.97	16.10	15.39	14.66	14.06	13.52	13.01	12.11	11.33	10.64	10.21	9.711	9.212	8.680	8.208
21.35	20.13	19.03	18.04	17.22	16.39	15.72	15.09	14.52	13.49	12.60	11.82	11.32	10.76	10.19	9.588	9.053

DEFLECTION IN DEGREES "f" FOR ONE COIL, AT THE TORQUE AND DESIGN STRESS SHOWN ABOVE; USING THE INSIDE DIAMETER IN THE LEFT COLUMN AND THE WIRE DIAMETER SHOWN AT THE TOP. TO DETERMINE THE TOTAL NUMBER OF COILS "F" REQUIRED, DIVIDE THE DEFLECTION SPECIFIED, BY THE VALUE FOR "f" SHOWN IN THE BODY OF THE TABLE. INTERMEDIATE SIZES MAY BE INTERPOLATED.

DEFLECTION IN DEGREES: $f = \frac{392\ S\ D}{E\ d}$

TORQUE "T" $= .0982\ S\ d^3$

MUSIC WIRE — ← E=28,500,000 → — MUSIC WIRE

22 .049	23 .051	24 .055	25 .059	26 .063	27 .067	28 .071	29 .075	30 .080	31 .085	32 .090	33 .095	34 .100	35 .106	36 .112	37 .118	1/8 .125
183,	182,	180,	178,	176,	174,	173,	171,	169,	167,	166,	164,	163,	161,	160,	158,	156,
2.114	2.371	2.941	3.590	4.322	5.139	6.080	7.084	8.497	10.07	11.88	13.81	16.00	18.83	22.07	25.49	29.92
16.67	16.03	14.88	13.88	13.00	12.44	11.81	11.17	10.50	9.897	9.418	8.934	8.547	8.090	7.727	7.353	6.973
18.25	17.53	16.26	15.15	14.18	13.56	12.85	12.15	11.40	10.74	10.21	9.676	9.248	8.743	8.341	7.929	7.510
19.83	19.04	17.64	16.42	15.36	14.67	13.90	13.13	12.31	11.59	11.00	10.42	9.948	9.396	8.955	8.504	8.046
21.40	20.55	19.02	17.70	16.54	15.79	14.95	14.11	13.22	12.43	11.80	11.16	10.65	10.05	9.569	9.080	8.583
22.98	22.06	20.40	18.97	17.72	16.90	15.99	15.09	14.13	13.28	12.59	11.90	11.35	10.70	10.18	9.655	9.119
24.56	23.56	21.79	20.25	18.90	18.02	17.04	16.07	15.04	14.12	13.38	12.64	12.05	11.35	10.80	10.23	9.655
26.14	25.07	23.17	21.52	20.08	19.14	18.09	17.05	15.94	14.96	14.17	13.39	12.75	12.01	11.41	10.81	10.19
27.71	26.58	24.55	22.80	21.26	20.25	19.14	18.03	16.85	15.81	14.97	14.13	13.45	12.66	12.03	11.38	10.73

MUSIC WIRE ← E=28,500,000 → | ← E=28,500,000 OIL TEMPERED MB

30 .080	31 .085	32 .090	33 .095	34 .100	35 .106	36 .112	37 .118	1/8 .125	10 .135	9 .1483	5/32 .1563	8 .162	7 .177	3/16 .1875	6 .192	5 .207
169,	167,	166,	164,	163,	161,	160,	158,	156,	161,	158,	156,	154,	150,	149,	146,	143,
8.497	10.07	11.88	13.81	16.00	18.83	22.07	25.49	29.92	38.90	50.60	58.44	64.30	81.68	96.45	101.5	124.6
17.76	16.65	15.76	14.87	14.15	13.31	12.64	11.96	11.26	10.93	9.958	9.441	9.064	8.256	7.856	7.565	7.015
18.67	17.50	16.55	15.61	14.85	13.97	13.25	12.53	11.80	11.44	10.42	9.870	9.473	8.620	8.198	7.891	7.312
19.58	18.34	17.35	16.35	15.55	14.62	13.87	13.11	12.34	11.95	10.87	10.30	9.882	8.984	8.539	8.218	7.609
20.48	19.19	18.14	17.10	16.25	15.27	14.48	13.68	12.87	12.47	11.33	10.73	10.29	9.348	8.881	8.545	7.906
21.39	20.03	18.93	17.84	16.95	15.92	15.10	14.26	13.41	12.98	11.79	11.16	10.70	9.713	9.222	8.872	8.202
22.30	20.88	19.72	18.58	17.65	16.58	15.71	14.83	13.95	13.49	12.25	11.59	11.11	10.08	9.564	9.199	8.499
23.21	21.72	20.52	19.32	18.36	17.23	16.32	15.41	14.48	14.00	12.71	12.02	11.52	10.44	9.905	9.526	8.796
24.12	22.56	21.31	20.06	19.06	17.88	16.94	15.99	15.02	14.52	13.16	12.44	11.92	10.81	10.25	9.852	9.093

←E=28,500,000→ OIL TEMPERED MB—ASTM A229 CLASS 1

1/8 .125	10 .135	9 .1483	5/32 .1563	8 .162	7 .177	3/16 .1875	6 .192	5 .207	7/32 .2188	4 .2253	3 .2437	1/4 .250	9/32 .2813	5/16 .3125	11/32 .3438	3/8 .375
156,	161,	158,	156,	154,	150,	149,	146,	143,	142,	141,	140,	139,	138,	137,	136,	135,
29.92	38.90	50.60	58.44	64.30	81.68	96.45	101.5	124.6	146.0	158.3	199.0	213.3	301.5	410.6	542.5	700.0
16.09	15.54	14.08	13.30	12.74	11.53	10.93	10.51	9.687	9.208	8.933	8.346	8.125	7.382	6.784	6.292	5.880
17.17	16.57	15.00	14.16	13.56	12.26	11.61	11.16	10.28	9.766	9.471	8.840	8.603	7.803	7.161	6.632	6.189
18.24	17.59	15.91	15.02	14.38	12.99	12.30	11.81	10.87	10.32	10.01	9.333	9.081	8.225	7.537	6.972	6.499
19.31	18.62	16.83	15.88	15.19	13.72	12.98	12.47	11.47	10.88	10.55	9.827	9.559	8.647	7.914	7.312	6.808
20.38	19.64	17.74	16.74	16.01	14.45	13.66	13.12	12.06	11.44	11.09	10.32	10.04	9.069	8.291	7.652	7.118
21.46	20.67	18.66	17.59	16.83	15.18	14.35	13.77	12.66	12.00	11.62	10.81	10.52	9.491	8.668	7.993	7.427
22.53	21.69	19.57	18.45	17.64	15.90	15.03	14.43	13.25	12.56	12.16	11.31	10.99	9.912	9.045	8.333	7.737
23.60	22.72	20.49	19.31	18.46	16.63	15.71	15.08	13.84	13.11	12.70	11.80	11.47	10.33	9.422	8.673	8.046
24.67	23.74	21.41	20.17	19.28	17.36	16.40	15.74	14.44	13.67	13.24	12.30	11.95	10.76	9.799	9.013	8.356
25.75	24.77	22.32	21.03	20.10	18.09	17.08	16.39	15.03	14.23	13.78	12.79	12.43	11.18	10.18	9.353	8.665
26.82	25.79	23.24	21.89	20.91	18.82	17.76	17.04	15.63	14.79	14.31	13.28	12.91	11.60	10.55	9.693	8.975
27.89	26.82	24.15	22.74	21.73	19.55	18.44	17.70	16.22	15.35	14.85	13.78	13.38	12.02	10.93	10.03	9.284

FIG. 135. Design Formulas for Torsion Springs

Property	Round Wire	Square Wire	Rectangular Wire
d = Wire dia.	$\sqrt[3]{\frac{10.18\ T}{S_b}}$	$\sqrt[3]{\frac{6\ T}{S_b}}$	$t = \sqrt{\frac{6\ T}{S_b\ b}}$
	$\sqrt[4]{\frac{4000\ T\ N\ D}{E\ F°}}$	$\sqrt[4]{\frac{2375\ T\ N\ D}{E\ F°}}$	$b = \frac{2375\ T\ N\ D}{E\ t^3\ F°}$
S_b = Stress, bending	$\frac{10.18\ T}{d^3}$	$\frac{6\ T}{d^3}$	$\frac{6\ T}{b\ t^2}$
	$\frac{E\ d\ F°}{392\ N\ D}$	$\frac{E\ d\ F°}{392\ N\ D}$	$\frac{E\ t\ F°}{392\ N\ D}$
N = Active Coils	$\frac{E\ d\ F°}{392\ S_b\ D}$	$\frac{E\ d\ F°}{392\ S_b\ D}$	$\frac{E\ t\ F°}{392\ S_b\ D}$
	$\frac{E\ d^4\ F°}{4000\ T\ D}$	$\frac{E\ d^4\ F°}{2375\ T\ D}$	$\frac{E\ b\ t^3\ F°}{2375\ T\ D}$
F° = Deflection	$\frac{392\ S_b\ N\ D}{E\ d}$	$\frac{392\ S_b\ N\ D}{E\ d}$	$\frac{392\ S_b\ N\ D}{E\ t}$
	$\frac{4000\ T\ N\ D}{E\ d^4}$	$\frac{2375\ T\ N\ D}{E\ d^4}$	$\frac{2375\ T\ N\ D}{E\ b\ t^3}$
T = Torque (Also = P x R)	$0.0982\ S_b\ d^3$	$0.1666\ S_b\ d^3$	$0.1666\ S_b\ b\ t^2$
	$\frac{E\ d^4\ F°}{4000\ N\ D}$	$\frac{E\ d^4\ F°}{2375\ N\ D}$	$\frac{E\ b\ t^3\ F°}{2375\ N\ D}$
ID_1 = Inside Diameter After Deflection	$\frac{N\ (ID\ free)}{N + \frac{F°}{360}}$	$\frac{N\ (ID\ free)}{N + \frac{F°}{360}}$	$\frac{N\ (ID\ free)}{N + \frac{F°}{360}}$

$$d = \sqrt[3]{\frac{10.18T}{S_b}} = \sqrt[3]{\frac{10.18 \times 4}{160000}} = \sqrt[3]{0.000\,254\,5} = 0.0\,634 \text{ in.}$$

(Note that T = torque = P X R). The nearest standard wire size is 0.063 in., the same as found my Method 1.

4. The stress is

$$S_b = \frac{10.18T}{d^3} = \frac{10.18 \times 4}{0.063^3} = 162\,849 \text{ psi}$$

(This is nearly the same as by Method 1.) The correct mean diameter D = 0.750 + 0.063 = 0.813 in.

5. The number of active coils (Fig. 135) is

$$N = \frac{EdF}{392 S_b D} = \frac{28\,700\,000 \times 0.063 \times 200}{392 \times 162\,849 \times 0.813} = 6.968 \text{ (say 7)}$$

The answer is the same as found by Method 1. The body length, total stress, reduction in ID, etc., are made exactly as shown in Method 1.

Method 3: Metric

Torque in the metric system is in newton metres (N·m). Formerly it was customary to use kilogram centimetres, but the new term should be used. However, it is simpler to use N·mm, remembering that 1 N·m = 1000 N·mm.

Conversion factors follow (also see Fig. 215):

1 oz-in. = 0.007 061 552 N·m = 7.061 552 N·mm

1 lb-in. = 0.112 98 N·m = 112.98 N·mm

1 N·m = 8.8507 lb-in. = 141.6119 oz-in. = 1000 N·mm

Strict conversion in the SI system is supposed to have all dimensions in metres on the assumption that millimetres and centimetres might eventually become obsolete, but this is problematical because small diameters of wire can be better visualized when expressed in millimetres, as in the following example. Transposing the previous Problem 4 into metric SI units:

d	= 0.063 in. X 25.4 = 1.6002 mm	(Say 1.6 mm)
ID	= 0.750 in. X 25.4 = 19.05 mm	(Say 19 mm)
T	= 4 in.-lb X 112.98 = 451.92 N·mm	(Say 450 N·mm)
S_b	= 162 849 psi X 0.006 894 757 = 1 122.8 MPa	(Say 1 123 MPa)
D	= ID + d = 19 + 1.6 = 20.6 mm	(Say 20.6 mm)
E	= 28 700 000 psi; from table, E = 197 900 MPa	(Fig. 112)

If tables of characteristics are not available, it would be necessary to assume a safe stress S_b and a mean diameter D as before.

The design formulas (Fig. 135) using the even units are

$$d = \sqrt[3]{\frac{10.18T}{S_b}} = \sqrt[3]{\frac{10.18 \times 450}{1123}} = \sqrt[3]{4.079} = 1.598 \text{ mm (say 1.6 mm)}$$

$$S_b = \frac{10.18T}{d^3} = \frac{10.18 \times 450}{1.6^3} = 1118.4 \text{ MPa}$$

S_b is quite close to the converted figure, the slight difference being dur to rounding off unwieldy figures.

$$N = \frac{EdF}{392 S_b D} = \frac{197\,900 \times 1.6 \times 200}{392 \times 1\,118.4 \times 20.6} = 7.012 \text{ (say 7 as before)}$$

$$\text{Body length} = 7 + 1 \times d = 8 \times 1.6 = 12.8 \text{ mm}$$

Allow, say, 0.35 mm between coils $\times$ 6 spaces = 6 $\times$ 0.35 = 2.1 mm

$$\text{Body length} = 12.8 + 2.1 = 14.9 \text{ (say 15 mm)}$$

$$\text{ID after deflection} = \frac{N(\text{ID free})}{N + F/360} = \frac{7 \times 19}{7 + 200/360} = 17.60 \text{ mm}$$

The curvature correction factor (Fig. 114) and other dimensions are determined exactly as shown in Method 1, but using the metric values.

Torsion Springs—Design Hints

1. *Proportions.* A spring index, inside diameter divided by wire diameter between 4 and 14 is best. Larger ratios require more than average tolerance. Ratios of 3 or less often cannot be coiled on automatic spring coiling machines because of arbor breakage. Springs with smaller or larger indexes do not give results exactly as the design formulas.
2. *Spring Index.* The spring index must be used with caution. In design formulas it is D ÷ d. For shop measurement it is OD ÷ d. For arbor design it is ID ÷ d. Conversions are easily performed by adding or subtracting 1 from D ÷ d.
3. *Total Coils.* Torsion springs with less than 3 coils buckle and are difficult to test accurately. When 30 or more coils are used, light loads will not deflect all the coils simultaneously due to friction with the supporting rod. Shop

men usually prefer to have the total coils specified to the nearest fraction in eighths or quarters such as $5^1/_8$, $5^1/_4$, $5^1/_2$, whenever possible.

4. *Rods.* Torsion springs should be supported by a rod running through the center. If unsupported, or if held by clamps or lugs, they buckle and reduce their torque and are subjected to additional stresses.
5. *Diameter Reduction.* The inside diameter reduces during deflection. This should be computed and clearance provided over the supporting rod. Also, allowances should be considered for normal spring diameter tolerances.
6. *Winding.* The coils of a spring may be closely or loosely wound, but they should not be wound with the coils pressed tightly together. Tight-wound springs with initial tension in the coils do not deflect uniformly and are difficult to test accurately. A slight space between the coils of about 20 to 25 percent of the wire thickness is desirable. Square and rectangular sections should be avoided whenever possible as they are difficult to wind, expensive, and are not always available.
7. *Hand.* The hand or direction of coiling should be specified and the spring designed so deflection causes the spring to wind up and have more coils. This increase in coils and in overall length should be allowed for during design. Deflecting the spring in an unwinding direction causes high stresses and may cause early failure. When a spring is sighted down the longitudinal axis, it is "right hand" when the direction of the wire into the spring takes a clockwise direction or if the angle of the coils follows an angle similar to the threads of a standard bolt or screw, otherwise it is "left hand." A spring must be coiled right hand to engage the threads of a standard machine screw.
8. *Arm Length.* All the wire in a torsion spring is active between the points where the forces are applied. Deflection of long extended arms can be calculated by allowing one third of the arm length, from the point of force contact to the body of the spring, converted into coils. If the length of arm is equal or less than one half the length of one coil, it can be safely neglected in most applications.
9. *Bends.* Arms should be as straight as possible. Bends are difficult to produce, often are made by secondary operations, and are expensive. Sharp bends are stress raisers that cause early failure. Bend radii should be as large as practicable. Hooks tend to open during deflection; their stresses can be calculated in the same way as for extension springs.
10. *Double Torsion.* These consist of a left-hand-wound series of coils and a series of right-hand-wound coils connected at the center. They are difficult to manufacture and are expensive. It often is better to use two separate springs. For torque and stress calculations, each series is calculated separately as individual springs; then the torques are added together, but the deflections are not added.

14 Special Springs

General

In addition to the regular helical springs, there are many special types that are used occasionally.

FLAT SPRINGS

Flat springs are used to keep within restricted space conditions in various products such as switches and relays. Several materials are used including flat high-carbon spring steel, phosphor-bronze, beryllium-copper, stainless steel, nickel-silver, and a few others. Formulas for flat springs (Fig. 136) are based on standard beam formulas where the deflections are small—never more than 25 percent of the spring length.

The bending stress in any member of a machine or mechanical product, regardless of shape or material, is equal to the bending moment divided by the section modulus. This simple formula often comes in handy when designing beams, castings, springs, lever arms, and other parts of a machine:

$$S_b = \frac{B_m}{S_m}$$

The standard beam formulas used by spring designers, their bending moments, deflections, section modulus, moment of inertia, and elements of sections are found in many engineering handbooks.

FIG. 136. Formulas for Flat Springs (based upon standard beam formulas where the deflection is small, not exceeding 25 percent of L)

PROPERTY	P F L PLAN b	L P F PLAN b	L P F b	L P F b $\frac{b}{4}$
Deflection F	$\frac{P\,L^3}{4\,E\,b\,t^3}$	$\frac{4\,P\,L^3}{E\,b\,t^3}$	$\frac{6\,P\,L^3}{E\,b\,t^3}$	$\frac{5.22\,P\,L^3}{E\,b\,t^3}$
	$\frac{S_b\,L^2}{6\,E\,t}$	$\frac{2\,S_b\,L^2}{3\,E\,t}$	$\frac{S_b\,L^2}{E\,t}$	$\frac{0.87\,S_b\,L^2}{E\,t}$
Force or Load P	$\frac{2\,S_b\,b\,t^2}{3\,L}$	$\frac{S_b\,b\,t^2}{6\,L}$	$\frac{S_b\,b\,t^2}{6\,L}$	$\frac{S_b\,b\,t^2}{6\,L}$
	$\frac{4\,E\,b\,t^3\,F}{L^3}$	$\frac{E\,b\,t^3\,F}{4\,L^3}$	$\frac{E\,b\,t^3\,F}{6\,L^3}$	$\frac{E\,b\,t^3\,F}{5.22\,L^3}$
Stress, Bending S_b	$\frac{3\,P\,L}{2\,b\,t^2}$	$\frac{6\,P\,L}{b\,t^2}$	$\frac{6\,P\,L}{b\,t^2}$	$\frac{6\,P\,L}{b\,t^2}$
	$\frac{6\,E\,t\,F}{L^2}$	$\frac{3\,E\,t\,F}{2\,L^2}$	$\frac{E\,t\,F}{L^2}$	$\frac{E\,t\,F}{0.87\,L^2}$
Thickness t	$\frac{S_b\,L^2}{6\,E\,F}$	$\frac{2\,S_b\,L^2}{3\,E\,F}$	$\frac{S_b\,L^2}{E\,F}$	$\frac{0.87\,S_b\,L^2}{E\,F}$
	$\sqrt[3]{\frac{P\,L^3}{4\,E\,b\,F}}$	$\sqrt[3]{\frac{4\,P\,L^3}{E\,b\,F}}$	$\sqrt[3]{\frac{6\,P\,L^3}{E\,b\,F}}$	$\sqrt[3]{\frac{5.22\,P\,L^3}{E\,b\,F}}$

Courtesy U.S. Department of Defense, MIL-STD-29A.

Notes:

1. In the inch-pound system, F, t, b, and L are in inches, P is in pounds, and S_b is in pounds per square inch.
2. In the metric system, F, t, b, and L are in millimetres, P is in newtons, and S_b is in megapascals.
3. Recommended design stresses: for flat cold-rolled spring steel use the music wire stresses for tension springs (Fig. 79). For other materials use the torsion spring stresses for that material, such as stainless steel, beryllium copper, etc.

LEAF SPRINGS

The design of leaf springs (Fig. 137) is thoroughly covered in the SAE Handbook, Supplement 7, called *Manual on Design and Application of Leaf Springs.* A wide variety of types of leaf springs for use in the automotive industry and for railway use are available.

Many different arrangements for the assembly of the leaves for laminated springs exist, including half elliptic symmetrical and asymmetrical; quarter elliptic; with or without helper springs to absorb shock. Depending upon the type of loading, springs can be designed for either progressive or variable rates with short or long leaves.

ASTM A 147 covers limiting stresses, heat treatment, sizes, mechanical testing, test loads, and similar data.

ASTM A 68 covers carbon steel for leaf springs.

ASTM A 552 covers several alloy steels.

FIG. 137. Variable-rate leaf spring.

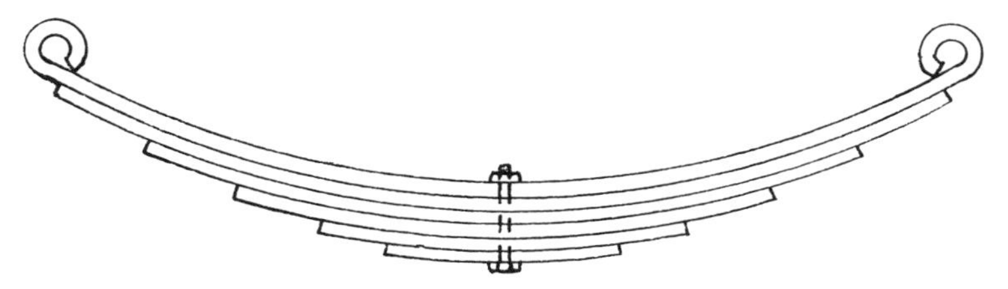

FIG. 138. Spiral spring.

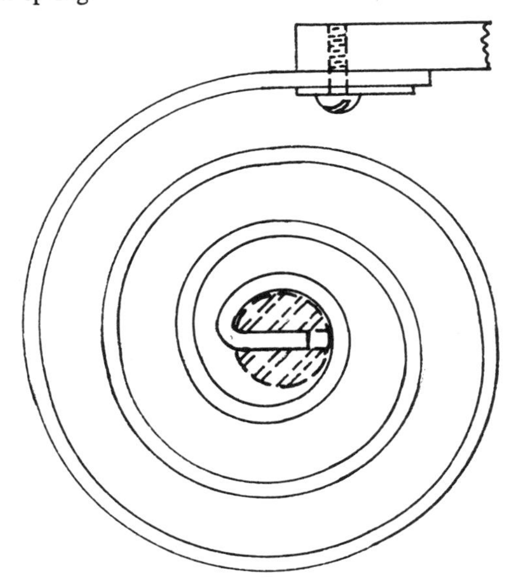

FIG. 139. Motor or power spring.

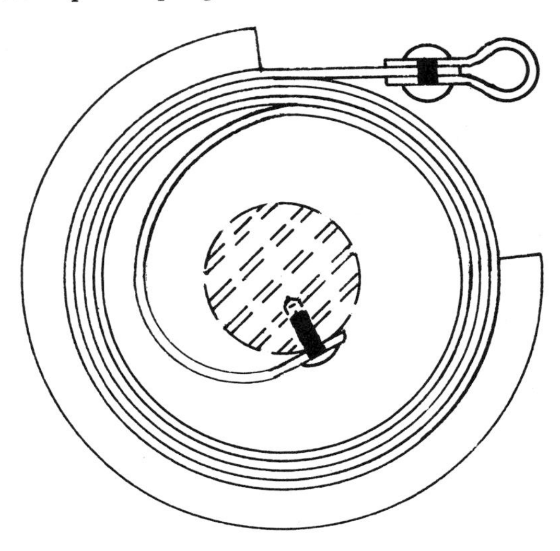

FIG. 140. Spiral power spring formulas.

Stress in bending	Active length	Thickness
$S_b = \dfrac{6T}{bt^2}$	$L = \dfrac{\pi EUt}{S_b}$	$t = \sqrt{\dfrac{6T}{S_b b}}$
Deflection	**Torque**	**Revolutions**
$F^0 = \dfrac{114.6 S_b L}{Et}$	$T = \dfrac{S_b bt^2}{6}$	$U = \dfrac{S_b L}{\pi Et}$

$$\text{ID of case} = \sqrt{2.55tL + (\text{arbor diam.})^2}$$

SPIRAL SPRINGS

If a spiral spring (Fig. 138) with two or more active coils is wound with a space between the coils and the outer end is clamped or securely fastened, the regular formulas (Fig. 140) for power springs provide accurate results. If the outer end is held with a hook or is free to rotate, more angular deflection will occur, as much as 20 percent for four coils, but proportionately less with more coils. Always specify the developed length (L) and allow for end formations and inactive portions. Usually about one half to three fourths of coil is inactive due to contact with the rotating shaft.

POWER OR MOTOR SPRINGS

Power or motor springs (Fig. 139) cannot be calculated with great accuracy. However, when the ratio of length to thickness and the area of spring to area of case meet certain requirements, it is possible to compute forces and deflections with reasonable accuracy. Either the case or the arbor may revolve.

Several spring manufacturers keep accurate records of a large number of motor springs, listing the forces and deflections, and willingly provide such data upon request. Formulas for spiral power springs are given in Fig. 140.

Requirements for Design

1. The arbor diameter should be 12 to 25 times the thickness (t) of the material.
2. The length (L) should be 5 000 to 15 000 times the thickness (t).
3. The area of the spring (L $\times$ t) should equal half the available area of the case ±10 percent. (Available area of case equals the area inside the case minus the area of the arbor.)
4. The bending stress S_b for tempered polished and blued SAE 1095 clock spring steel should not exceed 225 000 psi (1550 MPa) for thicknesses up to 0.040 in. (1 mm); 200 000 psi (1380 MPa) for 0.041 to 0.060 in. (1.1 to 1.5 mm), and 175 000 psi (1200 MPa) for thicknesses over 0.060 in. (1.5 mm). These stresses should be reduced approximately 20 percent for stainless steel and 50 percent for spring-quality nonferrous materials for average service.

E = 30 000 000 for flat spring steel; for other materials see Fig. 112.

b = breadth or width, t = thickness, T = torque = P $\times$ R, where P = force and R = distance to center line of spring. Values can be in inches and pounds per square inch or in millimetres and megapascals.

The torque (T) versus deflection (F^0) does not follow a straight-line curve as in compression springs. A spring unwound half way (50 percent) after being fully wound exerts 82 percent of its total force, not 50 percent as might be presumed (Fig. 141).

A table (Fig. 142) can be used to determine an appropriate thickness and width needed to exert a torque required. The table is based on blue-tempered spring steel SAE 1095, with a width of 1 in. (25.4 mm) for a spring fully wound up. A width one half as wide would exert one half the torque, and a width twice as wide would double the torque. The torque for any width is by direct proportion, for average service.

PROBLEM 5

What is the thickness (t) required, the stress S_b, length L, and ID of case for a spiral motor spring (Fig. 139) 1 in. wide, on a 3/8-in.-diam. arbor to exert a torque (T) of 14 in.-lb when fully wound up with 12 revolutions (U)? Also, what is the torque when the spring is unwound 6 revolutions?

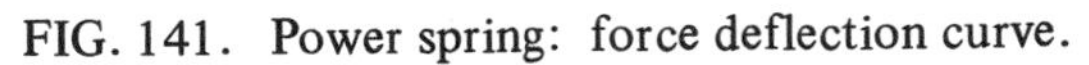
FIG. 141. Power spring: force deflection curve.

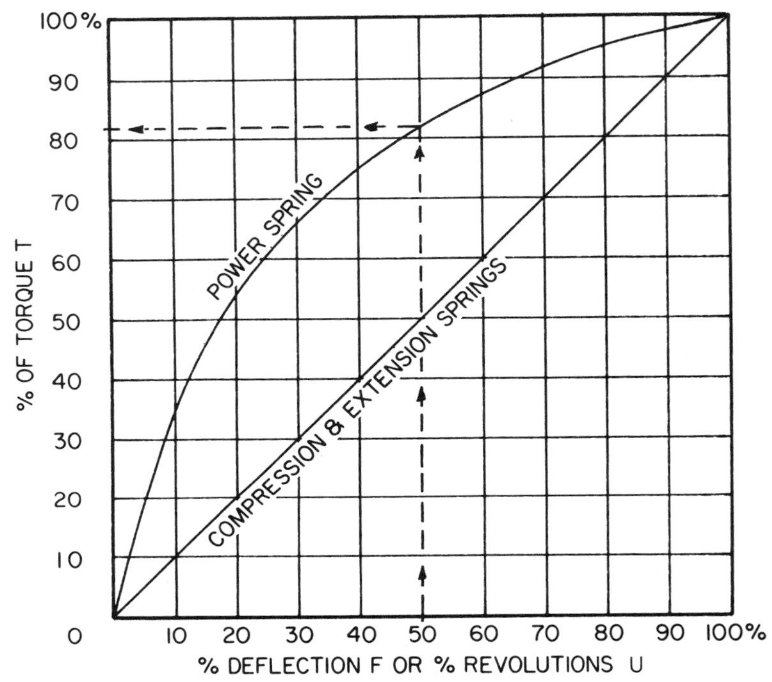

Courtesy Spring Manufacturers Institute.

FIG. 142. Torque for 1-in. (25.4-mm) width.

Thickness t		Torque T		Thickness t		Torque T	
(in.)	(mm)	(in.-lb)	(N·m)	(in.)	(mm)	(in.-lb)	(N·m)
0.008	0.20	2.40	0.27	0.040	1.00	60.0	6.78
0.010	0.25	3.75	0.42	0.045	1.14	67.5	7.63
0.012	0.30	5.40	0.61	0.050	1.27	84.4	9.54
0.015	0.38	8.45	0.95	0.060	1.52	120	13.56
0.018	0.46	12.15	1.37	0.070	1.78	143	16.16
0.020	0.51	15.00	1.69	0.080	2.03	186	21.02
0.025	0.64	23.40	2.64	0.090	2.29	236	26.66
0.030	0.76	33.70	3.81	0.100	2.54	292	32.99
0.035	0.89	46.00	5.20	0.125	3.18	456	51.52

Method 1: By Inch-Pounds

From the torque table (Fig. 142), a torque of 15 in.-lb (which is close to the requirement) requires a thickness (t) of 0.020 in. for a 1-in. width. Using this thickness, in the formulas, all requirements can be found:

$$\text{stress } S_b = \frac{6T}{bt^2} = \frac{6 \times 14}{1 \times 0.020^2} = 210\,000 \text{ psi}$$

$$\text{length } L = \frac{\pi EUt}{S_b} = \frac{3.1416 \times 30000000 \times 12 \times 0.020}{210\,000} = 107.7 \text{ in. (say 108 in.)}$$

$$\text{ID of case} = \sqrt{2.55tL + (\text{arbor diam.})^2} = \sqrt{2.55 \times 0.020 \times 108 + (0.375)^2}$$

$$= \sqrt{5.648} = 2.376 \text{ in. (say } 2\,{}^3/_8 \text{ in.)}$$

The torque when the spring is unwound 6 revolutions is found from Fig. 141, as follows:

The percentage of unwinding = 6 ÷ 12 × 100 = 50 percent. From the bottom of the curve, the 50 percent figure shown by the dashed line intersects the curve at 82 percent. Therefore the torque = 82 percent of 14 in. lb = 11.48 in.-lb.

Method 2: By Metrics

The same problem with dimensions transposed to the nearest even SI units appears as follows: width b = 25 mm, arbor = 9.5 mm, torque T = 1.58 N·m, revolutions U = 12.

From the torque table, Fig. 142, for a torque (T) of 1.69 N·m (which is close enough to 1.58) a thickness of 0.51 mm is found (use 0.50 mm).

The same formulas are used as before (Note: The torque T in newton metres must be multiplied by 1000 to have it conform with dimensions in millimetres.

$$\text{stress } S_b = \frac{6T}{bt^2} = \frac{6 \times (1.58 \times 1000)}{25 \times 0.50^2} = 1517 \text{ MPa}$$

$$\text{length } L = \frac{\pi EUt}{S_b} = \frac{3.1416 \times 206\,800 \times 12 \times 0.50}{1\,517} = 2569.6 \text{ mm}$$

(2569.6 mm × 0.3937 = 101.17 in. which is quite close to the 108 in. found in the previous solution, thereby proving the correctness. The slight difference is caused by rounding off the SI units.)

FIG. 143. Constant force spring.

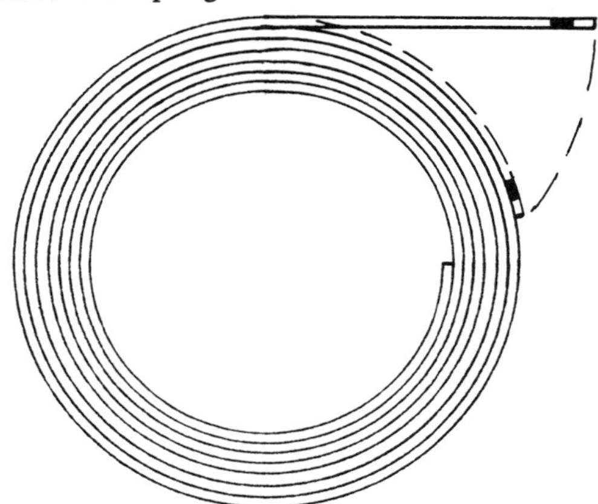

Note: The outer end usually lays curved against the body.

$$\text{ID of case} = \sqrt{2.55tL + (\text{arbor diam.})^2} = \sqrt{2.55 \times 0.50 \times 2569.6 + (9.5)^2}$$

$$= \sqrt{3366.49} = 58.021 \text{ mm (say 60 mm)}$$

(60 mm × 0.03 937 = 2.36 in. which is quite close to the $2^3/_8$ in. in the previous example.)

Other Types of Springs

CONSTANT FORCE SPRINGS

Constant force springs (Fig. 143) have an appearance similar to spiral power or motor springs, but the inner end is usually not fastened, thus leaving it free to rotate. The coils of flat strip are formed to remain in a coiled position without expanding; retaining rings to hold the coils are not required. Such springs can have the outer end pulled outward, and the force exerted to uncoil the spring can be uniformly constant for each increment of deflection. Most constant force springs are fitted snugly over rollers with the roller diameter about 15 to 20 percent larger than the ID of the inner coil to assure a snug fit. Various types of mountings to obtain different loadings can be assembled. Flat blue-tempered SAE 1095 spring steel strip and stainless steel types 301 and 302 strip, with Rockwell hardness of C48 to C52 and tensile strengths above 250,000 psi (1725 MPa) are the most popular materials used. Registered trade names are often used, such as "negator" and "spirator."

The complicated design procedures can be eliminated by selecting a standard design from the thousands now available from several spring manufacturers.

FIG. 144. Belleville washers.

For Stacks of Springs in Parallel

Total Deflection=

Deflection of single spring

Total Force=

Force on one spring x number of springs

For Stacks of Springs in Series

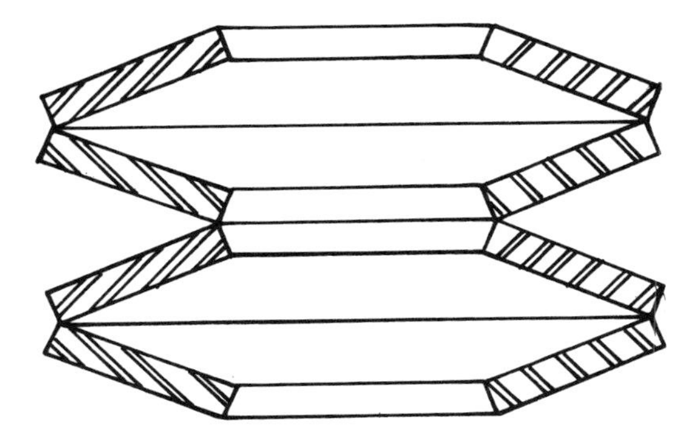

Total Deflection=

Deflection per spring x number of springs in stack

Total Force=

Force on one spring

A constant force spring on a storage drum with the outer end fastened to an output drum becomes a constant torque motor spring. The spring is reversely wound against its natural curvature to store energy. On releasing the output drum, the spring unwinds and returns over its storage drum.

BELLEVILLE WASHERS

Dished washers sustain relatively heavy forces with small deflections. The deflections can be increased with the same applied force by stacking the washers as shown in the lower portion of Fig. 144. Other variations in deflections and forces are obtainable by varying the stacking method. Each washer should be treated as one coil of a compression spring to analyze the deflections and forces exerted by different stackings.

Calculations for stress analysis, forces, and deflections are quite complicated, and several complete sets of calculations are often necessary. This is because the wide variations in ratios of OD to ID, height to thickness, and other factors require several formulas and constants obtained from more than 24 curves.

Several companies specialize solely in the manufacture of Belleville washers, curved, and wave-type washers. A few spring manufacturers also make such washers. Fortunately, they now stock these items in many sizes and willingly supply lists showing sizes, forces, and prices; designers are urged to obtain such lists rather than devote hours of calculations to obtain characteristics of washers that have already been standardized.

VOLUTE SPRINGS

Heavy sizes, made from thick, annealed steel bars or plates and then hardened and tempered, have been used as shock absorbers on Army tanks and heavy field artillery. They are seldom used in industry because of their high cost, long production time, difficulties in forming, and lack of proper materials in the sizes required. Small, light volute springs formerly used in hand tools have often been replaced with standard compression springs at much lower cost.

TORSION BARS

The automotive industry has found several applications where a straight bar can be twisted to provide a force. Such bars can be used to counterbalance the weight of hoods, absorb road shocks, and so on.

The formulas in Fig. 145 can be used to calculate the torsional stress and the degrees of deflection in a straight rod subject to a twisting action. The torque T is usually known.

FIG. 145. Formulas for torsion bars.

Section	Stress S	Deflection F^0	d (or t)
Round	$\frac{5.09T}{d^3}$	$\frac{584TL}{Gd^4}$	$d = \sqrt[3]{\frac{5.09T}{S}}$
Square	$\frac{4.8T}{t^3}$	$\frac{407TL}{Gt^4}$	$d = \sqrt[3]{\frac{4.8T}{S}}$

Note: use the recommended design stresses for compression springs for the material selected (see Chapter 10).

GARTER SPRINGS

Garter springs are long, closely coiled extension springs with the ends fastened together and then used in the form of a ring to hold mechanical seals on a shaft, or to hold round segments together, or as driving belts such as are often found in cameras, projection machines, and similar devices. Several methods of fastening the ends are used: (1) half hooks on each end can be hooked together; (2) a small screw can be inserted tightly in one end and the other end of the spring is reverse twisted so it winds itself onto the projecting end of the screw; (3) several coils on one end can be wound at a reduced diameter—the spring is then reverse twisted and the reduced end is threaded into the other end.

A familiar application of a garter spring is for an oil seal used on a shaft. The pressure per lineal inch exerted by the spring on the shaft is determined by the amount of contraction exerted by the spring. This is often difficult to calculate precisely due to the many variables involved including initial tension, rate, friction, effect of end fasteners, etc., but an approximation of this pressure follows:

First determine the deflection of the spring.

If the ID of the connected spring ring is stretched to its loaded position, the longitudinal deflection $F = \pi$ (shaft diam. minus ID of free spring ring). The rate of the spring should be calculated by the usual formulas (Fig. 118); this rate times the deflection equals the spring tension. The total compressive force on the shaft equals twice this tension, and the compressive force per inch of circumference on the shaft equals this force divided by the shaft diameter.

If the spring has initial tension, which is customary, the force on each inch of circumference on the shaft is obtained by the following formula:

$$\text{P per inch} = 2\pi r - \frac{2\pi \text{ ID } r}{\text{shaft diam.}} + \frac{2 \text{ IT}}{\text{shaft diam.}}$$

where r = rate and ID = inside diameter of the spring ring with ends fastened together in the free contracted position before stretching over the shaft.

The pressure of each coil is P divided by the number of coils per inch.

SNAP RINGS

Retaining rings, usually made from square or rectangular wire, have been used as housing shoulders to hold ball bearings in place. Most sizes have been standardized, including commercial tolerances and allowable variations. This data is available from ball bearing manufacturers. In addition, a wide variety of retaining rings, both internal and external types, are available and stocked by several manufacturers.

STRANDED WIRE SPRINGS

Compression springs made from stranded music wire (somewhat similar to wire cable) have been used to dampen the migratory waves that traverse a spring under shock loading. They have been used successfully in machine guns, small weapons, and stapling guns to dampen the high-velocity displacement of the coils. Usually three wires about 1/16 in. (1.6 mm) in diameter or smaller are twisted together. Each individual wire may be given a helical twist just prior to being joined in the stranding operation to assure a tight contact in the strand.

It is essential that the hand of the strand be opposite in direction to the hand of the spring, otherwise the torsional stresses will tend to loosen the strand. The ends of the strand or the end coil should be soldered, brazed, or a touch of weld can be used to prevent unraveling of the strand. Shotpeening is not recommended, as the small shot lodges tightly between the strands and is difficult to detect and remove.

The design procedure is similar to that used for regular compression springs, using the same formulas (Fig. 118) except that each strand carries its proportional share of the load. Also, a spring made from three-stranded wires with the area of the three diameters equal to 68 percent of the diameter of a single wire customarily used for a spring will exert the same loads and deflections, have the same rate, stress, solid height, and pitch as an equal size conventional spring.

Stranded wire is difficult to obtain in some sizes, and such springs are quite expensive when compared to comparable compression springs.

CLUTCH SPRINGS

Clutch springs, also called wrap springs, are a form of close-wound torsion spring made of square wire. They are accurately made with unusually close tolerances on their inside diameters to fit snugly over rotating shafts. If placed over two shafts with ends abutting, the spring becomes an overrunning clutch allowing both shafts to revolve in one direction, but on reversal, the inside diameter slightly enlarges, permitting only one shaft to revolve. They can be used as positive action on-off clutches, so precise acting as to permit one revolution or a partial revolution, and one may be combined with another acting as a brake, to start and stop rotation in any desired position with a minimum of cumulative error.

Interesting Problems

Spring manufacturers have long been known to employ engineers with special abilities to solve complicated problems. Their customers often call upon them for help even though the problems may be only remotely connected to the subject of springs.

For the mathematician, many interesting problems arise in the spring industry, such as: How fast does a compressed spring expand? How long, in seconds, does it take for a compressed spring to return to its original length? What is the time required for a spring to move a weight? What force should a spring exert to propel a 1-lb ball 20 ft vertically in 1 sec, and what spring deflection is required? How long does it take a spring to move a trip hammer weighing 2 lb through a required angle and strike an anvil with a force of 40 lb?

These and many similar problems are solved by knowing that the energy of a spring equals one half the total load times the distance moved. Energy also equals one half the mass times the square of the velocity, $E = \frac{1}{2} mv^2$. Work equals force times distance.

It is also interesting to note that it takes a spring exactly the same time to return to zero force from any deflected position. A compression spring deflected half way to its solid height takes as long to return to normal as it does if deflected to solid. Therefore, to get a fast-operating spring, it should be deflected from one loaded position to a lesser loaded position.

A few problems of general interest follow.

PROBLEM 6

A sailboat (Fig. 146) uses a cable (as a halyard) consisting of 7 strands of 0.025-in.-diam. stainless steel wire to hoist a sail up to the top of its mast. The sail and its support rod (gaff) weigh 60 lb. An extension spring that will

FIG. 146. Spring and cable for sailboat.

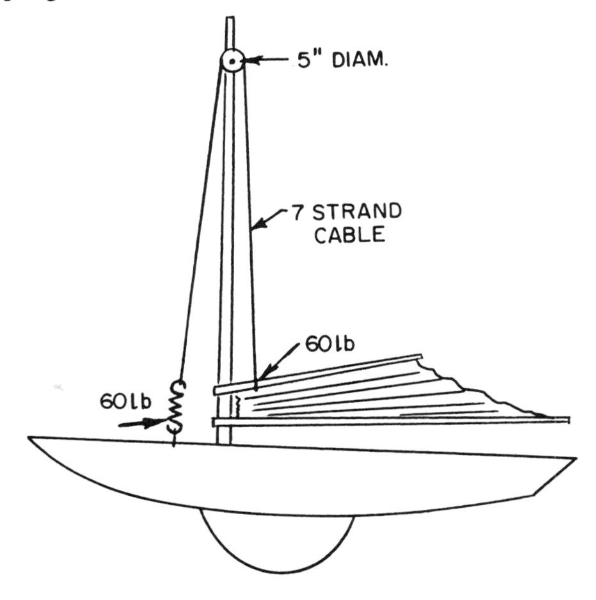

deflect an additional inch in case of overload, is needed to hold the bottom of the cable. The cable passes over a 5-in.-diam. pulley at the top of the mast. Design a proper spring and check the stresses in the cable to see if it is safe.

Analysis

A study of the problem indicates that (1) a compression spring fitted with drawbars (or one with a long bolt through the center) should be used to prevent overstretching; (2) a spring made from stainless steel type 302 should be used to avoid corrosion; (3) any reasonable diameter and length of spring would be satisfactory.

Part 1

The maximum force on the spring would be 60 lb plus 30 lb for the extra inch of deflection, which equals 90 lb at 3 in. deflection.

Using the formulas (Fig. 118) and assuming a stress of 80 000 psi (Fig. 90), select an arbitrary D of, say, 1 in.; G is 10 000 000 (Fig. 112) and the wire diameter d is

$$d = \sqrt[3]{\frac{2.55PD}{S}} = \sqrt[3]{\frac{2.55 \times 90 \times 1}{80\,000}}$$

$$= \sqrt[3]{0.00286875} = 0.142 \text{ in.}$$

The number of active coils is

$$N = \frac{Gdf}{\pi SD^2} = \frac{10\,000\,000 \times 0.142 \times 3}{3.1416 \times 80\,000 \times 1^2} = 16.95 \text{ (say 17)}$$

The balance of the design can be calculated the same as was done in Problem 6, but there is no need to do it at this time.

Part 2

Each strand of wire in the cable is subjected to a tensile force of 90 lb ÷ 7 = 12.86 lb and is also subjected to a tensile stress when being bent around the 5-in.-diam. pulley.

The area of the wire = $0.7854\ d^2 = 0.7854 \times 0.025^2 = 0.000491$. The stress due to the force is

$$S_t = \frac{P}{A} = \frac{12.86}{0.000491} = 26\,190 \text{ psi}$$

This is a low tensile stress, but it must be added to the tensile stress caused by the pulley.

Part 3

The tensile stress due to bending the wire strand as it goes around the pulley is

$$S_b = \frac{dE}{D + d}$$

E is 28 000 000 psi (Fig. 112), so

$$S_b = \frac{0.025 \times 28\,000\,000}{5 + 0.025} = 139\,300 \text{ psi}$$

The total tensile stress is 26 190 + 139 300 = 165 490 psi. Checking this stress with the curves (Fig. 91), it is seen that the stress is below the lowest curve and therefore is satisfactory.

PROBLEM 7

Design a spring for a key-operated electric switch requiring an open-wound torsion spring that must first be compressed ¾ in. when the key is inserted and then be rotated 90° when the key is turned. A light compressive force of ¾ lb is required, and the torque at 90° should be more than 1/8 in.-lb. The OD is 5/8 in. Expected life is 1 000 000 cycles. Determine the wire diameter, number of coils, and all stresses.

Analysis

The object is to have a switch that requires the operator to hold the key in the contact position, otherwise the key will be ejected from the switch. Music wire could be used, as it is tough. The forces are low and the stresses in compression and torsion must be low because the combination of stresses are hard on any material. The end coils should have short arms to hold the coils during the twisting operation, but a sketch or figure is not required at this stage.

Part 1

Compression. From the tables (Fig. 115) to the right of 5/8-in. OD in the first box, a load of 1.782 lb and a deflection per coil of 0.331 in. is found and at the top of that column, a wire diameter of 0.030 in. is located.

The 0.331-in. deflection is for oil-tempered MB wire with a modulus of elasticity G of 11 200 000 psi (Fig. 112), but 0.030-in.-diam. music wire has a G value of 12 000 000 psi (Fig. 112), so the 0.331 in. must be reduced in proportion to the G values as follows:

$$f = 0.331 \times \frac{11\,200\,000}{12\,000\,000} = 0.3089 \text{ in. (say 0.309 in.)}$$

Using the percentage method, the force percent = ¾ lb ÷ 1.782 × 100 = 42 percent.

The stress S_t at ¾ lb = 42 percent of 100 000 = 42 000 psi.

The deflection per coil f = 42 percent of 0.309 = 0.1297 (say 0.130 in.).

The number of active coils N = 0.750 ÷ 0.130 = 5.77 (say 5¾).

The solid height = 5¾ + 1 × 0.030 = 0.203 in.

The free height = the solid height, 0.203 in., plus the deflection, 0.75 in., plus about 20 percent for extra deflection, say 0.15 in., = 1.103 in. (say 1 7/64 in.). This design will be satisfactory if it can also exert the torque required.

Part 2: Torsion

The turning action of the key causes a torque T and a tensile or bending stress S_b.

The spring with 5¾ coils of 0.030-in.-diam. music wire and 5/8-in. OD will be twisted 90°. The mean diameter D = OD − d = 0.625 − 0.030 = 0.595 in. The torque is

$$T = \frac{Ed^4F}{4\,000ND} = \frac{29\,500\,000 \times 0.030^4 \times 90}{4\,000 \times 5.75 \times 0.595}$$

$$= 0.157 \text{ in.-lb.}$$

This torque is satisfactory, as it is more than the 1/8 in.-lb specified.

The stress is

$$S_b = \frac{10.18T}{d^3} = \frac{10.18 \times 0.157}{0.030^3} = 59\,195 \text{ psi}$$

And now the question arises: Are the combined stresses safe? According to Timoshenko, the equivalent torsion stress is

$$S_{bt} = \tfrac{1}{2}\sqrt{S_b{}^2 + 4S_t{}^2} = \tfrac{1}{2}\sqrt{59\,195^2 + 4(42\,000)^2}$$

$$= \tfrac{1}{2}\sqrt{3\,504\,048\,025 + 4(1\,764\,000\,000)}$$

$$= \tfrac{1}{2}\sqrt{10\,560\,048\,025} = 50\,138 \text{ psi in torsion}$$

The tension equivalent is twice this or 102 762 psi.

Comparison of these stresses with the curves (Figs. 78 and 79) show that the 50 138 psi is below the lowest curve (Fig. 78) and the 102 762 psi is also below the lowest curve (Fig. 79). Therefore, a long life can be expected. If these stresses are multiplied by their curvature correction factors (Fig. 114), they are still quite safe and provide a safety factor so that additional deflection and rotation can be made if desired.

PROBLEM 8

Design a compression spring for a shock absorber with three forces (Fig. 147), suitable for 1 000 000 deflections, with no permanent set at a compressed height of 2¾ in.

FIG. 147. Shock absorber spring.

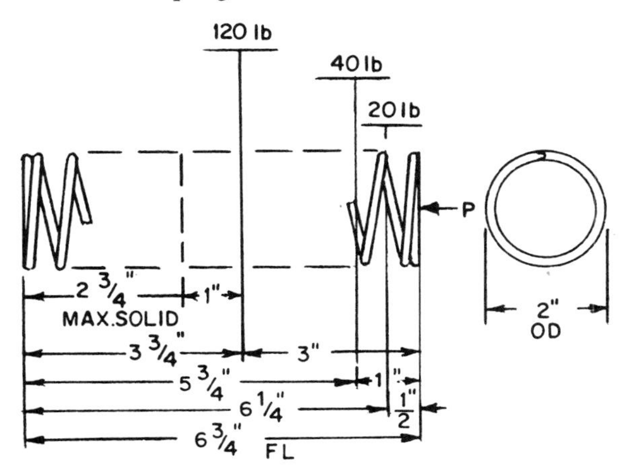

Analysis

A study of the design indicates that chromium-vanadium wire, ASTM A231, could be selected due to its good resistance to shock; the end coils should be closed and ground to reduce buckling; the rate is 120 ÷ 3 = 40 lb/in., the force at 2¾ in. is 40 × 4 = 160 and, from the tables (Fig. 115), using an OD of 2 in., a force of 194.3 lb and a deflection per coil of 0.436 in. will be found in the column headed by no. 5 (0.207-in.) wire.

By using these data, the stress and number of coils can be determined as follows:

Using the percentage method, the force percent is 120 ÷ 194.3 × 100 = 61.76 percent.

Therefore, the stress at 120 lb = 61.76 percent of 100 000 = 61 760 psi.

The deflection per coil = 61.76 percent of 0.436 = 0.269 in.

The number of active coils = 3 ÷ 0.269 = 11.15 (say 11).

The total number of coils = 11 + 2 = 13.

The solid height = 13 × 0.205 – 2.665 in., which is less than the 2¾-in. maximum solid height and therefore is satisfactory.

The spring index = OD ÷ d – 1 = 2 ÷ 0.205 – 1 = 8.75.

The correction factor K = 1.17 (Fig. 114).

The total stress at 120-lb load = 1.17 × 61 760 = 72 259 psi.

The total stress at 2¾-in. length = 61 760 ÷ 3 × 4 = 82 346 psi, and corrected = 1.17 × 82 346 = 96 345 psi.

From the fatigue strength curves (Fig. 86), it will be seen that both stresses are below the lower curve and therefore are quite safe. Also, by comparing the stresses with the endurance limit curves (Fig. 110), it will be seen that the spring will have an exceptionally long life. If the shock absorber is subjected to the dynamic force of a falling weight, additional calculations are required (Fig. 122). Also refer to Problem 9.

FIG. 148. Portable diving board.

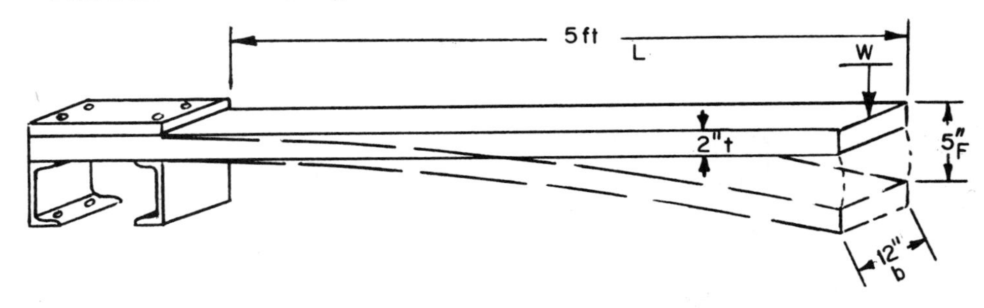

PROBLEM 9

How far will a steel spring deflect when a falling weight of 2 lb drops 10 in. upon it? The rate of the spring is known to be ½ lb/in.; to what force would it be subjected?

Analysis

From the formula (Fig. 122) and solving for F:

$$F_1{}^2 = \frac{2P(h + F)}{r} = \frac{2 \times 2\,(10 + F)}{\frac{1}{2}} = \frac{40 + 4F}{\frac{1}{2}} = 80 + 8F$$

Therefore,

$$F_1{}^2 = 80 + 8F \quad \text{or} \quad F_1{}^2 - 8F - 80 = 0$$

This equation is solved by the quadratic formula:

$$x = \frac{-b \pm \sqrt{b^2 - 4ac}}{2a}$$

where a = 1, b = –8, c = –80, and x = F.

Solving for the maximum value of deflection:

$$F = \frac{8 + \sqrt{64 + 4 \times 1 \times 80}}{2 \times 1} = 13.798 \text{ in. (say 13.8 in.)}$$

The force would then equal the rate times this deflection, or 0.5 × 13.8 = 6.9 lb.

PROBLEM 10

A manufacturer wanted to make a wooden diving board (Fig. 148). The board is to be of oak and will be subjected to a maximum deflection of 5 in. (He therefore considered it to be a spring!). What is the weight of the board, the total stress, and the force at 5 in. deflection?

Analysis

Various handbooks give widely different values for wood. The weights usually are not listed, but specific gravities at various moisture contents are listed. For select-quality, seasoned white oak, with 70 percent moisture content, the specific gravity averages 0.59, the modulus of elasticity E varies from 2 200 000 to 780 000 (average = 1 490 000), and the proportional limit varies from 4700 to 8200 psi (average = 6450 psi).

Specific gravity is the relationship between the weight of a substance and the weight of water. Water weighs 62.4 lb/ft^3, therefore the weight of this white oak is 0.59 X 62.4 = 36.8 lb/ft^3.

The number of cubic feet in the board = $^2/_{12}$ X 1 X 5 = 0.833. Therefore the weight = 36.8 X 0.833 = 30.65 lb.

The weight of the board may be considered as concentrated at its center of gravity, which in this case is 2.5 ft (30 in.) from either end. The area of maximum stress is at the support.

The stress due to the weight of the board alone (Fig. 136) is

$$S_1 = \frac{6PL}{bt^2} = \frac{5 \times 30.65 \times 30}{12 \times 2^2} = 95.8 \text{ psi}$$

The stress due to deflection of 5 in. (Fig. 136) is

$$S_2 = \frac{3EtF}{2L^2} = \frac{3 \times 1\,490\,000 \times 2 \times 5}{2 \times 60^2} = 6\,208 \text{ psi}$$

The total stress = 95.8 + 6208 = 6303.8 psi, which is below the average proportional limit.

The load supported by the board at 5 in. deflection (Fig. 136) is

$$P = \frac{Ebt^3F}{4L^3} = \frac{1\,490000 \times 12 \times 2^3 \times 5}{4 \times 60^3} = 827.77 \text{ lb}$$

The rate r = P/F = 827.77/5 = 165.55 lb/in.

Note: The deflection and rate therefore are also sufficient to take care of the dynamic loading of a man weighing 165 lb. With static loading the board will deflect 1 in. with a suddenly applied force, 2 in.; and with a bouncing force, 5 in. (Fig. 122), provided that he doesn't bounce too high!

SPRING MANUFACTURE

15 Methods

Introduction

The spring manufacturing industry, consisting of only a few hundred plants scattered throughout the United States, is small when compared with the automobile and machine tool industries, but it is one of the most important producers of parts that are vital to practically all products and equipment made.

Before World War I, many machine builders made their own springs, but they found that design and manufacture was difficult, time consuming, and expensive. Automatic coiling machines were practically unknown, but soon they were on their way and in the 1920s many spring manufacturing plants were established and other manufacturers gladly gave up trying to make their own springs. However, there will always be some product manufacturers producing a few of their own specialized springs economically.

The design of automatic equipment for high-production coiling, grinding, testing, looping, and the application of electronic equipment and numerical control, has brought more qualified engineers into the industry so that the future looks bright in this highly competitive field.

Manufacture

Some of the more simply designed compression, extension, and torsion springs may be made by only one coiling operation, but this is unusual, as most designs require several operations. The usual sequence of operations for high-quality compression springs, in proper order, are (1) checking quality of wire, (2) automatic coiling, (3) heating to remove residual stresses caused by the coiling operation, (4) pressing to remove permanent set, (5) measuring to segregate into groups for better grinding, (6) grinding end coils to the proper degree, (7) testing for forces, deflections, and solid height, (8) shotpeening to increase fatigue life, (9) electroplating for rust resistance, (10) baking to remove hydrogen embrittlement, (11) final inspection for appearance and quality control, and finally, (12) counting, packaging, shipping, and billing.

In addition to the operations above, it may also be necessary to include others, such as deburring or chamfering the end coils, magnaflux testing for seams, stamping or engraving part numbers, coloring for identification, 100 percent checking of dimensions and forces and fatigue testing to assure the required fatigue life. Extension and torsion springs may require additional operations, including looping, bending, coning, assembling with special hooks, pulling out ends, and forming to specifications. Flat springs may require piercing, clipping, bending, slotting, or threading. And, all along the way, general inspection and quality control methods may be required on many operations. Heat treatment alone is a complex operation and must be done with extreme care to obtain the best possible results.

Coiling

LATHE COILING

Prior to 1915 only a few dozen spring companies existed, and they coiled practically all of their springs on lathes. Even today, lathe coiling is often done where the quantity is low and the wire size is above $^{1}/_{8}$ in. (3 mm) diameter. The lathes now used often are back-geared for slowly revolving the arbor, and the long worm gear shaft that pushes the carriage is specially made with heavy teeth spaced at a wide pitch. The carriage has a heavy bronze split nut that can be engaged or withdrawn from the worm gear shaft quickly. A series of springs starting with ends closed, then properly pitched, then ends closed again, and so on can be coiled to a length of several feet, say 1 to 2 m. The several springs are then chopped off in a power press or cut with high-speed, thin grinding wheels on abrasive cut-off machines.

Another method of lathe coiling that is often used eliminates the carriage completely and uses a helios pitch tool (Fig. 149). This tool can wind compression springs with a uniform pitch, then close the ends, then again wind with a pitch in the same manner performed by the specially arranged carriage described in the preceding paragraph. The tool is available in three sizes. Size I has a wire capacity of 0.008 to 0.060 in. (0.20 to 1.5 mm) for use with arbor diameters running from 0.020 to 0.550 in. (0.50 to 14 mm). Size II uses wires from 0.020 to 0.200 in. (0.50 to 5 mm) with arbors from 0.120 to 1.60 in. (3 to 40 mm). Size III uses wires from 0.080 to 0.600 in. (2 to 15 mm) with arbors from 0.40 to 3.5 in. (10 to 90 mm) in diameter.

FIG. 149. A helios pitch tool used for coiling springs on a lathe for small production runs. (Photo courtesy The Carlson Co.)

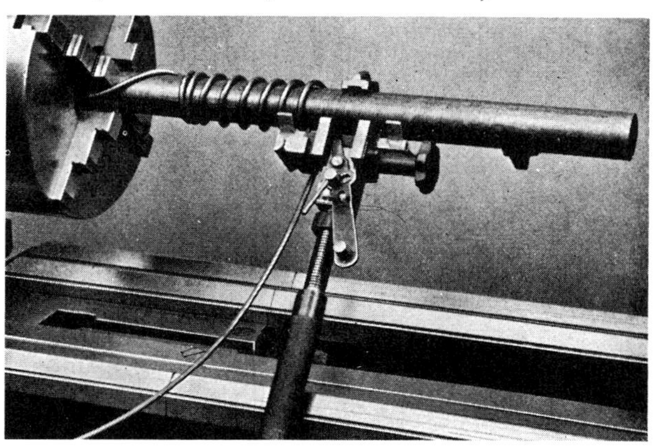

FIG. 150. A specially built lathe-type coiling machine that automatically sets the pitch and closes the end coils.

A third method of lathe coiling uses a coiled wire to act as a spacer between the coils, but this older method has been practically superseded by the methods described above.

A new lathe type of coiling machine (Fig. 150), made in England and first marketed in 1974, takes much of the labor away from lathe coiling. Instead of using a standard lathe with a carriage, this completely new machine

FIG. 151. A hand-operated machine for coiling all types of small springs, for samples and small production lots. (Photo courtesy The Carlson Co.)

uses the principle of a lathe combined with an automatic pitch control device and a pneumatic cut-off tool. This coiler, for both small and reasonably large quantities of springs, has an infinitely variable speed control, and the preset electric counters automatically coils the correct number of coils at a predetermined pitch.

The machines are made in two sizes: the smaller one coils wire up to $^5/_{16}$ in. (8 mm), with lengths up to 35 in. (890 mm) and outside diameters up to 6 in. (152 mm); the larger model coils wire from $^3/_{16}$ in. (5 mm) to $^3/_4$ in. (19 mm) and outside diameters up to 8 in. (200 mm) with lengths up to 67 in. (1700 mm).

HAND COILING

Hand-operated machines (Fig. 151) make accurate compression, extension, and torsion springs quickly in lots from 1 to 500 or more. Inexperienced operators can easily make compression springs with an exact pitch and with closed coils on both ends automatically. Extension springs can be wound tight, close, or open, and torsion springs can be made with long or short arms on both ends.

Easily adjustable cams set the pitch and length. Adjustable rack stops determine the number of coils. Arbors determine the diameters. These coilers make 2 or 2000 springs all exactly alike. 250 to 400 compression or 200 to 300 extension or torsion springs can be coiled per hour. The set-up time is usually less than 5 min.

These machines are made in two sizes. The smaller machine coils wire from 0.004 to 0.063 in. (0.10 to 1.6 mm) with outside diameters up to 1 in. (25 mm) and lengths up to 4 in. (100 mm).

The larger size coils wire diameters from 0.041 to 0.156 in. (1 to 4 mm) with outside diameters up to 2 in. (950 mm) and lengths up to 8 in. (200 mm).

AUTOMATIC COILING

The introduction of automatic spring coiling machines that could coil compression springs with both ends closed and do it quickly revolutionized the spring manufacturing industry. Although the first patent on an automatic coiler was issued in 1908, it was not until ten years later, in 1918, that the first universal coiler as we know it today was patented. And the next truly significant advancement occurred in 1939 with the introduction of a torsion attachment that could be mounted on a spring coiler to produce a torsion spring with an arm on both ends. Since that time, many patents have been granted on various types of spring machinery, including automatic coilers that can loop both ends of extension springs, coilers with double action to double the output, numerically controlled coilers, coilers that are programmable, duplex coilers, torsion spring coilers, four-slide, vertical-slide, multislide, radial tooling, grinders, electronic length measuring gauges with pitch adjustment devices, and a multitude of spring testers.

Automatic coilers are made for certain purposes and are known in accordance with their general principle of operation as follows.

Segment. Segment-type coilers (Fig. 152) take their name from the large segment of a gear which oscillates and drives the gears that turn the wire feed rolls. The distance the segment moves is adjustable, and the wire feed is quite accurate. This type is the workhorse of the spring industry and is used for short runs, usually 100 and up, and for long production runs of many millions. It coils about 50 to 200 springs per minute, depending upon the wire feed length. Several patented accessories are adaptable to increase the feed, to adjust the spring lengths, and to provide secondary operations. Several sizes are made to coil wire from 0.003 to 0.750 in. (0.075 to 19 mm). In 1976 a Japanese firm marketed an automatic coiler to coil annealed wire up to 1 in. (25 mm) in diameter.

Clutch. Clutch-type coilers (Fig. 153) use a large clutch to replace the gear segment to drive the wire feed rolls. They are used to produce long springs

FIG. 152. An automatic segment-type coiler showing tooling, adjustments, and a spring being coiled. (Photo courtesy Sleeper and Hartley Corp.)

such as garage door lifting springs, where the length of wire is far beyond the limited lengths available with segment coilers. The speed of coiling is determined by the wire feed speed, which usually runs between 100 to 200 ft (30 to 60 m) per minute, although faster feeding has been used. These coilers are not ordinarily used for short springs because fast operation of the clutch causes slippage during the engaging and disengaging operation of both clutch and brake mechanism, which causes some inaccuracies in spring lengths due to variations in the amount of wire feed. Several sizes are made.

Escapement. Escapement-type coilers use an escapement mechanism that allows the wire feed rolls to revolve continuously and then moves a roller into position under cams attached to the upper wire feed rolls to lift them a few thousandths of an inch (about 0.10 mm) above the wire. This lifting stops the wire feed so

FIG. 153. A clutch-type coiler for making long springs with many coils. (Photo courtesy Torin Corp.)

that the cut-off tool can cut off the completed spring without stopping the feed rolls. The wire feed is more than triple that of the segment coiler, and small springs can be coiled twice as fast. Several sizes are made.

Double Action. Double-action coilers (Fig. 154) use a double-acting rack to replace the gear segment, so that springs are coiled on both the forward and reverse stroke, thereby doubling production. The European system of using two coiling points improves the accuracy of spring diameters. Other improvements consist of automatic gear changing by adjusting push rods; placing adjustable controls on the front of the coiler for speed control, pitch adjustment, diameter control, stop and start buttons; and a combined output counter and production output tachometer that shows the quantity of springs being coiled per minute while the coiler is running. A programmable coiler (Fig. 155) simplifies the set-up procedure. Several sizes of these machines made in England are available.

Popcorn. Popcorn coilers (Fig. 156) are so-called because they sound like a corn popping machine, a noise caused by the noisy operation of the cams and cam follower and the cut-off tool. This type is used for high-quantity production of short springs with only a few coils, as the principal limitation is in the short wire feed. The wire feed rolls are grooved for about 85 percent of

FIG. 154. An automatic double-action machine for high-speed coiling. (Photo courtesy Bennett Tools Ltd.)

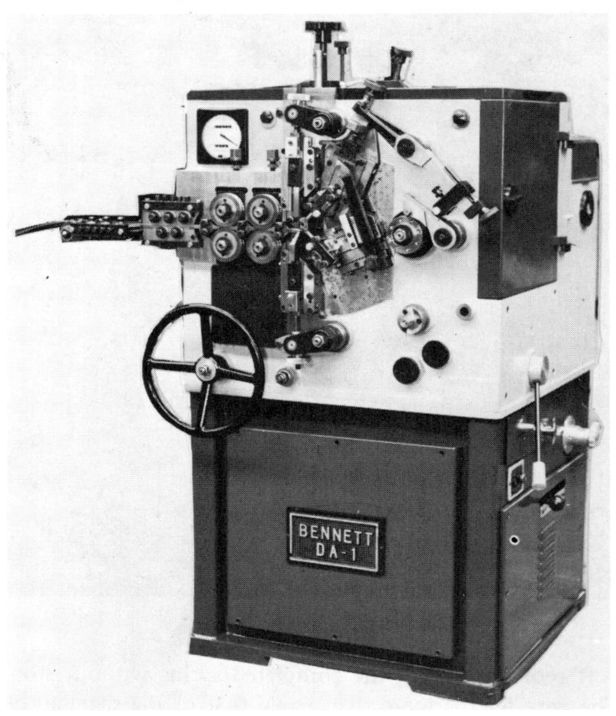

their periphery, and one spring is produced for each revolution. These machines are made in two sizes. The smaller machine has a wire feed of 8 in. (200 mm) with a production output from 10 000 to 18 000 springs per hour. The larger model has a wire feed up to 20 in. (500 mm) with a production output from 5 000 to 12 000 springs per hour.

AUTOMATED MACHINES

Several methods are used for looping the ends of extension springs right on the coiling machine (Fig. 157). Some machines coil, loop, heat, test, separate, and count, and package the springs entirely automatically. These machines are made in West Germany.

TORSION COILERS.

Torsion coilers (Figs. 158, 159, and 160) coil springs over retractable reciprocating arbors that move up and down. Several adjustable slides hold tools to form special bends. The arbor rotates to form the coils and retracts to produce

FIG. 155. A programable coiler with simplified adjustments, showing tooling and a spring being coiled. (Photo courtesy Sleeper and Hartley Corp.)

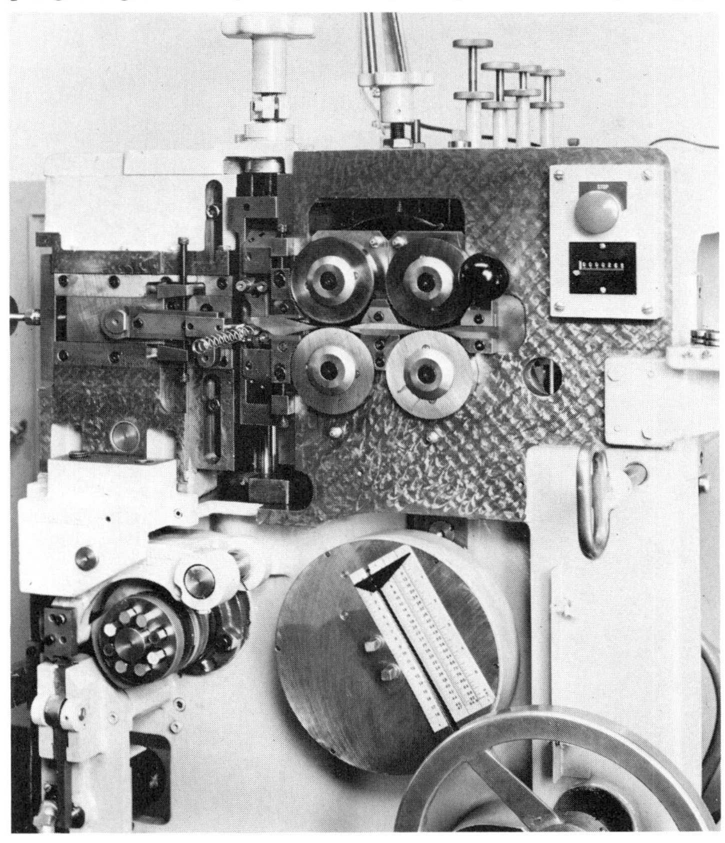

the desired coil spacing, then unwinds and the completed spring with all bends drops off the arbor. The machines operate continuously and do not require much attention after set-up. However, special tooling is usually required for each type of spring produced. Several types and sizes of these machines are available.

SPECIAL COILERS

Several types of special coilers are obtainable for coiling garter springs, flexible shafts, casings, sewer cleaning augers, rings, lock washers, rectangular springs, bed springs, and similar items.

HOT COILING

Springs using wire diameters about ½ in. (12 mm) and larger (often even

FIG. 156. A popcorn-type coiler for high-speed production of short springs. (Photo courtesy E. A. Samuel Machine Co.)

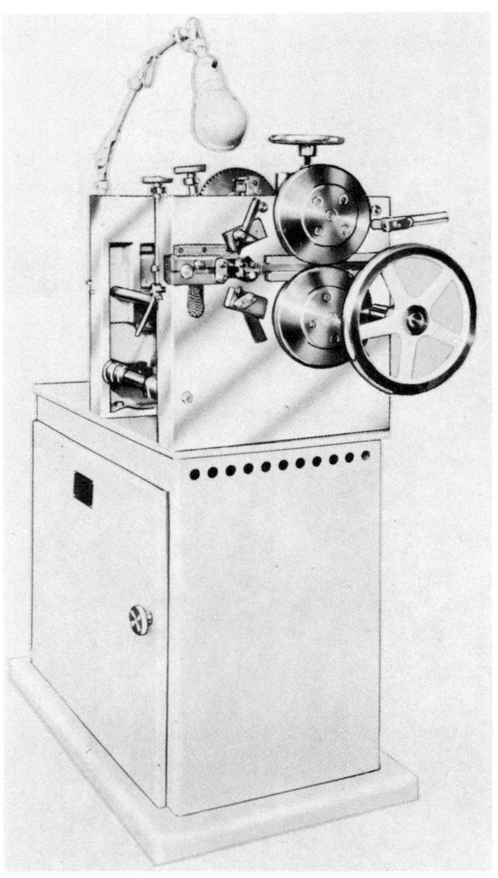

lighter) are usually coiled by the hot-coiling method whereby rods are heated red hot and then coiled on specially built arbor-type coilers. These machines hold large rotating arbors around which the hot wire or bars are coiled. The pitch is obtained by guiding the bar through a slot cut spirally into a pitch tool (or shaft) which revolves just above the arbor. It is necessary to have a number of these pitch tools, and it is necessary to cut special ones for some springs. Hydraulic actuated pitch tools on modern coilers eliminate the necessity of cutting special pitch tools, as they contain adjustable pitch mechanisms.

FIG. 157. An automated machine that coils extension springs, forms hooks on both ends, tests for force and length, and then separates springs into three categories, too weak, OK, and too strong, all automatically! (Photo courtesy Wafios Machine Corp.)

SLIDE MACHINES

A wide variety of four-slide, multislide, and verticalslide machines (Figs. 161 to 163) are available for making hooks, clips, small assemblies, and special springs. They are versatile, can form both wire and flat strip, but require special tooling. They are used only for large-quantity production, usually from 20 000 pieces and upward.

Pressing

The removal of permanent set from compression springs (Fig. 164) so that they will meet requirements of specifications calling for "no permanent set" consists of compressing the springs to the solid position one or more times, usually after heating. Such springs may be coiled a little longer than the specified free

FIG. 158. A duplex wire-working machine for coiling torsion springs and making special end formations. (Photo courtesy Sleeper and Hartley Corp.)

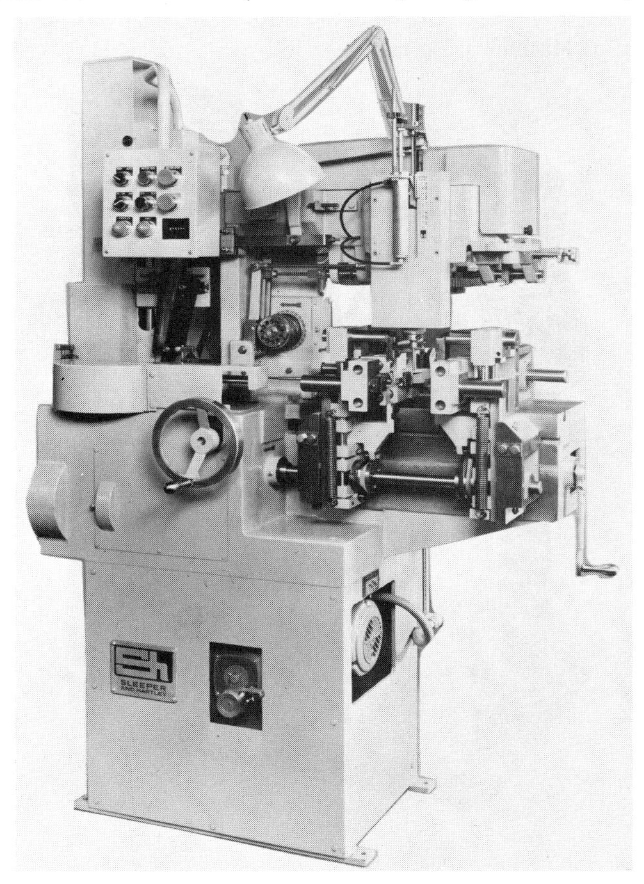

height so that after the set has been removed, the specified free height is obtained. Pressing also traps beneficial stresses in the springs, thereby increasing the fatigue life and endurance limit.

Several methods are used depending upon the size of the spring, the force required to compress it, and the quantity involved. Light springs in small quantities may be compressed by hand over close-fitting arbors—usually after heating and before grinding. Kick presses and small air presses can be used for small springs where large quantities are involved. Power presses, specially built motorized units, and hydraulic presses are used for heavier springs.

FIG. 159. A numerically controlled torsion spring coiler with sensors. (Photo courtesy Torin Corp.)

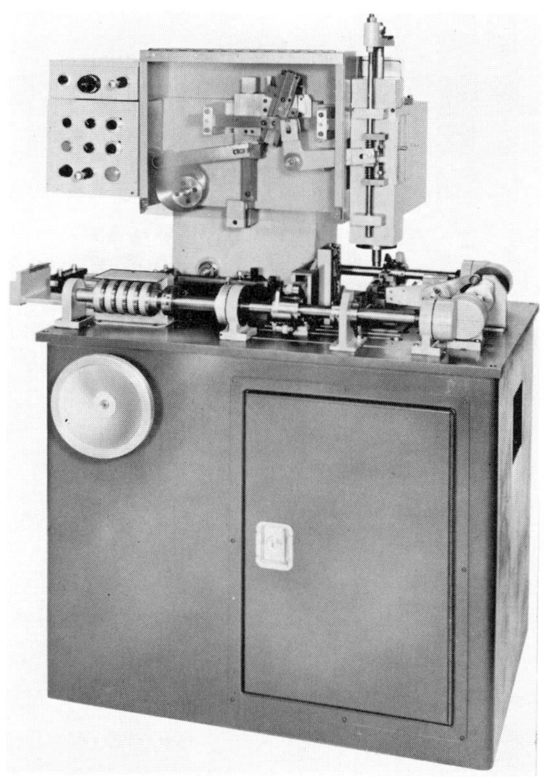

Measuring

One hundred percent checking of the free length manually is a time-consuming, expensive operation. It is occasionally necessary due to extremely small tolerances on forces and lengths of some specifications. High-speed coiling combined with slight variations in wire hardness and wire diameters cause variation in spring lengths that affect the forces and deflections. Small compression springs are placed over a vertical arbor with two grooves indicating length limits; springs too long are compressed and short springs are extended so that they return to the specified length. The corrected springs are then heated to stabilize and maintain the corrected lengths.

Fortunately, several types of electronic length measuring gauges were developed during the 1960s and 1970s. These gauges have practically eliminated manual gauging. These gauges are mounted directly onto the automatic

FIG. 160. A torsion spring coiler with extra slides to carry several tools. (Photo courtesy Wafios Machine Corp.)

spring coilers and measure the length of each spring as it is coiled and then separates the springs into three lots: too short, OK, and too long. The short springs often can be saved by heating them to extend their lengths. The long springs also can be saved occasionally by pressing to shorten them. The cost of saving the rejected springs must be carefully considered, as it may be higher than the cost of coiling more springs. Some gauges also reset the pitch tool so that when the coiler starts making springs that are out of tolerance, the gauge automatically brings the lengths back into tolerance, thus eliminating large quantities of rejects.

Grinding

End coils of compression springs are often ground so the springs will (1) obtain a good seat against a mating part, (2) reduce buckling, (3) stand straight,

FIG. 161. A four-slide machine for making wire forms and flat spring clips. (Photo courtesy A. H. Nilson Machine Co.)

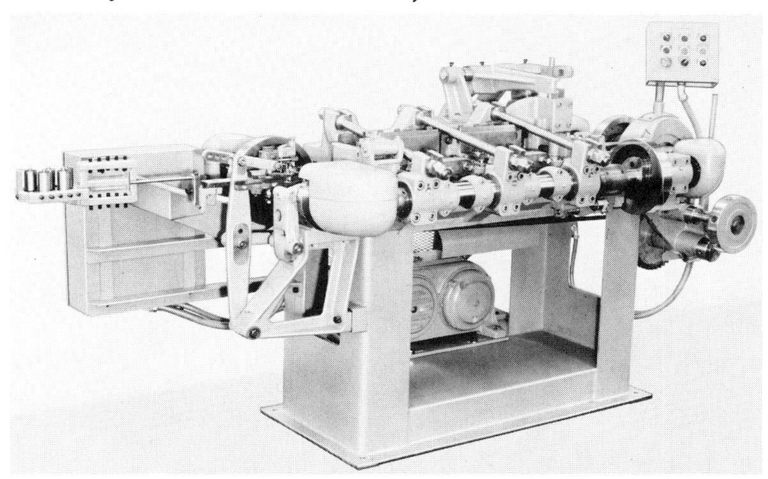

FIG. 162. A multislide machine for making spring clips and small assemblies. (Photo courtesy U. S. Baird Corp.)

and (4) exert a more uniform pressure against a diaphragm or mating part. Grinding is a slow, expensive operation compared to other operations, and should be avoided whenever it is practicable to do so, especially on light springs with wire diameters under $^1/_{32}$ in. (0.80 mm) and where the spring index (OD ÷ d) = 13 or larger. Several types of disk-grinding machines are available.

FIG. 163. A vertical-slide machine for easier installation of tools. (Photo courtesy Torin Corp.)

Small bench-mounted grinders (Fig. 165) that hold from 10 to 100 springs or more in paddles are useful for small production runs from 100 to 5000 or more. They can grind springs with wire diameters from 0.010 to 0.125 in. (0.25 to 3 mm) and larger with lengths up to 4½ in. (114 mm). Two 1-horsepower motors with horizontal spindles are used with 10-in. (254-mm) diameter grinding wheels.

Small springs with lengths up to 6 in. (150 mm) in large quantities are ground economically on double vertical spindle spring grinders (Fig. 166) which use two 5-horsepower motors with 18-in. (457-mm) wheels. The springs are inserted in holes drilled into removable plates, and a motor-driven turntable carries the springs between the two grinding wheels.

Two grinders can be mounted on a single bed plate (Fig. 167) with a large power-driven turntable that carries the springs between four grinding wheels. These machines and some heavier-duty machines are used to grind larger springs. They are available with grinding wheels up to 3 ft (1 m) in diameter and with four motors up to 25 horsepower. Even larger grinders with vertical turntables are used.

FIG. 164. Pressing springs solid to remove permanent set on a hydraulic press. Presses with turntables are often used. (Photo courtesy The Carlson Co.)

FIG. 165. A small, double-wheel bench grinder for grinding both ends of small compression springs. (Photo courtesy The Carlson Co.)

FIG. 166. An automatic grinder with turntable to grind both ends of small compression springs, for high-quantity production. (Photo courtesy The Bendix Corporation)

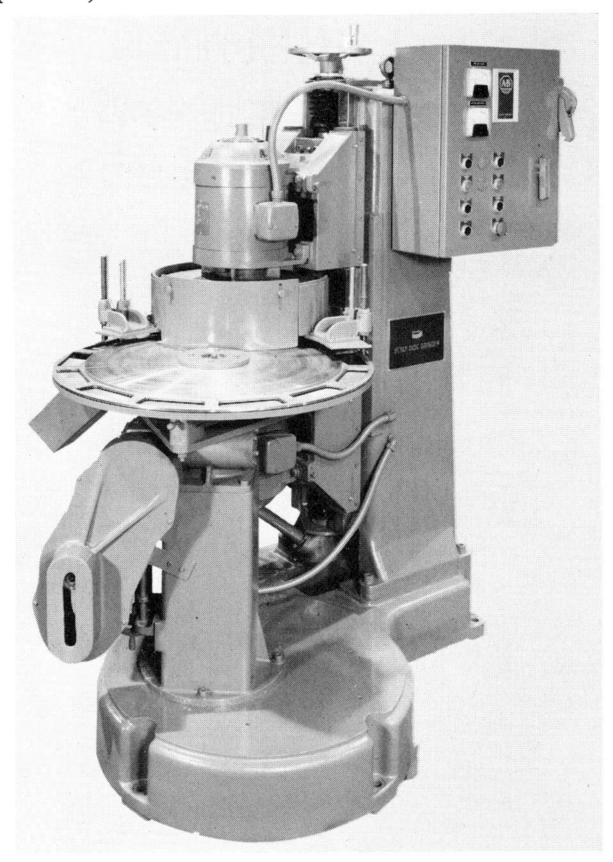

In addition to the grinders described, there are special machines to grind large, heavy, hot-rolled springs and single-spindle pedestal types for heavy springs. Small, light springs in small quantities are often ground on simple bench grinders with 6-in. (150-mm) wheels used for small tool sharpening.

Looping

Small extension springs with wire diameters up to 1/16 in. (1.5 mm) in small lots up to about 1000 pieces can be economically looped with special looping pliers. Larger lots up to 20 000 or more are looped with special looping tools in kick presses. Larger quantities can be looped using special automatic

FIG. 167. A double grinding machine using a large turntable to bring compression springs between two sets of grinding wheels. The first set usually does rough grinding and the second set produces a smooth finishing grind. (Photo courtesy Gardner Machine, Division of Litton Industries, (nc.)

loopers (Fig. 168 and 169). These are usually hopper fed, but are restricted to light wire diameters up to 0.072 in. (91.83 mm) maximum. Larger springs are often looped with heavier looping tools in power presses.

Testing

The need for high-precision testing of small springs used in computers, missiles, instruments, and electronic components requires extremely accurate, sensitive spring testers, often with capacities up to only 200 g (7 oz). Larger testers are available for checking forces up to 1000 lb (454 kg or 4450 N). Heavier springs are often tested on ordinary tensile-testing machines. Between these two forces, a large variety of spring testers must be used in order to obtain test results within proper tolerances. Obviously a 1000-lb (450-kg) capacity tester should not be used for testing springs requiring a test for only a few pounds. For this reason, testers are available in the capacities shown in Fig. 170 and in some other capacities also.

Electronic testers (Fig. 171) are available in several capacities for testing forces in grams or newtons and using dial gauges to check lengths in thousandths of an inch or in millimetres. These testers are unusually accurate and sensitive. The 200- and 2000-g capacity testers are the most popular.

FIG. 168. A hopper-fed machine that loops both ends of small extension springs. (Photo courtesy Sleeper and Hartley Corp.)

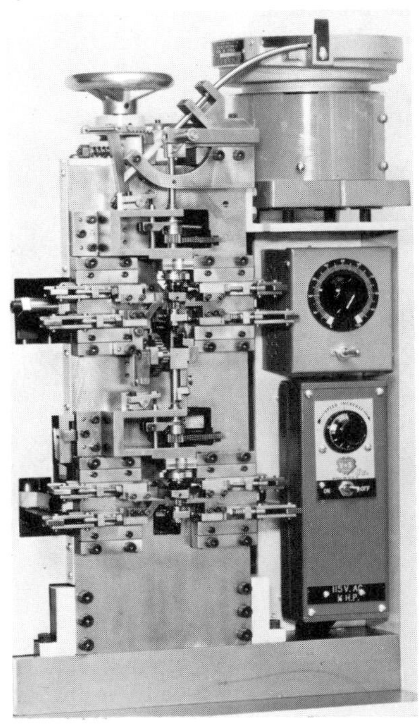

Light force testers (Fig. 172), economically priced, with capacities usually up to 5 lb (22 N) and 25 lb (111 N) are in great demand for small springs. These are balance-beam types of testers using dead weights in ounces, pounds, grams, and newtons and also using dial indicators for checking lengths.

Medium force testers (Fig. 173) with capacities up to 200 lb (890 N) have been the world's most popular testers for many years. They are used in almost every country where springs must be tested, because of their versatility and accuracy. They are of the balance-beam type with a 1-to-10 ratio, so that only a 1-lb weight is needed to check 10 lb (1 N to check 10 N). They can use flat weights in all systems, including ounces, pounds, grams, kilograms, and newtons. Length scales up to 12 in. (300 mm) and dial indicators in thousandths of an inch and millimetres are available, so the testers can be used for any weighing system.

Several accessories are available for these testers, including squareness-under-load fixtures, zero-setting devices, light force testing devices, dial indicators in several capacities, scales in inches or millimetres, and many types of weights in various denominations.

FIG. 169. A machine for looping large extension springs with wire diameters up to 0.072 in. (1.83 mm). (Photo courtesy Torin Corp.)

FIG. 170. Capacities of spring testers.

Pounds	Kilograms	Newtons
0.44	0.200	2.0
4.4	2.000	20.0
5.0	2.268	22.0
25.0	11.340	110.0
44.0	20	200.0
100.	45	450.0
200.	90	900.0
500.	227	2225.
1000.	454	4450.

FIG. 171. Electronic spring testers using solid state equipment and integrated circuits came of age during the 1970s. They are widely used for small, lightly loaded springs. Several sizes are available. (Photo courtesy The Carlson Co.)

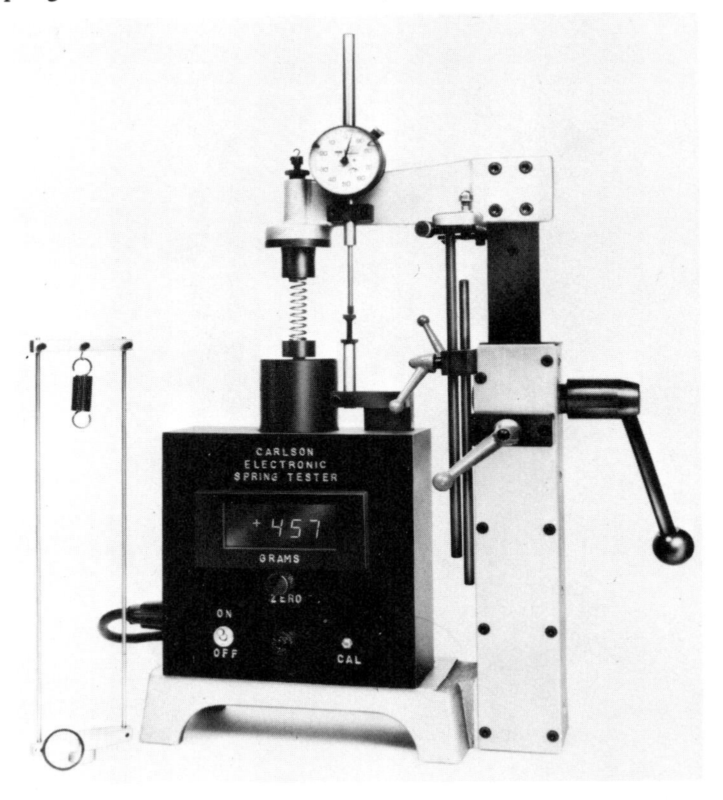

High-power spring testers (Fig. 174) are used to check heavy springs, usually from 100 to 1000 lb (450 to 4450 N) with lengths up to 24 in. (600 mm). These testers show the force on a dial-type meter, and some are available with digital read-outs. Other types with large dials using balance beams are also available. High-power spring testers with capacities up to 10 000 lb (44 500 N) are useful for testing heavy hot-rolled railway springs, although regular tensile testers are often used for that purpose.

Checking compression springs for squareness under force (Fig. 175) is often necessary for springs exerting a force against a diaphragm as in regulating valves, because squarely ground springs may not exert the same force at all points. This tester shows the amount of tipping up to 4°.

Several testers are available for testing torque and deflection of torsion

FIG. 172. A balance-beam tester, using flat weights, for light springs; often used because of low cost, excellent accuracy, and reliability. (Photo courtesy The Carlson Co.)

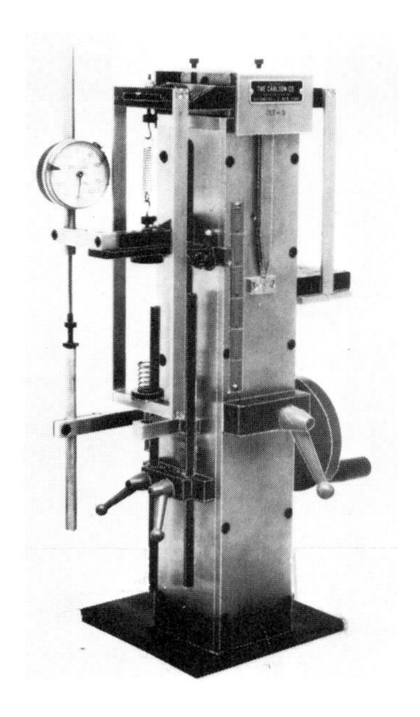

springs. The simple balance-beam type (Fig. 176) is widely used as it is easily adaptable to a large variety of different sizes and types of torsion springs having either short or long lengths. Weights can be in ounces, pounds, grams, kilograms, or newtons, and the forces expressed in inch-pounds, newton-metres, gram-centimeters, etc.

All torsion testers must be used with care and with an understanding of the forces that affect torsion springs. Allowance for, or a determination of, the friction between the spring and the test arbor over which it rides should be made; this is the problem that causes the most trouble. Another is positioning the spring initially and setting the protractor to the zero position correctly. Also, if the coils rub against each other, or if initial tension exists, proper testing is difficult and often impossible with repeatable results.

Light torque testers up to 4 in.-lb (0.45 N · m) and medium torque testers up to 80 in.-lb (9 N · m) are readily available. Heavier torque testers are usually built specially to order.

TWIST TESTING

Testing the torsional quality of wire is a new and important practice. It has

FIG. 173. A balance-beam tester with a beam ratio of 1 to 10; used for both light and medium springs up to 200 lb (890 N). This is the world's most popular spring tester because of its versatility, excellent accuracy, and low cost. (Photo courtesy The Carlson Co.)

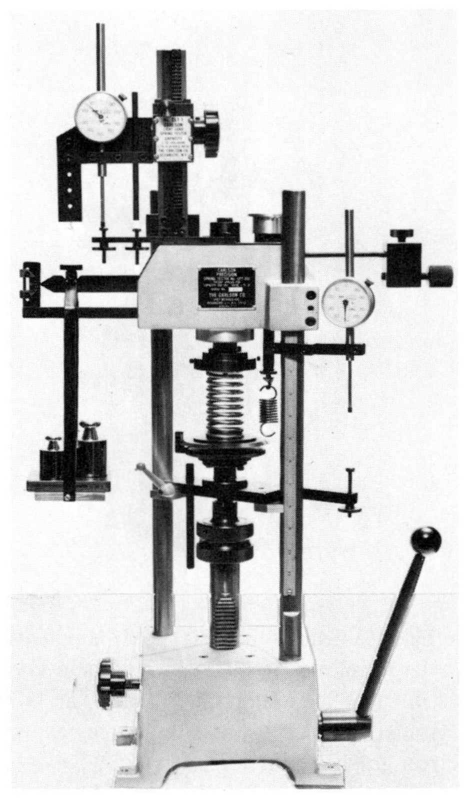

often been observed that two wires with identical tensile strengths and hardness would produce springs with widely different fatigue lives. Some springs lasted for only 10 000 cycles and others up to 1 000 000 cycles. The difference is due to variations in the orientation of the grain structure, the ratio of pearlite to ferrite, and the rotation of crystals to line up in preferred directions. Regulated patenting in the wire mills produces excellent torsional qualities by heating the wire to 1650°F (900°C), quenching in lead at 990°F (530°C), and then after drawing through the final die using a stress-relieving heat of 600°F (315°C) for only 3 min. This treatment improves the torsional quality so much that it doubles the number of twists to produce a fracture! All wire for important springs should be twist tested to check the quality and torsional ability of the wire.

FIG. 174. A high-power spring tester of the electronic type; used for both compression and extension springs with loads up to 1000 lb (4450 N). (Photo courtesy The Carlson Co.).

ISO-136 is the number of the International Organization Standard for torsion testing wire accepted in 31 countries. DIN 51212 is the German standard. ASTM E 558 is the American standard.

Several types of machines are available for twist testing wire to check quality and torsional ability. One model (Fig. 177) can test wires up to $^{3}/_{16}$ in. (4.75 mm) in diameter. The wire is twisted slowly and uniformly in one direction until it breaks. Music wire, hard-drawn MB, and beryllium-copper usually require more than 25 complete twists. Oil-tempered MB and alloy steels require more than 12.

Good-quality wire breaks with a straight-faced, smooth fracture at 90° to the longitudinal axis, with a reasonable number of twists. Brittle wire often breaks in less than five twists. Seams and inclusions in the wire are readily discernible, and delamination often found in the stainless 300 series, or in overdrawn wire, are reasons for rejection.

FIG. 175. A special tester used to check squareness under load of compression springs, with a capacity of 500 lb (2225 N). (Photo courtesy Link Engineering Co.).

FATIGUE TESTING

Testing springs under various forces and deflections to determine their fatigue life is important because the Department of Defense and some manufacturers often specify that certain springs must withstand 100 000 or 1 000 000 or more cycles without failure. Fatigue testing also provides data needed for the determination of recommended design stress curves, the beneficial effects of shotpeening, speed of cycling, and quality of material.

Fatigue testers are usually built specially to order depending upon the size, force, and deflection of the springs to be tested. The standard tester

FIG. 176. A torsion spring tester to measure torque in inch-pounds or newton-metres and angular rotation in degrees. (Photo courtesy The Carlson Co.)

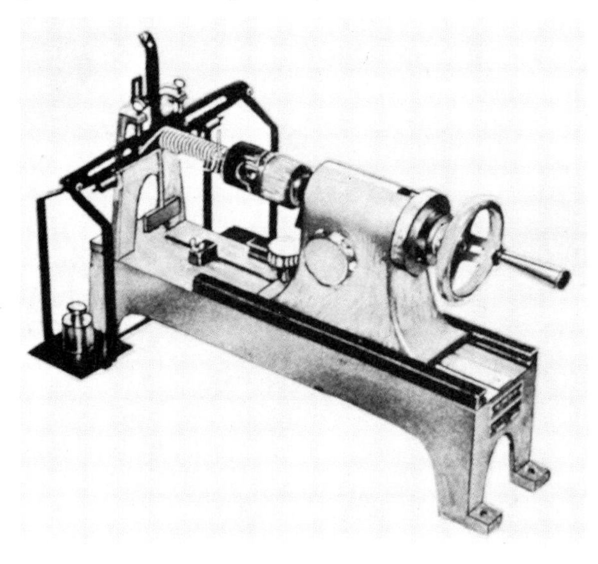

FIG. 177. A twist tester to check torsional ability and quality of wire. (Photo courtesy The Carlson Co.)

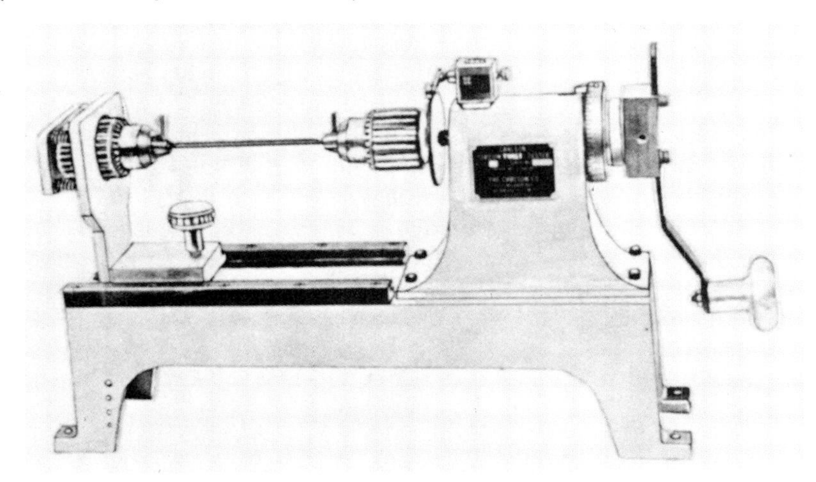

FIG. 178. A fatigue tester to check the number of cycles to break a spring. (Photo courtesy The Carlson Co.)

(Fig. 178) has a wide range of features including cycling from 70 to 420 cycles per minute, although 300 to 350 with 400 maximum are best for most springs. Higher speeds cause the springs to become quite hot. It has an infinitely variable stroke from 0 to 2 in. (0 to 50 mm) and can test springs from 0 to 6 in. (150 mm) with forces up to 1000 lb (4450 N) depending upon the stroke and speed of cycling.

Several other types, made in Europe or Japan, can be obtained for testing heavy coil springs such as are used in automobiles and for testing leaf springs.

Figure 179 shows tools customarily used on automatic spring coilers.

Shotpeening

Shotpeening is one of the greatest advances in the spring industry as it increases the fatigue life of compression springs at least 30 percent and has often increased the life from 2 to 10 times! Shotpeening may be given to all highly stressed compression springs made from any material over $^{1}/_{16}$ in. (1.5 mm) in diameter and some smaller sizes. Extension and closely wound torsion springs are difficult to shotpeen because the tiny steel shot, about $^{1}/_{64}$ in.

FIG. 179. Wire guides, coiling points, arbors, feed rolls, pitch spacers, and cut-off tools used on automatic spring coilers are often made from carbide to provide long tool life. (Photo courtesy Wesson, Division of Fansteel.)

(0.4 mm) in diameter, is often trapped between the coils and is difficult to remove. Closely wound springs also do not get shotpeened on the inner area of the coils where the stress is highest and where shotpeening does the most good.

Two types of machines are available. One type hurls the shot from a paddle wheel at a high centrifugal force onto the springs being revolved in a container or on an endless belt below the paddle wheel. Another type uses compressed air to hurl the shot. Each machine has certain advantages over the other, but both perform adequately.

The longer fatigue life due to shotpeening is accomplished by the following combination of effects:

1. Minute surface irregularities are hammered smooth.
2. The wire surface is cleaned, and sharp burrs are dulled.

3. The wire surface is slightly hardened, which raises the tensile strength and elastic limit.
4. Most important, beneficial compressive stresses are trapped near the surface which must be overcome by the destructive tensile stresses that cause fatigue failure before breakage can occur.

Note: most authorities claim that all heating and stress-relieving operations should be done before shotpeening because heating above 500°F (260°C) counteracts much of the beneficial effects of the trapped compressive stresses which are relieved by heating. Other authorities claim that a light stress-relieving heat is not detrimental. However, if shotpeened springs are electroplated, they should be heated immediately thereafter to relieve hydrogen embrittlement.

Shotpeening springs made from oil-tempered wire is an efficient method to remove the scale and clean the wire to prepare it for electroplating. The slightly roughened surface may not take a bright, glossy finish, but this is usually not considered objectionable. Shotpeening springs made of stainless steel must be done with care to prevent the steel shot from adhering to the surface, as such shot could rust and pit the wire. Wire brushing, tumbling, or passivating are recommended to overcome this objection.

Many spring companies have their own shotpeening and dust-collecting systems; others use commercial shotpeening facilities.

Finishing

Many types of finishes are applied to springs. A simple oil dip is often sufficient. Paint, lacquer, and enamel are occasionally used, but some types are brittle. Black japanning is often used, as it is flexible and has a bright, shiny appearance. Nickel and chromium plating are often specified, but these are not flexible and tend to chip. Zinc, although flexible, is not often specified due to its lack of brightness.

Cadmium plating is the best finish for small springs; it is flexible, highly corrosion resistant, is easiest to apply, and has a pleasing color.

Wire also may be purchased with a cadmium coating, but its mechanical properties should be checked as it may not meet requirements for tensile strength and torsion abilities.

Phosphate, chromate, and oxide coatings are occasionally used, but they provide only a limited amount of corrosion resistance. These finishes also can pick up hydrogen in the coating bath, which can cause failure due to embrittlement, but heating immediately after finishing at 400°F (200°C) for an hour removes almost all tendency to failure.

Cleaning the springs before applying a finish is often necessary. Removing oil and dirt can be accomplished by dipping in commercial solvents, alkalies, and detergents or by using vaporizing degreasers. Heat-treating scale can be removed by tumbling, sand blasting, and shotpeening.

Hydrogen Embrittlement

Spring steels are susceptible to embrittlement during electroplating due to atomic hydrogen which is diffused along the grain boundaries. This adsorbed hydrogen interferes with the slip that ordinarily occurs between the crystal planes when deflection takes place–just as grains of sand placed between one's fingers would interfere with the movement and prevent sliding the fingers back and forth.

The higher the carbon content and the rougher the surface, the more atomic hydrogen is adsorbed. Hydrogen is released from the aqueous solution whenever an electric current is sent through a finishing tank. Springs with residual stresses are more susceptible; therefore all springs should be stress relieved before plating.

Stretching extension springs so electroplating can deposit a coating inside the spring and between the coils should not be done, as this causes a stressed condition which is highly inclined toward embrittlement. Thin steel strip is especially susceptible; the higher the carbon content, the more it tends to embrittlement.

Springs should not be dipped in strong acid cleaning solutions, as this causes a high evolution of atomic hydrogen, although chemical inhibitors in the solutions are helpful.

Dehydrogenation, or removing of hydrogen, restores the springs to their normal condition if done immediately. This is accomplished by baking the springs in an oven as soon as possible after the springs have been plated and washed. A delay of 5 hours may be too long in many cases. The correct temperature and time depends upon the type of steel. Nonferrous materials are immune to hydrogen embrittlement.

Authorities differ on the amount of heat and the length of time at heat because they prefer methods which have suited their particular requirements. Some recommend temperatures as high as 500°F (260°C), but temperatures above 425°F (218°C) may cause an objectionable brown color to appear on the coating, which could cause rejection. This brown coloring can be removed by any bright dip solution. The following values which do not cause discoloration and effectively drive out the hydrogen are often used:

For music wire: 400°F (200°C) for 45 to 60 min

For all other steels: 420°F (215°C) for 1 to 2 hours

Longer times at heat, such as 3 to 4 hours, are often used, although some specifications needlessly call for as much as 24 hours!

Nothing can be done to restore springs that have been proved defective due to hydrogen embrittlement; they should be scrapped. Even with the best regulated methods, occasional embrittlement may be encountered. This is the reason that most spring manufacturers will not assume responsibility for plated springs.

Torsion Springs—Manufacturing Problems

ARM LOCATIONS

Drawings of torsion springs often show the arms in a specified location. This requires an exact number of coils, otherwise the arms will not be located properly. Several methods to position the arms correctly are available as follows.

Example: Suppose that a drawing shows the arms 90° apart in the free position. This automatically requires that the number of coils end their total with ¾, such as 5¾. But if the design formula shows 5.6 coils are necessary to meet the torque and deflection, what should be done?

1. The theoretical approach is to redesign the spring to use slightly heavier wire so that more coils will be required in the hope that the design formula will result in 5¾ coils, or to redesign, perhaps again and again, with smaller spring diameters until finally a correct design is made. This method is time consuming, tedious, and often exasperating.

More practical solutions are available as follows:

2. Wind the spring with 5.6 coils and if the spring is made from music wire, hard-drawn MB, or any oil-tempered high-carbon steel, heating, to relieve residual stresses due to the coiling operation, will cause the spring diameter to reduce slightly and thereby increase the number of coils, probably to 5¾. Several trial heating temperatures and varying the time at heat may be necessary to determine the exact treatment, but it often can be accomplished!

3. If heating is not required, or if it does not provide sufficient coils, the spring can be wound with the 5¾ coils specified and the coiling department notified to coil the spring near the minimum ID and then the number of coils and the torque often will be within tolerance.

4. If the design formula showed that 5.9 coils were required, it would be necessary to coil the springs with 5¾ coils and near the maximum ID.

5. Stainless steels type 302, 304, and 316 react in an opposite manner. Heating causes the spring diameter to enlarge and to reduce the number of coils, so an opposite approach must be used to meet the specifications.

Allowances for changes in spring diameters which occur during heat treatment should be determined in the coiling and heating department, and several trial spring sizes are often required. These changes in spring diameters, plus small variations in wire diameters, combined with slight changes in hardness which affect spring-back, and the need for exact arm locations are the reasons for needing rather large tolerances on torsion springs.

16 Heat Treatment

A Simplified Explanation

This chapter contains data concerning heat treatment solely as it pertains to spring steel and other spring materials.

Steel is the most versatile metal known to man, principally because its various properties, such as hardness, strength, ductility, and forming ability, can be altered to suit requirements by heat treating. Spring steels are considered a "specialty product" of the steel industry and represent less than 2% of the total tonnage of steel made each year—yet they furnish more power and stored-up kinetic energy than the other 98 percent! To the metallurgist, steel is either hard or soft or at some stage in between, but to the layman steel is always hard because he compares it with aluminum, brass, plastic, or wood. A little thought, however, will readily deduce that to drill a hole in a steel plate the drill must be harder than the plate; to turn down a shaft on a lathe, or cut threads, the cutting tool must be hard; and even a hacksaw blade or common file must be hard. So, too, the springs used in clocks or typewriters and the formed wires in a fishhook must be of a prescribed hardness to obtain the desired spring properties.

Metallurgy is the science that deals with the constitution, structure, and properties of metals and alloys and makes materials suitable for use by all mankind.

The outstanding advantage of spring steel is its versatility. The mechanical properties of spring steel, such as hardness, tensile strength, elastic limit, ductility, and shaping abilities, can be controlled and changed at will by heat treatment. These properties reflect the constitution of the steel, its nature, and its usefulness.

Although heat treatment is a science and a distinct part of metallurgy, the purpose of this chapter is briefly to outline in nontechnical, simplified terms what happens to spring steel when it is hardened and tempered by heat treatment:

1. What happens and what takes place during heat treatment
2. The importance of proper heat-treating procedures
3. Metallurgical terms, explained simply

The metallurgist may find the briefness and simplified descriptions too elementary and too concise for his view of this comprehensive subject, but the person without this knowledge should find these data helpful in understanding how heat treatment affects spring steel.

Spring steels require much greater processing, more operations, and more controls than do ordinary steels. This is necessary because of the need for excellence and uniformity of product with regard to ductility, hardenability, accuracy of size, finish, and other properties necessary to provide proper resilience and resistance to fatigue. High endurance or long life under repeated flexure is of utmost importance along with proper formability, toughness and uniformity of structure.

Mills that produce spring steel are known as "specialty mills" because they manufacture special products of high-carbon steel. Such mills require specially selected steels having inherent characteristics for ductility obtained by the correct size and arrangement of the carbide particles and the ferrite areas in the steel. Fine-grain steels are tougher and preferable for thin stock, and a coarser grain is better for heavier sizes. Special practice and processing beyond normal commercial needs to obtain greater ductility, uniformity, and longer fatigue life are often used to meet a customer's requirements.

Many of the constitutional changes that occur due to the special practices and processes to obtain longer fatigue life of spring steels can be seen only under a microscope. Some processes for examination are eddy current techniques to determine imperfections, X-rays, electron diffraction, magnetic measurements, microscopic examination of polished and etched specimens, and so on.

The descriptive data apply principally to the high-carbon spring steels having a carbon content between 0.45 and 1.05 percent.

The classical method of heat treatment of spring steel involves:

1. Heating "red hot" to the proper hardening temperature, at a rate that will not set up high stresses.
2. Quenching in oil to obtain high hardness with high tensile strength.

3. Reheating (tempering) at an intermediate temperature to obtain toughness, flexibility, resilience, and a lower, more useful hardness with less brittleness.

This basic process has been known for several hundred years, but it is only in recent times that the full significance of these procedures has been understood. The strength, hardness, and other mechanical properties are dependent largely on the microstructure, that is, the composition, arrangement, and shape of the extremely small crystalline grains that collectively make up the steel and that can be seen only with a microscope.

Heat treatment affects the structure of the steel and changes it from one state to another. Spring steel may be about 98.8 percent iron with only $^{6}/_{10}$ths of 1 percent carbon and about ½ of 1 percent manganese with impurities of less than $^{1}/_{10}$ of 1 percent, yet the carbon content, small as it is, exercises a predominating influence and makes the steel amenable to hardening by heat treatment and causes the steel to have excellent spring properties.

Solid Solutions

Although carbon is present only in small quantities, it nevertheless has a tremendous significance, just as a little salt in a glass of water affects it considerably. Liquid solutions of salt and water are readily conceived, and everyone knows that the water can be evaporated until the salt crystallizes out of solution. A solid solution, that is, a solution of one solid with another, is less understood, nevertheless, such solid solutions are capable of existing at elevated temperatures without melting the solids! Spring steel goes into solid solution during the hardening heat treatment. As the steel cools, either the iron or the iron carbide may be made to crystallize out of solution, just as the salt crystallized out of its liquid solution.

Carbon exists in steel not as free carbon, but as a compound of carbide of iron having a formula Fe_3C. This compound is called "cementite," and it is extremely hard and has high tensile strength. If the carbon content is less than 0.85 percent and cooling occurs slowly, the iron (ferrite) will crystallize out of solution and thereby increase the concentration of the cementite. If the carbon content is over 0.85 percent, the cementite will crystallize out and thereby increase the ferrite; and at 0.85 percent carbon the ferrite and cementite crystallizes out in a very finely divided mixture, thereby establishing the eutectoid composition. The 0.85 percent figure in the eutectoid composition varies slightly depending upon the purity of the ores, the amounts of other chemicals, and impurities. Several authorities have listed a number of percentages all the way from 0.80 to 0.90 percent, but recent research shows that 0.85 percent is the most accurate one for general purposes.

Grain Structure

During heat treatment steel goes through various states or conditions, and the appearance of the grains under a microscope shows certain shapes, forms, and characteristics which over the years have been given identifying names, often after the name of the one who discovered them. These different grains are caused by the decomposition of austenite into one or more of its transformation products; the physical properties of the steel depend upon the nature of the grains.

Depending upon the heat treatment, the mixture of cementite and ferrite may exist in microscopical forms called pearlite, austenite, martensite, cementite, etc., and the spring properties depend largely upon which of these forms the mixture is in. Under the microscope the grains after polishing and etching have distinctive characteristics.

The metallurgical names for the appearance of the grain structure and other terms with their definitions follow in alphabetical order.

Austenite. Grain structure, (named after Roberts-Austen, 1897). When the whole of the carbide is in complete solution, the steel is "austenitic." This phase is a solid solution of carbon or iron carbide in gamma iron. A eutectoid steel of 0.85 percent carbon is completely transformed to austenite when heated to 1350°F (730°C). The carbide dissolves in the surrounding ferrite, forming a solution of carbon in iron known as austenite. It is stable only at high temperatures in the carbon steels. It transforms into martensite when the steel is quenched.

Carbide. Compounds of carbon with iron. Carbide is alpha iron in which an appreciable percentage of carbon is dissolved. The carbon in steel exists only as a carbide of iron (6.6 percent carbon and 93.4 percent iron in solid solution).

Carbon C. An element, crystalline in graphite, amorphous as coal, in a combined state with magnesite ores. Iron forms a carbide with carbon Fe_3C. It is the hardening element in carbon steels.

Steel having 0.85 percent carbon is 100 percent pearlite. If heated above 1350°F (730°C), the carbide is dissolved in the ferrite, thus forming 100 percent austenite.

If over 0.85 percent carbon (hypereutectoid), the free ferrite is gradually dissolved by the austenite progressively as the temperature is increased, until the structure is 100 percent austenite. The austenite transforms to martensite when the steel is quenched.

Cementite. Grain structure. This is a combination of iron and carbon; a carbide of iron. X-ray examination shows that the crystal has four molecules. Carbon in iron exists in this compound and increases the hardness of the steel.

Eutectoid. A simple carbon steel (iron-carbon alloy) having 0.85 percent carbon which has the lowest transformation temperature, 1350°F (730°C). It is 100 percent pearlite in the annealed condition. (Note: various authorities specify different values from 0.80 to 0.90 percent as the eutectoid composition).

Hypereutectoid. Having more carbon than the eutectoid composition, that is, more than 0.85 carbon. It also contains some free cementite.

Ferrite. Metallurgical term for pure unalloyed iron; it is that portion of steel that contains no carbon in solid solution. It is soft and ductile and forms in large crystals during cooling, but can be made to form in small crystals by rapid quenching; however, it does not harden on rapid cooling.

Martensite. Grain structure. A transition form between austenite and ferrite and cementite. It is the hard component of hardened carbon steels. Its needle-like structure is distinctive. It is the hardest constituent of steel of eutectoid composition and is formed by rapid cooling from quenching temperature. It is magnetic.

Patenting. A special annealing operation considerably above the critical temperature (1830°F or 1000°C) and quickly cooling in molten lead, resulting in a fine sorbitic structure having high strength and toughness that is good for wire drawing purposes. Salt mixtures too can be used for quenching, and the quench temperatures can vary considerably to obtain desired conditions.

Pearlite. Grain structure. A combination of ferrite and cementite wherein the cementite is in parallel lamellae—a eutectoid of cementite and crystallized iron. It resembles mother of pearl. It is a compound of iron and carbon as a result of the transformation of austenite into aggregations of ferrite and iron carbide.

Phase. The phases are states, physical and chemical, in which components exist and into which they pass. The phase rule is used to determine the number of phases that can exist in a system by connecting the number of components, degrees of freedom, and equilibrium.

A hypereutectoid steel of say 0.90 percent carbon in the annealed state has a grain structure of pearlite and cementite which is relatively soft.

When heated above the critical temperature, the orientation of the atomic structure changes from the body-centered cubic lattice (alpha iron) and transforms to the face-centered cubic form (gamma iron), the grain structure changes to austenite and cementite, and the steel is no longer magnetic.

When quenched from the above state, the austenite transforms first to pearlite, then to bainite, and finally to martensite, and the steel is too hard and too brittle for most applications (Rockwell C65). The transformation of austenite to martensite causes the steel to become hardened.

Tempering, by heating below the transformation temperature, ranging from 600 to 1000°F (315 to 538°C), reduces the hardness of the martensite to about Rockwell C42 to 46, thereby reducing the brittleness and producing a tough steel with high tensile strength combined with some ductility.

Austempering

Another method of heat treating spring steel to obtain improved qualities, particularly ductility, and resilience, and to raise the percentage of the elastic limit (although the tensile strength may be the same), is known as austempering.

Although introduced in the 1930s, it took 20 years of experimentation and research testing to standardize the most acceptable procedures.

Austempering differs from the usual heat-treating procedure for annealed steels such as heating, quenching, and tempering. It also requires special furnaces for decarburization and for heating to an austenitic structure and installed for a continuous operational sequence.

In austempering, the springs are not tempered by the annealing of the martensitic structure. The final structure is obtained by direct transformation of the austenite into a desired martensitic structure of proper hardness. After heating to the usual hardening temperature and customary soaking time, the springs are immersed in a salt bath at about 600°F (315°C). The transformation of the austenite at this temperature is not a sudden reaction like the usual martensite formation, but is connected with a definite time sequence at this temperature, such as 20 min. When transformation is complete, air or water cooling may be done.

Note: the temperature of the salt bath can be varied depending upon the composition, size, and hardness desired. In the United States temperatures often vary from 400 to 800°F (200 to 425°C). In Germany higher temperatures are used, from 575 to 850°F (300 to 450°C). The time at heat also varies widely from 15 min to several hours to obtain hardnesses from Rockwell 42 to 52 C.

Austempering is particularly useful for flat steel springs including Belleville washers, retaining rings, and spring clips.

Equilibrium Diagrams

Equilibrium diagrams, also referred to as iron-carbon or constitutional diagrams (Fig. 180), illustrate what states are normal under different temperatures and carbon contents.

An oil-tempered spring steel, ASTM A 229, with 0.60 percent carbon, if annealed would consist of pearlite and ferrite as shown in the equilibrium diagram (Fig. 180). When heated above the upper critical temperature and into the full annealing and hardening range, say 1500°F (815°C), the structure is austenitic. If allowed to cool slowly to room temperature, the structure reverts to its original condition, but if quickly quenched in oil the structure becomes martensite, characterized by a needle-like appearance when polished and etched and viewed under a microscope. The material is then quite hard and brittle and must be reheated, called tempering, to bring it to a desired hardness of Rockwell C42 to 46, with proper ductility and resilience.

Figure 181 gives general hints on heat treating springs.

FIG. 180. Equilibrium Diagram for Carbon Steels.

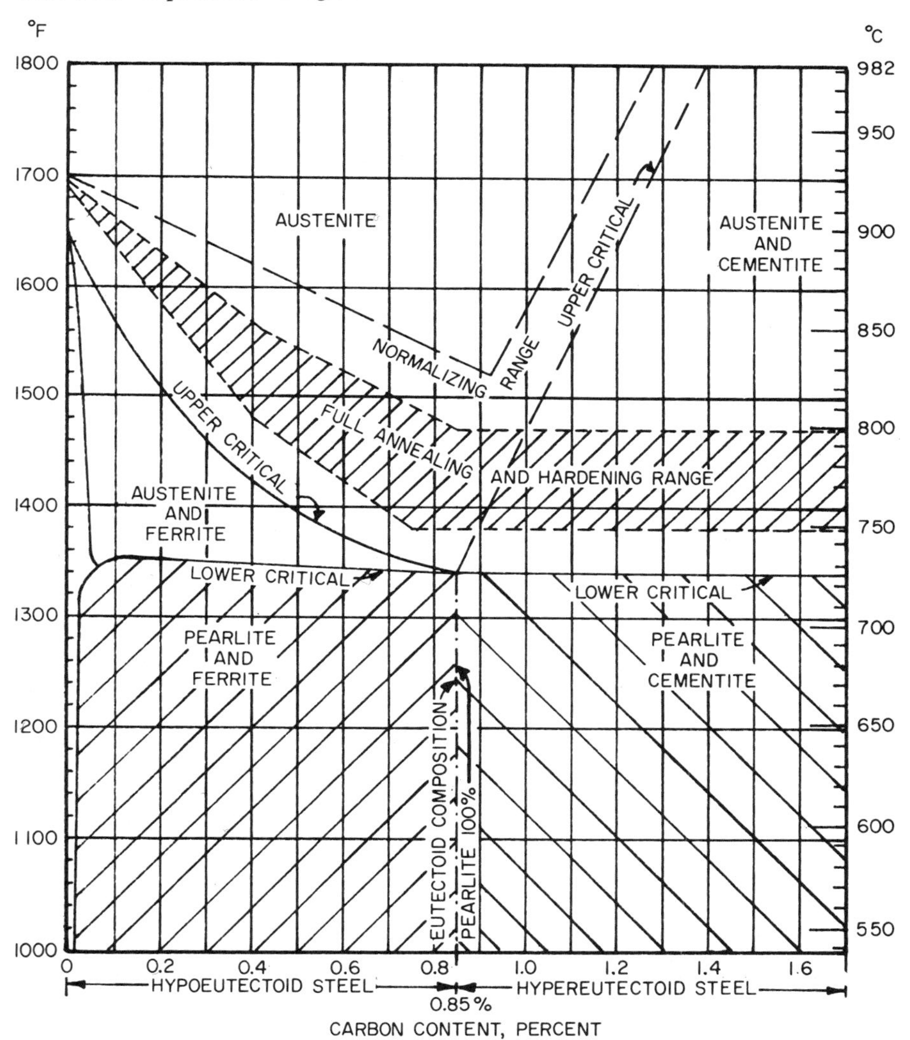

Coloring Springs Blue by Heat Treatment

Steel may be given an attractive blue, straw, or purple color by a light heat treatment called "blueing," "tempering," or baking. This coloring is a direct result of oxidation caused by uniting with oxygen at an elevated temperature. This blueing is also somewhat helpful in retarding corrosion for a short period of time, but it will not withstand a salt spray.

FIG. 181. General Hints on Heat Treating Springs.

1. *Compression springs* are baked after coiling (before setting) to relieve residual stresses and thus permit larger deflections before taking a permanent set. A light heat is sometimes given also after pressing to stabilize the material and to retard growth during transit.
2. *Extension springs* also are baked, but heat removes some of the initial tension. Allowance should be made for this loss. Baking at 500°F (260°C) for 30 min removes approximately 50 percent of the IT. The shrinkage in diameter however, will slightly increase the load and rate.
3. *Torsion springs* do not actually require baking, because coiling causes residual stresses in a direction that is helpful, but such springs frequently are baked so that jarring or handling will not cause them to lose position of ends.
4. *Outside diameters shrink* when springs of music wire, pretempered MB, and other carbon or alloy steels are baked. This slight reduction should be compensated for during coiling. Baking also slightly increases the free length. These two changes combined cause a little stronger force and increase the rate.
5. *Outside diameters expand* when springs of stainless steel (AISI 304) are baked. The free length also reduces slightly and these changes cause a little lighter force and decrease the rate.
6. *Inconel, monel, and nickel alloys* do not change much when baked.
7. *Beryllium-copper* shrinks and deforms when heated. Such springs usually are baked in fixtures or supported on arbors or rods during heating.
8. *Brass and phosphor-bronze* springs should be given a light heat only. Baking above 450°F (232°C) will soften the material. Do not heat in salt pots. (See Item 15.)
9. *Position of loops* will change with heat. Parallel hooks may change as much as 45° during baking. Torsion spring arms will alter position considerably. These changes should be compensated for during looping or forming.
10. *Quick heating* after coiling either in a high-temperature salt pot or by passing a spring through a gas flame is not good practice. Samples heated in this way will not conform with production runs that are properly baked. A small controlled-temperature oven should be used for samples and for small lot orders.
11. *Plated springs* should always be baked before plating to relieve coiling stresses and again after plating to relieve hydrogen embrittlement.
12. *Japanned springs* baked at about 400°F (204°C) for 30 min automatically provide a good baking treatment, and frequently this is the only heat necessary for some springs.

(continued)

FIG. 181 (continued)

13. *Hardness* values lower with high heat—but music wire, hard-drawn, and stainless will increase 2 to 4 points Rockwell C.
14. *Sharp bends* on loops or arms cause high stress concentrations; such springs should be baked as soon as possible after forming.
15. *Spring brass and phosphor-bronze* springs which are not very highly stressed and are not subject to severe operating use may receive adequate stress relieving after coiling by immersing them in boiling water for a period of 1 hour.

FIG. 182.

Temper Color of Carbon Steel	Temperature of Oven	
	Fahrenheit	Celsius
Light straw	420	215
Medium straw	450	232
Dark straw	475	245
Bronze (some purple)	500	260
Blue	540	282
Light blue	575	301
Gray	600	315

COLOR VARIATION

The amount of blue color, its uniformity, and evenness, depends principally upon using clean, bright steel, free from rust, oil, dirt, or coatings. The temperature of the oven determines the color, but the amount of air going into the oven, the type of steel, the type of heat (gas, oil, or electric) also affects the color at different temperatures (see Fig. 182). For these reasons, the temperatures listed in Fig. 182, are only approximate and may vary about 10°F.

TIME AT HEAT

The length of time at heat is not important, but the time should be long enough to permit the steel to be uniformly heated throughout; a longer time would not be economical.

Heat Treatment for Hard-Drawn and Pretempered Spring Materials

BAKING TEMPERATURE

Springs made from tempered wire, music wire, etc., do not require a hardening heat treatment, but they always should have a stress equalizing heat treatment after coiling wherever possible. Extension springs having initial tension and torsion springs do not always require stress relieving, but it is advisable to give them some heat, provided that the correct force and deflection can be maintained. In coiling, forming, or shaping springs, the wire is stressed beyond the elastic limit, thereby causing a slight plastic flow of the material to take place. Such springs are not stable; they have a strong tendency to grow in length if dropped, and should, where possible, be heated after coiling (preferably before setting). See Fig. 183. This treatment is called tempering, bluing, baking, stress relieving, seasoning, annealing, stress equalizing, etc.

BAKING TIME

The length of time that springs are allowed to remain at a stress-equalizing heat (after they reach the temperature recommended) is very important (Fig. 183). Readjustment of the structure takes time, and attempts to speed up the process by a higher temperature and shorter time should be avoided.

Hardening and Tempering Springs Made from Annealed Steel

Springs made from annealed or "soft" steels are hardened by heating above the critical range (the range of temperatures that cause a change in the internal molecular structure) and then immediately quenching in oil. This makes the steel brittle and hard (Rockwell C55 and above) and it is necessary to reduce this high quench hardness by heating the springs a second time at a lower temperature (This reheat is called tempering, drawing, blueing, baking, etc.) Both the "hardening" and "drawing" temperatures differ with each type of steel, as shown in Fig. 184.

Springs "soaked" at the hardening temperature for too long a time will be subject to decarburization (burning carbon out of the surface of the steel and causing a soft skin); this also causes an enlargement or growth in the size of the grainstructure which reduces the toughness, and causes brittleness without high hardness. Springs should be allowed to "soak" at the hardening temperature and allowed to remain at the drawing temperature in accordance with the wire size and the approximate values given in Fig. 184. Variations in carbon content affect time and temperature.

FIG. 183. Heat Treatment for Hard-Drawn and Pretempered Spring Materials.

Heat Treat. No.	Spring Material	ASTM No.	Baking Temperature					
			General service		Severe service		High-temp. service	
			°F	°C	°F	°C	°F	°C
HD-1	Hard-drawn wire	A227	450	230	500	260	Not used	
MB-1	Tempered MB	A229	450	230	550	290	650	345
HB-1	Tempered HB or XHB	–	450	230	550	290	650	345
CV-1	Chrome-vanadium	A231	500	260	600	315	700	370
SM-1	Silicon-manganese	–	500	260	600	315	700	370
CS-1	Chrome-silicon	A401	500	260	600	315	700	370
MW-1	Music wire[a]	A228	420	215	500	260	Not used	
MO-1	Monel 400	B164	500	260	600	315	700	370
PB-1	Phosphor bronze	B159	250	120	375	190	Not used	
SB-1	Spring brass	B134	250	120	375	190	Not used	
IN-1	Inconel 600	B166	750	400	850	455	900	480
IX-1	Inconel X 750	–	1250	675	1350	730	1350	730
SS-1	Stainless 18-8 & 16-2	A313	500	260	600	315	700	370
PH-1	17-7PH stainless	–	900	480	900	480	900	480
NC-1	Ni Span C 902	–	750°F (400°C) for 1 hr then 1200°F (650°C) for 2 hr.					

(continued)

FIG. 183 (continued)

Spring Material	Wire Diameter		Baking Time, min.[b]		
	inches	millimetres	General Service	Severe Service	High-temp. service
Music wire, stainless, & all other spring steels	Up to 0.015	Up to 0.38	10-15	15-20	20-30
	0.016-0.050	0.39-01.25	15-20	20-30	30-45
	0.051-0.120	1.26-3.00	20-25	30-40	45-60
	0.121-0.375	3.01-9.50	25-30	40-50	60-80
	0.375-0.500	9.51-12.70	30-45	50-60	60-90
Phosphor bronze & spring brass	Up to 0.025	Up to 0.65	20-30	30-45	Not used
	0.026-0.375	0.66-9.50	30-40	45-60	Not used
Monel and Inconel	Up to 0.025	Up to 0.65	30-45	50-60	1-2 hr
	0.026-0.375	0.66-9.50	45-60	60-80	2-3 hr
Inconel X	All Sizes		6 hr	8 hr	10 hr
17-7 PH Stainless	All Sizes		1 hr	4 hr	8 hr
Ni Span C 902	All Sizes		See above table, heat slowly, air cool		

[a]For light sections of music wire under 0.025 in. (0.65 mm) dia., use 350 to 400°F, 176 to 204°C.
[b]For recirculating air ovens–reduce 25 to 33% for liquid baths.

FIG. 184. Hardening and Drawing Temperatures

Heat Treat. No.	Spring Material Type	SAE Spec. No.	Carbon Content, %	Hardening Temp.		Drawing Temp.		Hardness, Rockwell "C" Scale
				°F	°C	°F	°C	
HD-2	Hard-drawn	1055	0.50-0.60	1500-1550	815-845	500-625	260-330	41-45
MB-2	"MB" grade	1065	0.60-0.75	1475-1525	800-830	500-625	260-330	42-46
HB-2	"HB" grade	1080	0.75-0.90	1500-1550	815-845	575-675	300-355	43-47
XH-2	"XHB" grade	1095	0.90-1.05	1500-1550	815-845	600-700	315-370	44-48
CV-2	Chrome-vanadium	6150	0.45-0.55	1600-1650	870-900	750-900	400-480	45-49
SM-2	Silicon-manganese	9260	0.55-0.65	1575-1625	860-885	600-700	315-370	43-47
CS-2	Chrome-silicon	–	0.50-0.60	1475-1525	800-830	700-800	370-425	46-51
SS-2	Stainless, type 420	51335	0.25-0.40	1825-1850	995-1010	750-850	400-455	46-51

Wire Diameter	Soak at Hardening	Drawing Time[a]
Up to 1/16" (1.58 mm)	2-3 min.	10-15 min.
Over 1/16" to 1/8" (1.58 to 3.17 mm)	3-5 min.	10-20 min.
Over 1/8 to 3/16" (3.17 to 4.76 mm)	5-8 min.	15-25 min.
Over 3/16" to 3/8" (4.76 to 9.52 mm)	8-12 min.	20-30 min.
Over 3/8" (9.52 mm)	10-15 min.	30-45 min.

[a]The drawing temperature and the length of time at drawing temperature determine the hardness. High drawing temperatures reduce the hardness. Alloy and stainless steels take more time at drawing than the carbon steels.

Note: Brinell hardness divided by two, times 1000, equals approximate tensile strength in pounds per square inch. Example: 350 Brinell ÷ 2 × 1000 = approx. 175,000 psi.

17 Corrosion

According to the dictionary, corrosion is the process of eating away, consuming, or wearing away by slow degrees.

The corrosion of metals may be defined as an attack of a metal by chemical or electrochemical action. An example of chemical attack is the eating away of metal by hydrochloric acid. Electrochemical attack may be by galvanic cell or by concentration cell corrosion.

Six types of corrosion are often noted: (1) rusting or tarnishing; (2) pitting or small indentations; (3) cracking, usually caused by a combination of stress and corrosion; (4) crevices under washers or gaskets; (5) particle—corrosion of one of an alloys constituents; (6) junction—where two different metals are joined.

Several types of corrosion such as stress corrosion or "season cracking" are described under copper-base alloys (chapter 7). Concentration cell corrosion, corrosion fatigue, and others are occasionally encountered with springs. However, atmospheric corrosion and galvanic corrosion give the most trouble.

Galvanic Corrosion

When two different metals widely separated in the galvanic series (Fig. 185) touch each other in water, in acid solutions, in a moist corrosive atmosphere, or in liquids capable of carrying current, one of the metals will corrode while the other is protected. This destructive change is caused by an electrochemical

FIG. 185. Galvanic Series

Anodic–Corroded End
Magnesium alloys
Zinc
Aluminum
Cadmium
Spring steels
Alloy steels
Stainless steels
Lead and tin
Nickel alloys
Inconel and Inconel-X
Spring brass
Phosphor-bronze
Beryllium-copper
Monel alloys
Silver and gold
Platinum
Cathodic End–Protected

reaction that is promoted by the galvanic voltage set up in the system. The corroding metal acts as an anode, the protected metal as a cathode, and the corrosive influence as an electrolyte. Copper-base alloys are cathodic and protected while causing corrosion to occur on steel, stainless steel, aluminum, iron, and other alloys near the anodic end.

Galvanic corrosion between two metals that are near each other in the galvanic series, such as cadmium and steel, would be of little consequence. When two metals more widely separated, such as aluminum and copper are coupled in sea water, galvanic corrosion will soon affect the aluminum as it is nearer to the anodic end of the series, but the copper will be galvanically protected from corrosion. Several tables of galvanic series are available depending upon the electrolyte and types of alloys; Fig. 185 is most applicable to springs.

The severity of corrosion depends upon the relative anodic and cathodic areas, strength of electrolyte, temperature, and polarization effects. Variations in the composition of the corrosive media also affects the amounts of galvanic corrosion.

Springs made of copper-base alloys such as phosphor-bronze or beryllium-copper should not be used in conjunction with aluminum or zinc die castings in water or wet atmospheres. Springs of stainless steel should not be used in conjunction with bronze castings in wet atmospheres. A bad example was a stainless steel spring used in a cast bronze rain jet for watering lawns—it wasn't long before the spring failed, as it was nearer the anodic end.

Metals near to each other in the galvanic series, such as Monel and phosphor-bronze or spring steel and cadmium, are well protected from galvanic corrosion. Metals further apart, such as beryllium-copper and zinc, in a wet atmosphere will cause the zinc to disintegrate.

Methods to reduce the severity of galvanic corrosion include the following: (1) select combinations of materials as close together as possible in the galvanic series; (2) insulate dissimilar metals by a coating of bitumastic material, a layer of plastic, or even a coating of enamel or paint; (3) keep dissimilar metals as far apart as possible; (4) add chemical inhibitors to the electrolyte; and (5) avoid threaded connections such as small bolts (the threads corrode very quickly)—a brazed connection works quite well, using bronze as the brazing material.

In the galvanic series (Fig. 185) no mention is made concerning "passivity or active" alloys. Some materials (especially the stainless steels) in certain electrolytes act more like the noble metals near the cathodic end and do not corrode rapidly. However, conditions can arise to reduce this passivity, so it is safer in design work to consider such conditions as very special and perhaps requiring corrosion tests.

18 Tolerances

Another controversial problem in the spring industry concerns commercial tolerances. This is because most designers are familiar with allowances and tolerances for turned shafts and other fitted members and attempt to apply those same tolerances to springs. Spring tolerances must be much larger than machined surface, casting, or forging tolerances because of the unavoidable variations in wire size. Slight variations in wire diameters or in hardness of the wire cause a difference in the amount of "spring back" which occurs when wire is coiled. Wear on dies and tools used in automatic spring coilers should be considered as well. Some leeway on wire size, dimensions, and number of coils should always be given in order that springs may be made to meet the more important load and deflection requirements.

The solution of the problem of tolerances is in many cases left to the judgment of the spring maker. A great many drawings and specifications omit all reference to tolerances. In such cases, it is assumed that commercial tolerances are applicable or that the spring maker is familiar with the customer's usual requirements. If doubt exists regarding variations permitted, it is quite frequently necessary to obtain such information through correspondence, which causes delay and should be avoided as much as possible. Delays of this type may be avoided by adding a note to the drawing or specification stating that commercial tolerances are acceptable, or by specifying the tolerances shown in Figs. 186 to 200.

Even the commercial tolerances agreed upon by spring manufacturing companies require certain significant interpretations, and in some instances it is necessary to devote additional time to the computation of tolerances which would be applicable. Much of this time can be saved by placing the required data in simplified tabular form and referring to them when necessary. Commercial tolerances are especially significant if it is realized that any spring developed within the limits indicated in Figs. 186 to 200 may be produced at a reasonable commercial cost and in a comparatively short manufacturing time.

Reductions to these tolerances add considerably to the cost of the springs and to manufacturing time because individual testing and adjusting of each spring may be necessary.

In the data in Figs. 186 to 200 it will be observed that the spring index, the ratio of the mean diameter of a spring divided by its wire diameter, D/d, is a primary controlling factor in determining most tolerances. This relationship is very important in spring design and manufacture, for with small ratios less wire is used and less "spring back" occurs so that smaller tolerances are applicable. The number of coils per running inch of length, or the total number of coils, in a spring also affects tolerances, for with a large number of coils it is necessary to use larger tolerances.

When the spring index D/d exceeds 16, larger tolerances than those listed here are ordinarily required, but standards for springs of this type vary with each manufacturer and it is suggested that for conditions not covered by these figures the advice of a reputable spring maker be obtained.

In a large majority of spring applications, extreme accuracy of length and diameters need not be closely held to precision tolerances. In such cases it is far better to eliminate tolerances completely and add "approximate" on such dimensions or state "commercial tolerances are applicable." This would result in a reduction of manufacturing time, lower the cost, and help to speed delivery.

The tolerances are shown in Figs. 186 to 200 in the inch-pound system and in metric (SI) units in even figures rounded off.

FIG. 186. Compression Spring Free Length Tolerances, Squared and Ground Ends (Inch-Pound System).[a,c]

Number of active coils per inch	Spring index, D/d						
	4	6	8	10	12	14	16
	Length tolerance (± inch per inch of free length[b]						
0.5	0.010	0.011	0.012	0.013	0.015	0.016	0.016
1	0.011	0.013	0.015	0.016	0.017	0.018	0.019
2	0.013	0.015	0.017	0.019	0.020	0.022	0.023
4	0.016	0.018	0.021	0.023	0.024	0.026	0.027
8	0.019	0.022	0.024	0.026	0.028	0.030	0.032
12	0.021	0.024	0.027	0.030	0.032	0.034	0.036
16	0.022	0.026	0.029	0.032	0.034	0.036	0.038
20	0.023	0.027	0.031	0.034	0.036	0.038	0.040

[a]For springs with closed ends not ground, multiply the above tolerances by 1.7
[b]For springs less than 0.5 inch long, use the tolerances for 0.5 inch long springs.
[c]Courtesy Spring Manufacturers Institute.

FIG. 187. Extension Spring Free Length and End Tolerances, (Inch-Pound and Metric Systems[a,c]

Free length tolerances[b]		End tolerances for both systems	
Spring free length (in.)	Tolerance (in.)	Total number of coils	Angle between loop planes (degrees)
Up to 0.5	±0.020		
Over 0.5 to 1.0	±0.030	3-6	±25
Over 1.0 to 2.0	±0.040	7-9	±35
Over 2.0 to 4.0	±0.060	10-12	±45
Over 4.0 to 8.0	±0.093	13-16	±60
Over 8.0 to 16.0	±0.156	Over 16	Random
Over 16.0 to 24.0	±0.218		

[a]For springs with closed ends not ground, multiply the above tolerances by 1.7.
[b]For springs less than 0.5 inch long, use the tolerances for 0.5 inch long springs.
[c]Courtesy Spring Manufacturers Institute.

FIG. 188. Compression Spring Squareness Tolerance (Inch-Pound and Metric Systems)[a,c]

Slenderness ratio, FL/D	Spring index, D/d						
	4	6	8	10	12	14	16
	Squareness tolerance (±degrees)						
0.5	3.0	3.0	3.5	3.5	3.5	3.5	4.0
1.0	2.5	3.0	3.0	3.0	3.0	3.5	3.5
1.5	2.5	2.5	2.5	3.0	3.0	3.0	3.0
2.0	2.5	2.5	2.5	2.5	3.0	3.0	3.0
3.0	2.0	2.5	2.5	2.5	2.5	2.5	3.0
4.0	2.0	2.0	2.5	2.5	2.5	2.5	2.5
6.0	2.0	2.0	2.0	2.5	2.5	2.5	2.5
8.0	2.0	2.0	2.0	2.0	2.5	2.5	2.5
10.0	2.0	2.0	2.0	2.0	2.0	2.5	2.5
12.0	2.0	2.0	2.0	2.0	2.0	2.0	2.5

Springs with closed and ground ends, in the free position. Squareness tolerances closer than those shown require special process techniques which increase cost. Springs made from fine wire sizes, and with high spring indexes, irregular shapes, or long free lengths require special attention in determining appropriate tolerance and feasibility of grinding ends.
[a]For springs with closed ends not ground, multiply the above tolerances by 1.7.
[b]For springs less than 0.5 inch long, use the tolerances for 0.5 inch long springs.
[c]Courtesy Spring Manufacturers Institute.

FIG. 189. Compression and Extension Spring Coil Diameter Tolerances (Inch-Pound System)[a]

Wire diameter, d (in.)	Spring index, D/d						
	4	6	8	10	12	14	16
	Coil diameter tolerance (±in.)						
0.015	0.002	0.002	0.003	0.004	0.005	0.006	0.007
0.023	0.002	0.003	0.004	0.006	0.007	0.008	0.010
0.035	0.002	0.004	0.006	0.007	0.009	0.011	0.013
0.051	0.003	0.005	0.007	0.010	0.012	0.015	0.017
0.076	0.004	0.007	0.010	0.013	0.016	0.019	0.022
0.114	0.006	0.009	0.013	0.018	0.021	0.025	0.029
0.171	0.008	0.012	0.017	0.023	0.028	0.033	0.038
0.250	0.011	0.015	0.021	0.028	0.035	0.042	0.049
0.375	0.016	0.020	0.026	0.037	0.046	0.054	0.064
0.500	0.021	0.030	0.040	0.062	0.080	0.100	0.125

[a](Apply to OD or ID).

FIG. 190. Compression Spring Normal Force Tolerances (Inch-Pound System)[a,b]

Length tolerance (±in.)	Deflection (in.)							
	0.05	0.10	0.15	0.20	0.25	0.30	0.40	0.50
	Tolerance (±percent of force)							
0.005	12	7	6	5				
0.010		12	8.5	7	6.5	5.5	5	
0.020		22	15.5	12	10	8.5	7	6
0.030			22	17	14	12	9.5	8
0.040				22	18	15.5	12	10
0.050					22	19	14.5	12

0.060	25	22	17	14
0.070		25	19.5	16
0.080			22	18
0.090			25	20
0.100				22

Length tolerance	Deflection (in.)						
	0.75	1.00	1.50	2.00	3.00	4.00	6.00
0.020	5						
0.030	6	5					
0.040	7.5	6	5				
0.050	9	7	5.5				
0.060	10	8	6	5			
0.070	11	9	6.5	5.5			
0.080	12.5	10	7.5	6	5		
0.090	14	11	8	6	5		
0.100	15.5	12	8.5	7	5.5		
0.200		22	15.5	12	8.5	7	5.5
0.300			22	17	12	9.5	7
0.400				21	15	12	8.5
0.500				25	18.5	14.5	10.5

[a]The force tolerances for compression springs are based upon the predetermined length tolerance and the deflection from free length to the loaded position. The length tolerance should be determined first.

[b]Courtesy Spring Manufacturers Institute.

FIG. 191. Compression Spring Force Tolerances (Metric System)[a]

Length tolerance (±mm)	Deflection from free length to force position (mm)							
	1.25	2.50	3.75	5.00	6.50	7.50	10.00	13.00
	Load tolerance (±percent of load)							
0.13	12	7	6	5				
0.25		12	8.5	7	6.5	5.5	5	
0.50		22	15.5	12	10	8.5	7	6
0.75			22	17	14	12	9.5	8
1.00				22	18	15.5	12	10
1.25					22	19	14.5	12
1.50					25	22	17	14
1.75						25	19.5	16
2.00							22	18
2.25							25	20
2.50								22

Length tolerance (±mm)	Deflection from free length to force position (mm)						
	20	25	40	50	75	100	150
	Load tolerance (±percent of load), continued						
0.50	5						
0.75	6	5					
1.00	7.5	6	5				
1.25	9	7	5.5				
1.50	10	8	6	5			
1.75	11	9	6.5	5.5			
2.00	12.5	10	7.5	6	5		
2.25	14	11	8	6.5	5.3		
2.50	15.5	12	8.5	7	5.5		
5.00		22	15.5	12	8.5	7	5.5
7.50			22	17	12	9.5	7
10.00				21	15	12	8.5
13.00				25	18.5	14.5	10.5

[a]First, the length tolerance should be predetermined.

FIG. 192. Extension Spring Force Tolerances (Inch-Pound and Metric Systems)[a,b]

Spring index, D/d		Wire diameter (in. & mm)										
	in.	0.015	0.022	0.032	0.044	0.062	0.092	0.125	0.189	0.250	0.375	0.437
	mm	0.40	0.50	0.80	1.10	1.60	2.35	3.15	4.75	6.35	9.50	11.0
	$\frac{FL}{F}$	Tolerance (±percent of force)										
4	12	20.0	18.5	17.6	16.9	16.2	15.5	15.0	14.3	13.8	13.0	12.6
	8	18.5	17.5	16.7	15.8	15.0	14.5	14.0	13.2	12.5	11.5	11.0
	6	16.8	16.1	15.5	14.7	13.8	13.2	12.7	11.8	11.2	9.9	9.4
	4.5	15.0	14.7	14.1	13.5	12.6	12.0	11.5	10.3	9.7	8.4	7.9
	2.5	13.1	12.4	12.1	11.8	10.6	10.0	9.1	8.5	8.0	6.8	6.2
	1.5	10.2	9.9	9.3	8.9	8.0	7.5	7.0	6.5	6.1	5.3	4.8
	0.5	6.2	5.4	4.8	4.6	4.3	4.1	4.0	3.8	3.6	3.3	3.2
6	12	17.0	15.5	14.6	14.1	13.5	13.1	12.7	12.0	11.5	11.2	10.7
	8	16.2	14.7	13.9	13.4	12.6	12.2	11.7	11.0	10.5	10.0	9.5
	6	15.2	14.0	12.9	12.3	11.6	10.9	10.7	10.0	9.4	8.8	8.3
	4.5	13.7	12.4	11.5	11.0	10.5	10.0	9.6	9.0	8.3	7.6	7.1
	2.5	11.9	10.8	10.2	9.8	9.4	9.0	8.5	7.9	7.2	6.2	6.0
	1.5	9.9	9.0	8.3	7.7	7.3	7.0	6.7	6.4	6.0	4.9	4.7
	0.5	6.3	5.5	4.9	4.7	4.5	4.3	4.1	4.0	3.7	3.5	3.4
8	12	15.8	14.3	13.1	13.0	12.1	12.0	11.5	10.8	10.2	10.0	9.5
	8	15.0	13.7	12.5	12.1	11.4	11.0	10.6	10.1	9.4	9.0	8.6
	6	14.2	13.0	11.7	11.2	10.6	10.0	9.7	9.3	8.6	8.1	7.6
	4.5	12.8	11.7	10.7	10.1	9.7	9.0	8.7	8.3	7.8	7.2	6.6
	2.5	11.2	10.2	9.5	8.8	8.3	7.9	7.7	7.4	6.9	6.1	5.6
	1.5	9.5	8.6	7.8	7.1	6.9	6.7	6.5	6.2	5.8	4.9	4.5
	0.5	6.3	5.6	5.0	4.8	4.5	4.4	4.2	4.1	3.9	3.6	3.5
10	12	14.8	13.3	12.0	11.9	11.1	10.9	10.5	9.9	9.3	9.2	8.8
	8	14.2	12.8	11.6	11.2	10.5	10.2	9.7	9.2	8.6	8.3	8.0
	6	13.4	12.1	10.8	10.5	9.8	9.3	8.9	8.6	8.0	7.6	7.2
	4.5	12.3	10.8	10.0	9.5	9.0	8.5	8.1	7.8	7.3	6.8	6.4
	2.5	10.8	9.6	9.0	8.4	8.0	7.7	7.3	7.0	6.5	5.9	5.5
	1.5	9.2	8.3	7.5	6.9	6.7	6.5	6.3	6.0	5.6	5.0	4.6
	0.5	6.4	5.7	5.1	4.9	4.7	4.5	4.3	4.2	4.0	3.8	3.7
12	12	14.0	12.3	11.1	10.8	10.1	9.8	9.5	9.0	8.5	8.2	7.9
	8	13.2	11.8	10.7	10.2	9.6	9.3	8.9	8.4	7.9	7.5	7.2
	6	12.6	11.2	10.2	9.7	9.0	8.5	8.2	7.9	7.4	6.9	6.4
	4.5	11.7	10.2	9.4	9.0	8.4	8.0	7.6	7.2	6.8	6.3	5.8
	2.5	10.5	9.2	8.5	8.0	7.8	7.4	7.0	6.6	6.1	5.6	5.2
	1.5	8.9	8.0	7.2	6.8	6.5	6.3	6.1	5.7	5.4	4.8	4.5
	0.5	6.5	5.8	5.3	5.1	4.9	4.7	4.5	4.3	4.2	4.0	3.3
14	12	13.1	11.3	10.2	9.7	9.1	8.8	8.4	8.1	7.6	7.2	7.0
	8	12.4	10.9	9.8	9.2	8.7	8.3	8.0	7.6	7.2	6.8	6.4
	6	11.8	10.4	9.3	8.8	8.3	7.7	7.5	7.2	6.8	6.3	5.9
	4.5	11.1	9.7	8.7	8.2	7.8	7.2	7.0	6.7	6.3	5.8	5.4
	2.5	10.1	8.8	8.1	7.6	7.1	6.7	6.5	6.2	5.7	5.2	5.0
	1.5	8.6	7.7	7.0	6.7	6.3	6.0	5.8	5.5	5.2	4.7	4.5
	0.5	6.6	5.9	5.4	5.2	5.0	4.8	4.6	4.4	4.3	4.2	4.0
16	12	12.3	10.3	9.2	8.6	8.1	7.7	7.4	7.2	6.8	6.3	6.1
	8	11.7	10.0	8.9	8.3	7.8	7.4	7.2	6.8	6.5	6.0	5.7
	6	11.0	9.6	8.5	8.0	7.5	7.1	6.9	6.5	6.2	5.7	5.4
	4.5	10.5	9.1	8.1	7.5	7.2	6.8	6.5	6.2	5.8	5.3	5.1
	2.5	9.7	8.4	7.6	7.0	6.7	6.3	6.1	5.7	5.4	4.9	4.7
	1.5	8.3	7.4	6.6	6.2	6.0	5.8	5.6	5.3	5.1	4.6	4.4
	0.5	6.7	5.9	5.5	5.3	5.1	5.0	4.8	4.6	4.5	4.3	4.1

[a] $\frac{FL}{F}$ = ratio of the spring free length (*FL*) to deflection (*F*).

[b] Courtesy Spring Manufacturers Institute.

FIG. 193. Compression and Extension Springs, Coil Diameter, Tolerance (Metric System)

Wire diameter (mm)	Spring index, D/d						
	4	6	8	10	12	14	16
	Coil diameter tolerance (±mm)						
0.40	0.05	0.05	0.08	0.10	0.15	0.15	0.20
0.60	0.05	0.08	0.10	0.15	0.20	0.20	0.25
0.90	0.05	0.10	0.15	0.20	0.25	0.30	0.35
1.30	0.08	0.15	0.20	0.25	0.30	0.40	0.45
2.00	0.10	0.20	0.25	0.30	0.40	0.50	0.55
3.00	0.15	0.25	0.30	0.45	0.50	0.60	0.75
4.50	0.20	0.30	0.40	0.60	0.70	0.85	0.95
6.00	0.30	0.40	0.50	0.70	0.90	1.00	1.25
9.50	0.40	0.50	0.65	0.90	1.15	1.40	1.60
13.00	0.50	0.75	1.00	1.50	2.00	2.50	3.00

[a]These tolerances should be applied to the OD or ID.
[b]Courtesy Spring Manufacturers Institute.

FIG. 194. Compression and Extension Springs, Number of Coils, Tolerances (Inch-Pound and Metric Systems)[a]

Compression springs	Extension Springs 3-5 coils	Extension Springs 6-8 coils	Extension Springs 9-12 coils	For additional coils, add-
±5%	±20%	±30%	±40%	1½% for each coil over 12

Torsion springs[a]

Number of coils	Tolerance	Number of coils	Tolerance
Up to 5	±5°	Over 10 to 20	±15°
Over 5 to 10	±10°	Over 20 to 40	±30°

[a]Courtesy of the Spring Manufacturers Institute.

FIG. 195. Extension spring hook and loop tolerances.

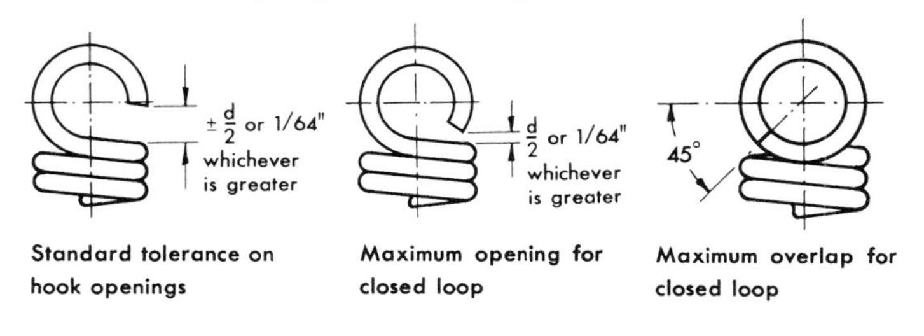

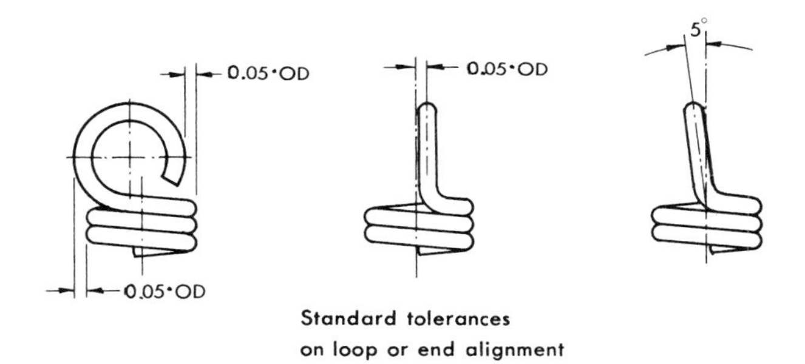

Standard tolerances
on loop or end alignment

FIG. 196. Torsion Spring Coil Diameter Tolerances (Inch-Pound System)[a]

Wire diameter d (in.)	Spring index, D/d						
	4	6	8	10	12	14	16
	Coil diameter tolerance (±in.)						
0.015	0.002	0.002	0.002	0.002	0.003	0.003	0.004
0.023	0.002	0.002	0.002	0.003	0.004	0.005	0.006
0.035	0.002	0.002	0.003	0.004	0.006	0.007	0.009
0.051	0.002	0.003	0.005	0.007	0.008	0.010	0.012
0.076	0.003	0.005	0.007	0.009	0.012	0.015	0.018
0.114	0.004	0.007	0.010	0.013	0.018	0.022	0.028
0.172	0.006	0.010	0.013	0.020	0.027	0.034	0.042
0.250	0.008	0.014	0.022	0.030	0.040	0.050	0.060

[a]Courtesy Spring Manufacturers Institute.

FIG. 197. Torsion Spring Coil Diameter Tolerances (Metric System)

Wire diameter, d (mm)	Spring index, D/d						
	4	6	8	10	12	14	16
	Coil diameter tolerance (±mm)						
0.40	0.05	0.05	0.05	0.05	0.08	0.08	0.10
0.60	0.05	0.05	0.05	0.08	0.10	0.12	0.15
0.90	0.05	0.05	0.08	0.10	0.15	0.18	0.25
1.25	0.05	0.08	0.12	0.18	0.20	0.25	0.30
2.00	0.08	0.12	0.18	0.25	0.30	0.40	0.45
3.00	0.10	0.18	0.25	0.35	0.45	0.55	0.70
4.00	0.15	0.25	0.35	0.50	0.70	0.85	1.00
6.00	0.20	0.35	0.55	0.75	1.00	1.25	1.50

FIG. 198. Torsion Spring Tolerance for Angular Relationship of Ends (Inch-Pound and Metric Systems)[a]

Number of coils, N	Spring index, D/d								
	4	6	8	10	12	14	16	18	20
	Free angle tolerance (±degrees)								
1	2	3	3.5	4	4.5	5	5.5	5.5	6
2	4	5	6	7	8	8.5	9	9.5	10
3	5.5	7	8	9.5	10.5	11	12	13	14
4	7	9	10	12	14	15	16	16.5	17
5	8	10	12	14	16	18	20	20.5	21
6	9.5	12	14.5	16	19	20.5	21	22.5	24
8	12	15	18	20.5	23	25	27	28	29
10	14	19	21	24	27	29	31.5	32.5	34
15	20	25	28	31	34	36	38	40	42
20	25	30	34	37	41	44	47	49	51
25	29	35	40	44	48	52	56	60	63
30	32	38	44	50	55	60	65	68	70
50	45	55	63	70	77	84	90	95	100

[a]Courtesy Spring Manufacturers Institute.

FIG. 199. Tolerances on Wire Diameters (Inch-Pound System)[a]

Music Wire	
Wire size range (in.)	Tolerance (±in.)
0.004-0.010	±0.0002
0.011-0.028	±0.0003
0.029-0.065	±0.0004
0.064-0.080	±0.0005
0.081-0.250	±0.001

Type 302 Stainless & 17-7 PH Stainless	
Wire size range (in.)	Tolerance (±in.)
0.007-0.0079	±0.0002
0.008-0.0119	±0.00025
0.012-0.0239	±0.0004
0.024-0.0329	±0.0005
0.033-0.0439	±0.00075
0.044 and larger	±0.001

Hard-Drawn MB and Oil-Tempered MB	
Wire size range (in.)	Tolerance (±in.)
0.010 -0.0199	±0.0005
0.020 -0.0347	±0.0006
0.0348-0.051	±0.0008
0.0511-0.075	±0.001
0.0751-0.109	±0.0015
0.1091-0.250	±0.002
0.2501-0.375	±0.0025
0.3751-0.625	±0.003

Chrome-Silicon	
Wire size range (in.)	Tolerance (±in.)
0.032 -0.072	±0.001
0.0721-0.375	±0.002

Chrome-Vanadium	
Wire size range (in.)	Tolerance (±in.)
0.020 -0.0275	±0.0008
0.0276-0.072	±0.001
0.0721-0.375	±0.002
0.3751-0.500	±0.003

[a]Courtesy Spring Manufacturers Institute.

FIG. 200. Tolerances on Wire Diameters (Metric System)

Music Wire	
Wire size range (mm)	Tolerance (±mm)
0.10-0.25	0.005
0.26-0.70	0.008
0.71-1.60	0.010
1.61-2.00	0.013
2.01-6.00	0.025

Type 302 and 17-7 PH Stainless	
Wire size range (mm)	Tolerance (±mm)
0.18-0.20	0.005
0.21-0.30	0.006
0.31-0.60	0.010
0.61-0.85	0.013
0.86-1.10	0.020
1.11 and larger	0.025

Hard-Drawn MB and Oil-Tempered MB	
Wire size range (mm)	Tolerance (±mm)
0.25-0.50	0.013
0.51-0.90	0.015
0.91-1.30	0.020
1.31-2.00	0.025
2.01-2.75	0.040
2.76-6.00	0.050
6.01-9.50	0.060
9.51-16.00	0.075

Chrome-Silicon	
Wire size range (mm)	Tolerance (±mm)
0.80-1.80	0.025
1.81-9.50	0.050

Chrome-Vanadium	
Wire size range (mm)	Tolerance (±mm)
0.50-0.70	0.020
0.71-1.80	0.025
1.81-9.50	0.050
9.50-13.0	0.075

19 Glossary of Spring Terminology*

Active coils. The coils which deflect under a load.

Age hardening. Also known as precipitation hardening. A hardening method for certain alloys whereby they are held at a medium high temperature for a long time, causing a precipitation of a constituent from a supersaturated solution. Examples are beryllium-copper and Monel K-500.

Alloy. A metal composed of two or more elements, such as chromium-silicon, stainless steel, or brass.

Arbor. Also called mandrel. A round rod, usually held in a chuck, over which wire is coiled to form a spring. On automatic coilers, it is used to determine the spring diameter and as an anvil for cutting off the spring.

Austempering. A hardening method whereby steel is quenched from above the transformation point to a quench temper slightly above the upper limit of martensite formation and held at that temperature until the austenite is transformed to bainite.

Austenite. A face-centered cubic solid solution of carbon or other elements in gamma iron (see equilibrium diagram, Fig. 180).

Bainite. A constituent of steel produced when austenite transforms at a temperature below that at which martensite is formed (as found in austempering).

*Additional terms relating to fatigue are described in Chapter 10. Leaf or laminated springs have a special terminology of their own consisting of over 50 terms; consult a leaf spring manufacturer or a spring consultant for details.

Baking. Heating to remove residual stresses or to remove hydrogen from electroplated springs.

Belleville springs. A dished washer, hardened to have resilient spring qualities.

Bend Test. A test to determine ductility by bending a specimen over a predetermined radius. See ASTM E 6.

Blueing. Heating bright steel in an oxidising environment to produce a blue color. Sometimes used to denote baking.

Brass. A copper-base alloy of copper and zinc.

Bronze. A copper-base alloy of copper and tin.

Buckling. Lateral movement of a spring when compressed.

Carbon steel. Steel which has no purposely added alloying elements to obtain certain mechanical properties.

Carburizing. Heating steel in contact with carbonaceous materials to add carbon to the surface for hardening.

Case hardening. Hardening the surface of steel. Several methods are used, such as carburizing, nitriding, induction hardening, and flame hardening.

Cast. The behavior of a single coil of wire when cut from a bundle, such as springing upward, outward, unwinding, or remaining at approximately the same diameter.

Coned disk spring. Same as a Belleville spring.

Creep. Plastic deformation of a material, proceeding slowly and continuously when stressed below the elastic limit, particularly at elevated temperatures, causing elongation and loss of load.

Decarburization. Loss of carbon from the surface of a steel caused by high temperature during rolling or heat treating.

Deflection. Displacement of the ends of a spring on the application of a load.

Ductility. Plastic permanent deflection of a material before fracture.

Elastic limit and endurance limit. See Chapter 10 on fatigue.

Fatigue life and fatigue strength. See Chapter 10 on fatigue.

Finish. For wire, the type of surface condition. For springs, the type of coating, such as paint, enamel, or electroplating. Sometimes used to denote the final checking, adjusting, or compliance of springs to specifications.

Force. Active power. A pressure due to a weight. Erroneously called "load."

Galvanizing. Applying zinc to a surface by passing the material through molten zinc. May also be applied by electroplating.

Gauge or Gage. A standard of thickness; several are used depending upon type of material and country of origin.

Gradient or Rate or Constant. Force per increment of deflection, such as pounds per inch, newtons per millimetre, torque per degree of deflection or newton-millimetres per degree.

Hand. The direction in which the coils of a spring are formed, such as left hand or right hand if fitted over a screw.

Heat setting or Hot pressing. Deflecting a spring in a fixture and heating higher than its operating temperature, to reduce permanent set or loss of load.

Helical or Helix. Springs of spiral or cylindrical shape such as compression, extension, or torsion springs in contrast to flat, cantilever, or motor springs.

Hydrogen Embrittlement. Brittleness in a material caused by adsorption of hydrogen during pickling or electroplating. See Chapter 15.

Hysteresis. The difference in forces at various deflections when loading and unloading a spring.

Index. A ratio of the mean coil diameter to the wire, D/d.

Initial tension. A force wound into an extension spring during coiling to hold the coils together.

Killed steel. Deoxidizing steel and permitting only a slight evolution of gases during solidification after pouring into molds, thereby producing more uniform chemical properties, especially suitable for wire.

Load. A force applied to a spring to cause deflection (specify force, not load).

Loop. A closed hook. Usually on extension springs to provide for fastening or for applying a load.

Mandrel. See arbor.

Mean coil diameter. A dimension: outside diameter minus thickness of wire, OD – d.

Mechanical properties. The properties of a metal associated with elastic or inelastic behavior when a force is applied or that relate to stress and strain. Such as tensile strength, elastic limit, elongation, ductility, and hardness.

Modulus of elasticity. See Chapter 10 and Fig. 112.

Moment. See torque and Chapter 13.

Natural frequency or Natural period of vibration. The rate or number of cycles per minute (or second) at which a spring will vibrate freely once it has been excited.

Normalizing. Heating and holding a material above the transformation range and cooling in air to refine the grain size and remove residual stresses.

Passivating. An acid-dipping treatment applied to stainless steel to remove contaminants and to restore the oxide film which may have been lost due to heat treatment or shotblasting, thereby improving its corrosion resistance.

Permanent set. The difference in length of a highly stressed spring after deflection to a specified position and released.

Pitch. The distance from center to center of the wire in two adjacent active coils. (Specify number of active coils rather than pitch.)

Precipitation hardening. See Age hardening.

Preset. Removal of set by deflecting a compression spring to solid.

Prestressing. Increasing the elastic limit of a highly stressed compression spring by coiling it slightly longer than required; heating to remove residual coiling stresses, then compressing solid to trap beneficial stresses so that the spring returns to the specified length.

Proportional limit. The maximum stress or load without deviating from a straight line curve for stress to strain as per Hooke's law. A special value of yield strength and elastic limit.

Rate. Force per unit of deflection. See Gradient.

Reel or Swift. A rotating device to hold coils of wire, permitting unwinding.

Residual stress. Stresses set up within a spring by coiling, compressing to remove set, shotpeening, cold working, or induced by heat treatment.

Resilience. Elastic ability of a material to recover or rebound to its original position or shape after deflection; possessing power of recovery.

Rod. A round, square, or shaped solid section of material furnished in straight lengths.

Scragging. Pressing to solid to remove set. Also denotes prestressing.

Set. Amount of permanent distortion when a spring is stressed higher than the elastic limit of the material; see Permanent set.

Shotpeen. Hurtling small steel shot against springs to clean surface and trap beneficial surface stresses, thereby increasing fatigue life. Shotblasting with sand or other shot is for cleaning only. See Chapter 15.

Slenderness ratio. FL/D, ratio of free length to mean diameter. If over 4, may cause buckling in compression springs, depending upon amount of deflection.

S/N curve. Fatigue strength curve, based on number of cycles to failure at various stresses. Also fatigue life curve.

Spiral. Winding around a center point and gradually receding from it. Usually, a clock or motor spring made from flat strip or one with a space between the coils.

Spring index. D/d, ratio of mean diameter to wire thickness. See Index. Best ratio is between 7 and 13. Troublesome if less than 4 or over 16.

Squareness. More accurately, deviation from squareness. The angular deviation, of a compression spring when standing upright, from the vertical.

Squareness under force. Same as Squareness, except with a compression spring under force.

Strain. The change or deformation produced by a stress as compared to the original size or shape.

Stress. The force divided by the area over which it acts. The intensity of internal force that tends to move, separate, or distort the grain structure.

Stress relieve. Low-temperature heat treatment to relieve residual stresses induced during coiling or forming so that springs will be more stable, more capable of withstanding higher stresses, and have longer fatigue life.

Tempering. Heating a hardened material to a medium or low temperature for a suitable time to modify the structure, thereby obtaining a desired hardness, toughness, and optimum mechanical properties. See Stress relieve.

Tensile strength. Often called ultimate tensile strength. The maximum tensile stress which a material is capable of sustaining. It equals the maximum force reached during a tensile test, divided by the original cross-sectional area. Values can be in pounds per square inch or megapascals.

Torque. A turning or twisting force in torsion springs tending to produce rotation. The force on a torsion spring required to cause a specified angular deflection. See Chapter 13.

Torsion test or Twist test. Twisting or rotating a wire in a testing instrument to determine the number of twists (in one direction only) until the wire breaks. See Chapter 15 and ASTM E 558.

Total coils. In compression springs, the sum of the active coils plus the dead coils forming the ends. In extension, torsion, spiral, and motor springs, equal to the active coils, and does not include the coils or material used for forming the ends.

Wahl factor. A curvature correction factor used to determine the total stress in a spring. In Europe, the Gohner correction factor, which is quite similar, is often used.

Wap. A European term denoting a single turn of wire in a coil.

20 Miscellaneous Tables

The figures in this chapter (Figs. 201 to 219) have been selected because of their usefulness to spring designers.

FIG. 201. Modulus of Elasticity: Frequently Used Conversions (psi to MPa)[a]

psi	MPa Exact	MPa Rounded	psi	MPa Exact	MPa Rounded
5 000 000	34 474	34,500	14 000 000	96 527	96 500
6 000 000	41 369	41 400	14 500 000	99 974	100 000
7 000 000	48 263	48 300	15 000 000	103 421	103 400
7 250 000	49 987	50 000	17 000 000	117 211	117 200
9 200 000	63 432	63 500	19 000 000	131 000	131 000
9 500 000	65 500	65 500	26 000 000	179 264	179 300
10 000 000	68 948	69 000	26 500 000	182 711	182 700
10 500 000	72 395	72 400	27 000 000	186 158	186 200
10 750 000	74 119	74 100	27 500 000	189 606	189 600
11 000 000	75 842	75 800	28 000 000	193 053	193 000
11 200 000	77 221	77 200	28 500 000	196 501	196 500
11 400 000	78 600	78 600	28 600 000	197 190	197 200
11 500 000	79 290	79 300	28 700 000	197 880	197 900
11 600 000	79 979	80 000	28 800 000	198 569	198 600
11 700 000	80 669	80 700	29 000 000	199 948	200 000
11 750 000	81 013	81 000	29 500 000	203 395	203 400
11 850 000	81 703	81 700	30 000 000	206 843	206 800
12 000 000	82 737	82 700	31 000 000	213 737	213 700

[a]psi × 0.006 894 757 = MPa. Also see Fig. 112.

FIG. 202. Modulus of Elasticity as Affected by Elevated Temperatures[a]

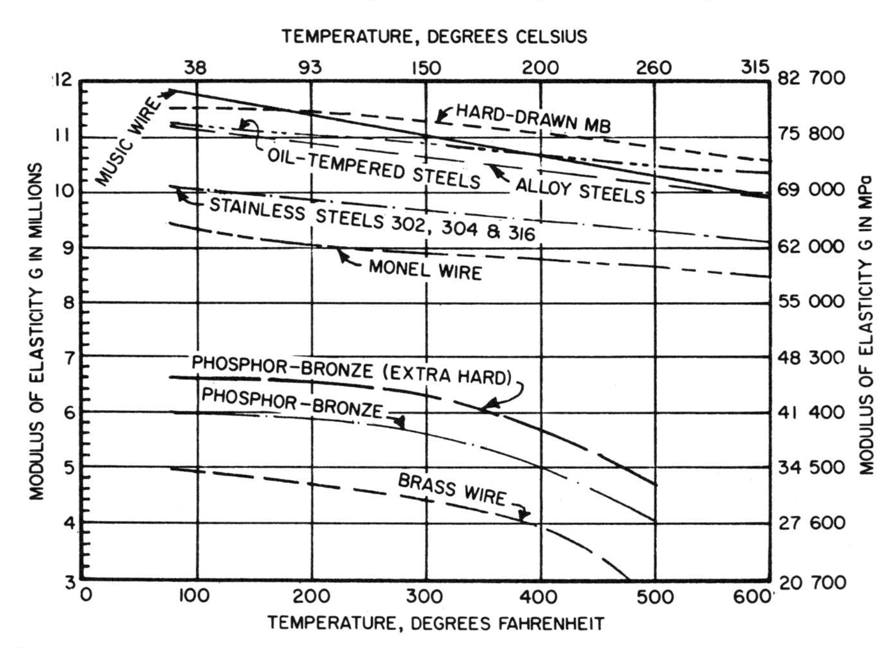

[a]The modulus of elasticity G lowers in value at elevated temperatures as shown by the curves. For example, the modulus of elasticity G for a spring made from oil-tempered MB steel drops from 11 200 000 psi (77 200 MPa) at room temperature to 10 600 000 psi (73 000 MPa) at 400°F (200°C), a drop of about 5½ percent. This drop in modulus means that the spring will deflect 5½ percent more before it exerts the same force as it did at room temperature. Conversely, if the compressed height remains constant, the spring will exert 5½ percent less force. This lowering in force should be allowed for when designing springs.

FIG. 203. Conversions of Weights Frequently Used with Spring Testers

1 oz. = 0.0625 lb = 28.349 52 g = 0.278 013 9 N
1 lb = 16 oz = 453.5924 g = 4.448 222 N
1 g = 0.035 273 97 oz = 0.002 204 622 lb = 0.009 806 652 N
1 N = 3.596 942 oz = 0.224 808 9 lb = 101.9716 g

(Values are rounded off.)

Ounces	Pounds	Grams	Newtons
1/64	0.001	0.443	0.0043
1/32	0.001 9	0.886	0.0087
1/16	0.003 9	1.772	0.0174
1/8	0.007 8	3.544	0.0348
1/4	0.015 6	7.087	0.0695
1/2	0.031 3	14.175	0.1390
1	0.062 5	28.350	0.278
2	0.125	56.699	0.556
3	0.187 5	85.049	0.834
4	0.250	113.398	1.112
5	0.312 5	141.748	1.390
10	0.625	283.495	2.780

Grams	Ounces	Pounds	Newtons
0.1	0.003 5	0.000 22	0.000 98
0.2	0.007 1	0.000 44	0.001 96
0.3	0.010 6	0.000 66	0.002 94
0.5	0.017 6	0.001 10	0.004 90
1	0.353 3	0.002 2	0.009 8
2	0.070 5	0.004 4	0.019 6
3	0.105 8	0.006 6	0.029 4
4	0.141 1	0.008 8	0.039 2
5	0.176 4	0.011 0	0.049 0
10	0.352 7	0.022 0	0.098 1
15	0.529 1	0.033 1	0.147 1
25	0.881 8	0.055 1	0.245 2

Pounds	Ounces	Grams	Newtons
0.1	1.6	45.359	0.4448
0.2	3.2	90.718	0.8896
0.3	4.8	136.08	1.334
0.5	8.0	226.80	2.224
1	16	453.59	4.448
2	32	907.18	8.896
3	48	1 360.78	13.345
4	64	1 814.37	17.793
5	80	2 267.96	22.241
10	160	4 535.92	44.482
15	240	6 803.89	66.723
25	400	11 339.81	111.206

Newtons	Ounces	Pounds	Grams
0.1	0.3597	0.022 5	10.20
0.2	0.7194	0.045 0	20.39
0.3	1.0791	0.067 4	30.59
0.5	1.7985	0.112 4	50.99
1	3.597	0.224 8	101.97
2	7.194	0.449 6	203.94
3	10.791	0.674 4	305.91
4	14.388	0.899 2	407.89
5	17.985	1.124 0	509.86
10	35.969	2.248 1	1 019.72
15	53.954	3.372 1	1 529.57
25	89.924	5.620 2	2 549.29

FIG. 204. Inches to Millimetres Conversion[a]

in.	mm.	in.	mm.	in.	mm.	in.	mm.	in.	mm.	in.	mm.	in.	mm.
·0010	·025	·0075	·191	·0140	·356	·040	1·02	·105	2·67	·170	4·32	·35	8·9
·0011	·028	·0076	·193	·0141	·358	·041	1·04	·106	2·69	·171	4·34	·36	9·1
·0012	·031	·0077	·196	·0142	·361	·042	1·07	·107	2·72	·172	4·37	·37	9·4
·0013	·033	·0078	·198	·0143	·363	·043	1·09	·108	2·74	·173	4·39	·38	9·7
·0014	·036	·0079	·201	·0144	·366	·044	1·12	·109	2·77	·174	4·42	·39	9·9
·0015	·038	·0080	·203	·0145	·368	·045	1·14	·110	2·79	·175	4·45	·40	10·2
·0016	·041	·0081	·206	·0146	·371	·046	1·17	·111	2·82	·176	4·47	·45	11·4
·0017	·043	·0082	·208	·0147	·373	·047	1·19	·112	2·84	·177	4·50	·50	12·7
·0018	·046	·0083	·211	·0148	·376	·048	1·22	·113	2·87	·178	4·52	·55	14·0
·0019	·048	·0084	·213	·0149	·378	·049	1·24	·114	2·90	·179	4·55	·60	15·2
·0020	·051	·0085	·216	·0150	·381	·050	1·27	·115	2·92	·180	4·57	·65	16·5
·0021	·053	·0086	·218	·0151	·384	·051	1·30	·116	2·95	·181	4·60	·70	17·8
0022	·056	·0087	·221	·0152	·386	·052	1·32	·117	2·97	·182	4·62	·75	19·1
·0023	·058	·0088	·224	·0153	·389	·053	1·35	·118	3·00	·183	4·65	·80	20·3
0024	·061	·0089	·226	·0154	·391	·054	1·37	·119	3·02	·184	4·67	·85	21·6
·0025	·064	·0090	·229	·0155	·394	·055	1·40	·120	3·05	·185	4·70	·90	22·9
·0026	·066	·0091	·231	·0156	·396	·056	1·42	·121	3·07	·186	4·72	·95	24·1
·0027	·069	·0092	·234	·0157	·399	·057	1·45	·122	3·10	·187	4·75	1·00	25·4
·0028	·071	·0093	·236	·0158	·401	·058	1·47	·123	3·12	·188	4·78	1·25	31·8
·0029	·074	·0094	·239	·0159	·404	·059	1·50	·124	3·15	·189	4·80	1·50	38·1
·0030	·076	·0095	·241	·0160	·406	·060	1·52	·125	3·18	·190	4·83	1·75	44·5
·0031	·079	·0096	·244	·0161	·409	·061	1·55	·126	3·20	·191	4·85	2·00	50·8
·0032	·081	·0097	·246	·0162	·412	·062	1·57	·127	3·23	·192	4·88	2·25	57·2
·0033	·084	·0098	·249	·0163	·414	·063	1·60	·128	3·25	·193	4·90	2·50	63·5
·0034	·086	·0099	·251	·0164	·417	·064	1·63	·129	3·28	·194	4·93	2·75	69·9
·0035	·089	·0100	·254	·0165	·419	·065	1·65	·130	3·30	·195	4·95	3·00	76·2
·0036	·091	·0101	·257	·0166	·422	·066	1·68	·131	3·33	·196	4·98	3·25	82·6
·0037	·094	·0102	·259	·0167	·424	·067	1·70	·132	3·35	·197	5·00	3·50	88·9
·0038	·097	·0103	·262	·0168	·427	·068	1·73	·133	3·38	·198	5·03	3·75	95·3
·0039	·099	·0104	·264	·0169	·429	·069	1·75	·134	3·40	·199	5·05	4·00	101·6
·0040	·102	·0105	·267	·0170	·432	·070	1·78	·135	3·43	·200	5·08	4·25	108·0
·0041	·104	·0106	·269	·0171	·434	·071	1·80	·136	3·45	·201	5·11	4·50	114·3
·0042	·107	·0107	·272	·0172	·437	·072	1·83	·137	3·48	·202	5·13	4·75	120·7
·0043	·109	·0108	·274	·0173	·439	·073	1·85	·138	3·51	·203	5·16	5·00	127·0
·0044	·112	·0109	·277	·0174	·442	·074	1·88	·139	3·53	·204	5·18	5·25	133·4

·0045	·114	·0110	·279	·0175	·445	·075	1·91	·140	3·56	·205	5·21	5·50	139·7
·0046	·117	·0111	·282	·0176	·447	·076	1·93	·141	3·58	·206	5·23	5·75	146·1
·0047	·119	·0112	·284	·0177	·450	·077	1·96	·142	3·61	·207	5·26	6·0	152·4
·0048	·122	·0113	·287	·0178	·452	·078	1·98	·143	3·63	·208	5·28	6·5	165·1
·0049	·124	·0114	·290	·0179	·455	·079	2·01	·144	3·66	·209	5·31	7·0	177·8
·0050	·127	·0115	·292	·0180	·457	·080	2·03	·145	3·68	·210	5·33	7·5	190·5
·0051	·130	·0116	·295	·0181	·460	·081	2·06	·146	3·71	·211	5·36	8·0	203·2
·0052	·132	·0117	·297	·0182	·462	·082	2·08	·147	3·73	·212	5·38	8·5	215·9
·0053	·135	·0118	·300	·0183	·465	·083	2·11	·148	3·76	·213	5·41	9·0	228·6
·0054	·137	·0119	·302	·019	·483	·084	2·13	·149	3·78	·214	5·44	9·5	241·3
·0055	·140	·0120	·305	·020	·508	·085	2·16	·150	3·81	·215	5·46	10·0	254·0
·0056	·142	·0121	·307	·021	·533	·086	2·18	·151	3·84	·216	5·49	10·5	266·7
·0057	·145	·0122	·310	·022	·559	·087	2·21	·152	3·86	·217	5·51	11·0	279·4
·0058	·147	·0123	·312	·023	·584	·088	2·24	·153	3·89	·218	5·54	11·5	292·1
·0059	·150	·0124	·315	·024	·610	·089	2·26	·154	3·91	·219	5·56	12·0	304·8
·0060	·152	·0125	·318	·025	·635	·090	2·29	·155	3·94	·220	5·59	12·5	317·5
·0061	·155	·0126	·320	·026	·660	·091	2·31	·156	3·96	·225	5·71	13·0	330·2
·0062	·157	·0127	·323	·027	·686	·092	2·34	·157	3·99	·230	5·84	13·5	342·9
·0063	·160	·0128	·325	·028	·711	·093	2·36	·158	4·01	·235	5·97	14·0	355·6
·0064	·163	·0129	·328	·029	·737	·094	2·39	·159	4·04	·240	6·10	14·5	368·3
·0065	·165	·0130	·330	·030	·762	·095	2·41	·160	4·06	·25	6·35	15	381·0
·0066	·168	·0131	·333	·031	·787	·096	2·44	·161	4·09	·26	6·60	16	406·4
·0067	·170	·0132	·335	·032	·813	·097	2·46	·162	4·12	·27	6·86	17	431·8
·0068	·173	·0133	·338	·033	·838	·098	2·49	·163	4·14	·28	7·11	18	457·2
·0069	·175	·0134	·340	·034	·864	·099	2·51	·164	4·17	·29	7·37	19	482·6
·0070	·178	·0135	·343	·035	·889	·100	2·54	·165	4·19	·30	7·62	20	58·00
·0071	·180	·0136	·345	·036	·914	·101	2·57	·166	4·22	·31	7·87	21	533·4
·0072	·183	·0137	·348	·037	·940	·102	2·59	·167	4·24	·32	8·13	22	558·8
·0073	·185	·0138	·351	·038	·965	·103	2·62	·168	4·27	·33	8·38	23	584·2
·0074	·188	·0139	·353	·039	·991	·104	2·64	·169	4·29	·34	8·64	24	609·6

[a]Courtesy International Nickel Co.

FIG. 205. Millimetres to Inches Conversion[a]

mm.	in.	mm.	in.	mm.	in.	mm.	in.	mm.	in.	mm.	in.	mm.	in
·010	·00039	·300	·0118	1·00	·0394	1·90	·0748	2·80	·1102	4·50	·177	17·5	·69
·011	·00043	·305	·0120	1·01	·0398	1·91	·0752	2·81	·1106	4·55	·179	18·0	·71
·012	·00047	·310	·0122	1·02	·0402	1·92	·0756	2·82	·1110	4·60	·181	18·5	·73
·013	·00051	·315	·0124	1·03	·0406	1·93	·0760	2·83	·1114	4·65	·183	19·0	·75
·014	·00055	·320	·0126	1·04	·0409	1·94	·0764	2·84	·1118	4·70	·185	19·5	·77
·015	·00059	·325	·0128	1·05	·0413	1·95	·0768	2·85	·1122	4·75	·187	20·0	·79
·016	·00063	·330	·0130	1·06	·0417	1·96	·0772	2·86	·1126	4·80	·189	20·5	·81
·017	·00067	·335	·0132	1·07	·0421	1·97	·0776	2·87	·1130	4·85	·191	21·0	·83
·018	·00071	·340	·0134	1·08	·0425	1·98	·0780	2·88	·1134	4·90	·193	21·5	·85
·019	·00075	·345	·0136	1·09	·0429	1·99	·0783	2·89	·1138	4·95	·195	22·0	·87
·020	·00079	·350	·0138	1·10	·0433	2·00	·0787	2·90	·1142	5·00	·197	22·5	·89
·021	·00083	·355	·0140	1·11	·0437	2·01	·0791	2·91	·1146	5·05	·199	23·0	·91
·022	·00087	·360	·0142	1·12	·0441	2·02	·0795	2·92	·1150	5·10	·201	23·5	·93
·023	·00091	·365	·0144	1·13	·0445	2·03	·0799	2·93	·1154	5·15	·203	24·0	·94
·024	·00094	·370	·0146	1·14	·0449	2·04	·0803	2·94	·1157	5·20	·205	24·5	·96
·025	·00098	·375	·0148	1·15	·0453	2·05	·0807	2·95	·1161	5·25	·207	25·0	·98
·026	·00102	·380	·0150	1·16	·0457	2·06	·0811	2·96	·1165	5·30	·209	25·5	1·00
·027	·00106	·385	·0152	1·17	·0461	2·07	·0815	2·97	·1169	5·35	·211	26·0	1·02
·028	·00110	·390	·0154	1·18	·0465	2·08	·0819	2·98	·1173	5·40	·213	26·5	1·04
·029	·00114	·395	·0156	1·19	·0469	2·09	·0823	2·99	·1177	5·45	·215	27·0	1·06
·030	·00118	·400	·0157	1·20	·0472	2·10	·0827	3·00	·1181	5·50	·217	27·5	1·08
·031	·00122	·405	·0159	1·21	·0476	2·11	·0831	3·01	·1185	5·55	·219	28	1·10
·032	·00126	·410	·0161	1·22	·0480	2·12	·0835	3·02	·1189	5·60	·221	29	1·14
·033	·00130	·415	·0163	1·23	·0484	2·13	·0839	3·03	·1193	5·65	·222	30	1·18
·034	·00134	·420	·0165	1·24	·0488	2·14	·0843	3·04	·1197	5·70	·224	31	1·22
·035	·00138	·425	·0167	1·25	·0492	2·15	·0846	3·05	·1201	5·75	·226	32	1·26
·036	·00142	·430	·0169	1·26	·0496	2·16	·0850	3·06	·1205	5·80	·228	33	1·30
·037	·00146	·435	·0171	1·27	·0500	2·17	·0854	3·07	·1209	5·85	·230	34	1·34
·038	·00150	·440	·0173	1·28	·0504	2·18	·0858	3·08	·1213	5·90	·232	35	1·38
·039	·00154	·445	·0175	1·29	·0508	2·19	·0862	3·09	·1217	5·95	·234	36	1·42
·040	·00157	·450	·0177	1·30	·0512	2·20	·0866	3·10	·1220	6·00	·236	37	1·46
·041	·00161	·455	·0179	1·31	·0516	2·21	·0870	3·11	·1224	6·05	·238	38	1·50
·042	·00165	·460	·0181	1·32	·0520	2·22	·0874	3·12	·1228	6·10	·240	39	1·54
·043	·00169	·465	·0183	1·33	·0524	2·23	·0878	3·13	·1232	6·15	·242	40	1·57
·044	·00173	·470	·0185	1·34	·0528	2·24	·0882	3·14	·1236	6·20	·244	41	1·61
·045	·00177	·475	·0187	1·35	·0532	2·25	·0886	3·15	·1240	6·25	·246	42	1·65
·046	·00181	·480	·0189	1·36	·0535	2·26	·0890	3·16	·1244	6·30	·248	43	1·69
·047	·00185	·485	·0191	1·37	·0539	2·27	·0894	3·17	·1248	6·35	·250	44	1·73
·048	·00189	·490	·0193	1·38	·0543	2·28	·0898	3·18	·1252	6·40	·252	45	1·77
·049	·00193	·495	·0195	1·39	·0547	2·29	·0902	3·19	·1256	6·45	·254	46	1·81
·050	·00197	·500	·0197	1·40	·0551	2·30	·0906	3·20	·1260	6·50	·256	47	1·85
·055	·00217	·51	·0201	1·41	·0555	2·31	·0909	3·21	·1264	6·6	·260	48	1·89
·060	·00236	·52	·0205	1·42	·0559	2·32	·0913	3·22	·1268	6·7	·264	49	1·93
·065	·00256	·53	·0209	1·43	·0563	2·33	·0917	3·23	·1272	6·8	·268	50	1·97
·070	·00276	·54	·0213	1·44	·0567	2·34	·0921	3·24	·1276	6·9	·272	51	2·01

·075	·00295	·55	·0217	1·45	·0571	2·35	·0925	3·25	·1280	7·0	·276	52	2·05
·080	·00315	·56	·0221	1·46	·0575	2·36	·0929	3·26	·1283	7·1	·280	53	2·09
·085	·00335	·57	·0224	1·47	·0579	2·37	·0933	3·27	·1287	7·2	·284	54	2·13
·090	·00354	·58	·0228	1·48	·0583	2·38	·0937	3·28	·1291	7·3	·287	55	2·17
·095	·00374	·59	·0232	1·49	·0587	2·39	·0941	3·29	·1295	7·4	·291	56	2·20
·100	·00394	·60	·0236	1·50	·0591	2·40	·0945	3·30	·1299	7·5	·295	57	2·24
·105	·0041	·61	·0240	1·51	·0594	2·41	·0949	3·31	·1303	7·6	·299	58	2·28
·110	·0043	·62	·0244	1·52	·0598	2·42	·0953	3·32	·1307	7·7	·303	59	2·32
·115	·0045	·63	·0248	1·53	·0602	2·43	·0957	3·33	·1311	7·8	·307	60	2·36
·120	·0047	·64	·0252	1·54	·0606	2·44	·0961	3·34	·1315	7·9	·311	61	2·40
·125	·0049	·65	·0256	1·55	·0610	2·45	·0965	3·35	·1319	8·0	·315	62	2·44
·130	·0051	·66	·0260	1·56	·0614	2·46	·0969	3·36	·1323	8·1	·319	63	2·48
·135	·0053	·67	·0264	1·57	·0618	2·47	·0972	3·37	·1327	8·2	·323	64	2·52
·140	·0055	·68	·0268	1·58	·0622	2·48	·0976	3·38	·1331	8·3	·327	65	2·56
·145	·0057	·69	·0272	1·59	·0626	2·49	·0980	3·39	·1335	8·4	·331	66	2·60
·150	·0059	·70	·0276	1·60	·0630	2·50	·0984	3·40	·1339	8·5	·335	67	2·64
·155	·0061	·71	·0280	1·61	·0634	2·51	·0988	3·41	·1343	8·6	·339	68	2·68
·160	·0063	·72	·0284	1·62	·0638	2·52	·0992	3·42	·1346	8·7	·343	69	2·72
·165	·0065	·73	·0287	1·63	·0642	2·53	·0996	3·43	·1350	8·8	·346	70	2·76
·170	·0067	·74	·0291	1·64	·0646	2·54	·1000	3·44	·1354	8·9	·350	71	2·80
·175	·0069	·75	·0295	1·65	·0650	2·55	·1004	3·45	·1358	9·0	·354	72	2·83
·180	·0071	·76	·0299	1·66	·0654	2·56	·1008	3·46	·1362	9·1	·358	73	2·87
·185	·0073	·77	·0303	1·67	·0657	2·57	·1012	3·47	·1366	9·2	·362	74	2·91
·190	·0075	·78	·0307	1·68	·0661	2·58	·1016	3·48	·1370	9·3	·366	75	2·95
·195	·0077	·79	·0311	1·69	·0665	2·59	·1020	3·49	·1374	9·4	·370	76	2·99
·200	·0079	·80	·0315	1·70	·0669	2·60	·1024	3·50	·1378	9·5	·374	77	3·03
·205	·0081	·81	·0319	1·71	·0673	2·61	·1028	3·55	·140	9·6	·378	78	3·07
·210	·0083	·82	·0323	1·72	·0677	2·62	·1032	3·60	·142	9·7	·382	79	3·11
·215	·0085	·83	·0327	1·73	·0681	2·63	·1035	3·65	·144	9·8	·386	80	3·15
·220	·0087	·84	·0331	1·74	·0685	2·64	·1039	3·70	·146	9·9	·390	81	3·19
·225	·0089	·85	·0335	1·75	·0689	2·65	·1043	3·75	·148	10·0	·394	82	3·23
·230	·0091	·86	·0339	1·76	0693	2·66	·1047	3·80	·150	10·5	·413	83	3·27
·235	·0093	·87	·0343	1·77	·0697	2·67	·1051	3·85	·152	11·0	·433	84	3·31
·240	·0094	·88	·0347	1·78	·0701	2·68	·1055	3·90	·154	11·5	·453	85	3·35
·245	·0096	·89	·0350	1·79	·0705	2·69	·1059	3·95	·156	12·0	·47	86	3·39
·250	·0098	·90	·0354	1·80	·0709	2·70	·1063	4·00	·157	12·5	·49	87	3·43
·255	·0100	·91	·0358	1·81	·0713	2·71	·1067	4·05	·159	13·0	·51	88	3·46
·260	·0102	·92	·0362	1·82	·0717	2·72	·1071	4·10	·161	13·5	·53	89	3·50
·265	·0104	·93	·0366	1·83	·0720	2·73	·1075	4·15	·163	14·0	·55	90	3·54
·270	·0106	·94	·0370	1·84	·0724	2·74	·1079	4·20	·165	14·5	·57	91	3·58
·275	·0108	·95	·0374	1·85	·0728	2·75	·1083	4·25	·167	15·0	·59	92	3·62
·280	·0110	·96	·0378	1·86	·0732	2·76	·1087	4·30	·169	15·5	·61	93	3·66
·285	·0112	·97	·0382	1·87	·0736	2·77	·1091	4·35	·171	16·0	·63	94	3·70
·290	·0114	·98	·0386	1·88	·0740	2·78	·1094	4·40	·173	16·5	·65	95	3·74
·295	·0116	·99	·0390	1·89	·0744	2·79	·1098	4·45	·175	17·0	·67	96	3·78

[a]Courtesy International Nickel Co.

FIG. 206. Fractions. Decimals. and Millimetres[a]

Inches		M M	Inches		M M	Inches		M M	Inches		M M
Fractions	Decimals		Fractions	Decimals		Fractions	Decimals		Fractions	Decimals	
—	.0004	.01	25/32	.781	19.844	—	2.165	55.	3-11/16	3.6875	93.663
—	.004	.10	—	.7874	20.	2-3/16	2.1875	55.563	—	3.7008	94.
—	.01	.25	51/64	.797	20.241	—	2.2047	56.	3-23/32	3.719	94.456
1/64	.0156	.397	13/16	.8125	20.638	2-7/32	2.219	56.356	—	3.7401	95.
—	.0197	.50	—	.8268	21.	—	2.244	57.	3-3/4	3.750	95.250
—	.0295	.75	53/64	.828	21.034	2-1/4	2.250	57.150	—	3.7795	96.
1/32	.03125	.794	27/32	.844	21.431	2-9/32	2.281	57.944	3-25/32	3.781	96.044
—	.0394	1.	55/64	.859	21.828	—	2.2835	58.	3-13/16	3.8125	96.838
3/64	.0469	1.191	—	.8661	22.	2-5/16	2.312	58.738	—	3.8189	97.
—	.059	1.5	7/8	.875	22.225	—	2.3228	59.	3-27/32	3.844	97.631
1/16	.062	1.588	57/64	.8906	22.622	2-11/32	2.344	59.531	—	3.8583	98.
5/64	.0781	1.984	—	.9055	23.	—	2.3622	60.	3-7/8	3.875	98.425
—	.0787	2.	29/32	.9062	23.019	2-3/8	2.375	60.325	—	3.8976	99.
3/32	.094	2.381	59/64	.922	23.416	—	2.4016	61.	3-29/32	3.9062	99.219
—	.0984	2.5	15/16	.9375	23.813	2-13/32	2.406	61.119	—	3.9370	100.
7/64	.109	2.778	—	.9449	24.	2-7/16	2.438	61.913	3-15/16	3.9375	100.013
—	.1181	3.	61/64	.953	24.209	—	2.4409	62.	3-31/32	3.969	100.806
1/8	.125	3.175	31/32	.969	24.606	2-15/32	2.469	62.706	—	3.9764	101.
—	.1378	3.5	—	.9843	25.	—	2.4803	63.	4	4.000	101.600
9/64	.141	3.572	63/64	.9844	25.003	2-1/2	2.500	63.500	4-1/16	4.062	103.188
5/32	.156	3.969	1	1.000	25.400	—	2.5197	64.	4-1/8	4.125	104.775
—	.1575	4.	—	1.0236	26.	2-17/32	2.531	64.294	—	4.1338	105.
11/64	.172	4.366	1-1/32	1.0312	26.194	—	2.559	65.	4-3/16	4.1875	106.363
—	.177	4.5	1-1/16	1.062	26.988	2-9/16	2.562	65.088	4-1/4	4.250	107.950
3/16	.1875	4.763	—	1.063	27.	2-19/32	2.594	65.881	4-5/16	4.312	109.538
—	.1969	5.	1-3/32	1.094	27.781	—	2.5984	66.	—	4.3307	110.
13/64	.203	5.159	—	1.1024	28.	2-5/8	2.625	66.675	4-3/8	4.375	111.125
—	.2165	5.5	1-1/8	1.125	28.575	—	2.638	67.	4-7/16	4.438	112.713
7/32	.219	5.556	—	1.1417	29.	2-21/32	2.656	67.469	4-1/2	4.500	114.300
15/64	.234	5.953	1-5/32	1.156	29.369	—	2.6772	68.	—	4.5275	115.
—	.2362	6.	—	1.1811	30.	2-11/16	2.6875	68.263	4-9/16	4.562	115.888
1/4	.250	6.350	1-3/16	1.1875	30.163	—	2.7165	69.	4-5/8	4.625	117.475
—	.2559	6.5	1-7/32	1.219	30.956	2-23/32	2.719	69.056	4-11/16	4.6875	119.063
17/64	.2656	6.747	—	1.2205	31.	2-3/4	2.750	69.850	—	4.7244	120.
—	.2756	7.	1-1/4	1.250	31.750	—	2.7559	70.	4-3/4	4.750	120.650
9/32	.281	7.144	—	1.2598	32.	2-25/32	2.781	70.6439	4-13/16	4.8125	122.238
—	.2953	7.5	1-9/32	1.281	32.544	—	2.7953	71.	4-7/8	4.875	123.825
19/64	.297	7.541	—	1.2992	33.	2-13/16	2.8125	71.4376	—	4.9212	125.
5/16	.312	7.938	1-5/16	1.312	33.338	—	2.8346	72.	4-15/16	4.9375	125.413
—	.315	8.	—	1.3386	34.	2-27/32	2.844	72.2314	5	5.000	127.000

21/64	.328	8.334	1-11/32	1.344	34.131	—	2.8740	73.	—	5.1181	130.
—	.335	8.5	1-3/8	1.375	34.925	2-7/8	2.875	73.025	5-1/4	5.250	133.350
11/32	.344	8.731	—	1.3779	35.	2-29/32	2.9062	73.819	5-1/2	5.500	139.700
—	.3543	9.	1-13/32	1.406	35.719	—	2.9134	74.	—	5.5118	140.
23/64	.359	9.128	—	1.4173	36.	2-15/16	2.9375	74.613	5-3/4	5.750	146.050
—	.374	9.5	1-7/16	1.438	36.513	—	2.9527	75.	—	5.9055	150.
3/8	.375	9.525	—	1.4567	37.	2-31/32	2.969	75.406	6	6.000	152.400
25/64	.391	9.922	1-15/32	1.469	37.306	—	2.9921	76.	6-1/4	6.250	158.750
—	.3937	10.	—	1.4961	38.	3	3.000	76.200	—	6.2992	160.
13/32	.406	10.319	1-1/2	1.500	38.100	3-1/32	3.0312	76.994	6-1/2	6.500	165.100
—	.413	10.5	1-17/32	1.531	38.894	—	3.0315	77.	—	6.6929	170.
27/64	.422	10.716	—	1.5354	39.	3-1/16	3.062	77.788	6-3/4	6.750	171.450
—	.4331	11.	1-9/16	1.562	39.688	—	3.0709	78.	7	7.000	177.800
7/16	.438	11.113	—	1.5748	40.	3-3/32	3.094	78.581	—	7.0866	180.
29/64	.453	11.509	1-19/32	1.594	40.481	—	3.1102	79.	—	7.4803	190.
15/32	.469	11.906	—	1.6142	41.	3-1/8	3.125	79.375	7-1/2	7.500	190.500
—	.4724	12.	1-5/8	1.625	41.275	—	3.1496	80.	—	7.8740	200.
31/64	.484	12.303	—	1.6535	42.	3-5/32	3.156	80.169	8	8.000	203.200
—	.492	12.5	1-21/32	1.6562	42.069	3-3/16	3.1875	80.963	—	8.2677	210.
1/2	.500	12.700	1-11/16	1.6875	42.863	—	3.1890	81.	8-1/2	8.500	215.900
—	.5118	13.	—	1.6929	43.	3-7/32	3.219	81.756	—	8.6614	220.
33/64	.5156	13.097	1-23/32	1.719	43.656	—	3.2283	82.	9	9.000	228.600
17/32	.531	13.494	—	1.7323	44.	3-1/4	3.250	82.550	—	9.0551	230.
35/64	.547	13.891	1-3/4	1.750	44.450	—	3.2677	83.	—	9.4488	240.
—	.5512	14.	—	1.7717	45.	3-9/32	3.281	83.344	9-1/2	9.500	241.300
9/16	.563	14.288	1-25/32	1.781	45.244	—	3.3071	84.	—	9.8425	250.
—	.571	14.5	—	1.8110	46.	3-5/16	3.312	84.1377	10	10.000	254.001
37/64	.578	14.684	1-13/16	1.8125	46.038	3-11/32	3.344	84.9314	—	10.2362	260.
—	.5906	15.	1-27/32	1.844	46.831	—	3.3464	85.	—	10.6299	270.
19/32	.594	15.081	—	1.8504	47.	3-3/8	3.375	85.725	11	11.000	279.401
39/64	.609	15.478	1-7/8	1.875	47.625	—	3.3858	86.	—	11.0236	280.
5/8	.625	15.875	—	1.8898	48.	3-13/32	3.406	86.519	—	11.4173	290.
—	.6299	16.	1-29/32	1.9062	48.419	—	3.4252	87.	—	11.8110	300.
41/64	.6406	16.272	—	1.9291	49.	3-7/16	3.438	87.313	12	12.000	304.801
—	.6496	16.5	1-15/16	1.9375	49.213	—	3.4646	88.	13	13.000	330.201
21/32	.656	16.669	—	1.9685	50.	3-15/32	3.469	88.106	—	13.7795	350.
—	.6693	17.	1-31/32	1.969	50.006	3-1/2	3.500	88.900	14	14.000	355.601
43/64	.672	17.066	2	2.000	50.800	—	3.5039	89.	15	15.000	381.001
11/16	.6875	17.463	—	2.0079	51.	3-17/32	3.531	89.694	—	15.7480	400.
45/64	.703	17.859	2-1/32	2.03125	51.594	—	3.5433	90.	16	16.000	406.401
—	.7087	18.	—	2.0472	52.	3-9/16	3.562	90.4877	17	17.000	431.801
23/32	.719	18.256	2-1/16	2.062	52.388	—	3.5827	91.	—	17.7165	450.
—	.7283	18.5	—	2.0866	53.	3-19/32	3.594	91.281	18	18.000	457.201
47/64	.734	18.653	2-3/32	2.094	53.181	—	3.622	92.	19	19.000	482.601
—	.7480	19.	2-1/8	2.125	53.975	3-5/8	3.625	92.075	—	19.6850	500.
3/4	.750	19.050	—	2.126	54.	3-21/32	3.656	92.869	20	20.000	508.001
49/64	.7656	19.447	2-5/32	2.156	54.769	—	3.6614	93.			

[a]Courtesy International Nickel Co.

FIG. 207. Metric Conversion Table. Inches and Fractions of an Inch to Millimetres. 39.37 in., U.S. Standard = 1 m = 100 cm = 1000-mm

Fractions	Inches										
	0	1	2	3	4	5	6	7	8	9	10
0	0.00	25.40	50.80	76.20	101.60	127.00	152.40	177.80	203.20	228.60	254.00
1/64	0.40	25.80	51.20	76.60	102.00	127.40	152.80	178.20	203.60	229.00	254.40
1/32	0.79	26.19	51.59	76.99	102.39	127.79	153.19	178.59	203.99	229.39	254.79
3/64	1.19	26.59	51.99	77.39	102.79	128.19	153.59	178.99	204.39	229.79	255.19
1/16	1.59	26.99	52.39	77.79	103.19	128.59	153.99	179.39	204.79	230.19	255.59
5/64	1.98	27.38	52.79	78.18	103.58	128.98	154.38	179.78	205.18	230.58	255.99
3/32	2.38	27.78	53.18	78.58	103.98	129.38	154.78	180.18	205.58	230.98	256.38
7/64	2.78	28.18	53.58	78.98	104.38	129.78	155.18	180.58	205.98	231.38	256.78
1/8	3.18	28.58	53.98	79.38	104.78	130.18	155.58	180.98	206.38	231.78	257.18
9/64	3.57	28.97	54.37	79.77	105.17	130.57	155.97	181.37	206.77	232.17	257.57
5/32	3.97	29.37	54.77	80.17	105.57	130.97	156.37	181.77	207.17	232.57	257.97
11/64	4.37	29.77	55.17	80.57	105.97	131.37	156.77	182.17	207.57	232.97	258.37
3/16	4.76	30.16	55.56	80.96	106.36	131.76	157.16	182.56	207.96	233.36	258.76
13/64	5.16	30.56	55.96	81.36	106.76	132.16	157.56	182.96	208.36	233.76	259.16
7/32	5.56	30.96	56.36	81.76	107.16	132.56	157.96	183.36	208.76	234.16	259.56
15/64	5.95	31.35	56.75	82.15	107.55	132.95	158.35	183.75	209.15	234.55	259.95
1/4	6.35	31.75	57.15	82.55	107.95	133.35	158.75	184.15	209.55	234.95	260.35
17/64	6.75	32.15	57.55	82.95	108.35	133.75	159.15	184.55	209.95	235.35	260.75
9/32	7.14	32.54	57.94	83.34	108.74	134.14	159.54	184.94	210.34	235.74	261.14
19/64	7.54	32.94	58.34	83.74	109.14	134.54	159.94	185.34	210.74	236.14	261.54
5/16	7.94	33.34	58.74	84.14	109.54	134.94	160.34	185.74	211.14	236.54	261.94
21/64	8.33	33.73	59.13	84.53	109.93	135.33	160.73	186.13	211.53	236.93	262.34
11/32	8.73	34.13	59.53	84.93	110.33	135.73	161.13	186.53	211.93	237.33	262.73
23/64	9.13	34.53	59.93	85.33	110.73	136.13	161.53	186.93	212.33	237.73	263.13
3/8	9.53	34.93	60.33	85.73	111.13	136.53	161.93	187.33	212.73	238.13	263.53
25/64	9.92	35.32	60.72	86.12	111.52	136.92	162.32	187.72	213.12	238.52	263.92
13/32	10.32	35.72	61.12	86.52	111.92	137.32	162.72	188.12	213.52	238.92	264.32
27/64	10.72	36.12	61.52	86.92	112.32	137.72	163.12	188.52	213.92	239.32	264.72
7/16	11.11	36.51	61.91	87.31	112.71	138.11	163.51	188.91	214.31	239.71	265.11
29/64	11.51	36.91	62.31	87.71	113.11	138.51	163.91	189.31	214.71	240.11	265.51
15/32	11.91	37.31	62.71	88.11	113.51	138.91	164.31	189.71	215.11	240.51	265.91
31/64	12.30	37.70	63.10	88.50	113.90	139.30	164.70	190.10	215.50	240.90	266.30
1/2	12.70	38.10	63.50	88.90	114.30	139.70	165.10	190.50	215.90	241.30	266.70
33/64	13.10	38.50	63.90	89.30	114.70	140.10	165.50	190.90	216.30	241.70	267.10
17/32	13.49	38.89	64.29	89.69	115.09	140.49	165.89	191.29	216.69	242.09	267.49
35/64	13.89	39.29	64.69	90.09	115.49	140.89	166.29	191.69	217.09	242.49	267.89
9/16	14.29	39.69	65.09	90.49	115.89	141.29	166.69	192.09	217.49	242.89	268.29
37/64	14.68	40.08	65.48	90.88	116.28	141.68	167.08	192.48	217.88	243.28	268.69
19/32	15.08	40.48	65.88	91.28	116.68	142.08	167.48	192.88	218.28	243.68	269.08
39/64	15.48	40.88	66.28	91.68	117.08	142.48	167.88	193.28	218.68	244.08	269.48
5/8	15.88	41.28	66.68	92.08	117.48	142.88	168.28	193.68	219.08	244.48	269.88
41/64	16.27	41.67	67.07	92.47	117.87	143.27	168.67	194.07	219.47	244.87	270.27
21/32	16.67	42.07	67.47	92.87	118.27	143.67	169.07	194.47	219.87	245.27	270.67
43/64	17.07	42.47	67.87	93.27	118.67	144.07	169.47	194.87	220.27	245.67	271.07
11/16	17.46	42.86	68.26	93.66	119.06	144.46	169.86	195.26	220.66	246.06	271.46
45/64	17.86	43.26	68.66	94.06	119.46	144.86	170.26	195.66	221.06	246.46	271.86
23/32	18.26	43.66	69.06	94.46	119.86	145.26	170.66	196.06	221.46	246.86	272.26
47/64	18.65	44.05	69.45	94.85	120.25	145.65	171.05	196.45	221.85	247.25	272.65
3/4	19.05	44.45	69.85	95.25	120.65	146.05	171.45	196.85	222.25	247.65	273.05
49/64	19.45	44.85	70.25	95.65	121.05	146.45	171.85	197.25	222.65	248.05	273.45
25/32	19.84	45.24	70.64	96.04	121.44	146.85	172.24	197.64	223.04	248.44	273.84
51/64	20.24	45.64	71.04	96.44	121.84	147.24	172.64	198.04	223.44	248.84	274.24
13/16	20.64	46.04	71.44	96.84	122.24	147.64	173.04	198.44	223.84	249.24	274.64
53/64	21.03	46.43	71.83	97.23	122.63	148.03	173.43	198.83	224.23	249.64	275.04
27/32	21.43	46.83	72.23	97.63	123.03	148.43	173.83	199.23	224.63	250.03	275.43
55/64	21.83	47.23	72.63	98.03	123.43	148.83	174.23	199.63	225.03	250.43	275.83
7/8	22.23	47.63	73.03	98.43	123.83	149.23	174.63	200.03	225.43	250.83	276.23
57/64	22.62	48.02	73.42	98.82	124.22	149.62	175.02	200.42	225.82	251.22	276.62
29/32	23.02	48.42	73.82	99.22	124.62	150.02	175.42	200.82	226.22	251.62	277.02
59/64	23.42	48.82	74.22	99.62	125.02	150.42	175.82	201.22	226.62	252.02	277.42
15/16	23.81	49.21	74.61	100.01	125.41	150.81	176.21	201.61	227.01	252.41	277.81
61/64	24.21	49.61	75.01	100.41	125.81	151.21	176.61	202.01	227.41	252.81	278.21
31/32	24.61	50.01	75.41	100.81	126.21	151.61	177.01	202.41	227.81	253.21	278.61
63/64	25.00	50.40	75.80	101.20	126.60	152.00	177.40	202.80	228.20	253.60	279.00

FIG. 207 (continued)

Inches										Fractions
11	12	13	14	15	16	17	18	19	20	
279.40	304.80	330.20	355.60	381.00	406.40	431.80	457.20	482.60	508.00	0"
279.80	305.20	330.60	356.00	381.40	406.80	432.20	457.60	483.00	508.40	1/64
280.19	305.59	330.99	356.39	381.79	407.19	432.59	457.99	483.39	508.80	1/32
280.59	305.99	331.39	356.79	382.19	407.59	432.99	458.39	483.79	509.19	3/64
280.99	306.39	331.79	357.19	382.59	407.99	433.39	458.79	484.19	509.59	1/16
281.39	306.79	332.19	357.59	382.99	408.39	433.79	459.19	484.59	509.99	5/64
281.78	307.18	332.58	357.98	383.38	408.78	434.18	459.58	484.98	510.38	3/32
282.18	307.58	332.98	358.38	383.78	409.18	434.58	459.98	485.38	510.78	7/64
282.58	307.98	333.38	358.78	384.18	409.58	434.98	460.38	485.78	511.18	1/8
282.97	308.37	333.77	359.17	384.57	409.97	435.37	460.77	486.17	511.57	9/64
283.37	308.77	334.17	359.57	384.97	410.37	435.77	461.17	486.57	511.97	5/32
283.77	309.17	334.57	359.97	385.37	410.77	436.17	461.57	486.97	512.37	11/64
284.16	309.56	334.96	360.36	385.76	411.16	436.56	461.96	487.36	512.76	3/16
284.56	309.96	335.36	360.76	386.16	411.56	436.96	462.36	487.76	513.16	13/64
284.96	310.36	335.76	361.16	386.56	411.96	437.36	462.76	488.16	513.56	7/32
285.35	310.75	336.15	361.55	386.95	412.35	437.75	463.15	488.55	513.95	15/64
285.75	311.15	336.55	361.95	387.35	412.75	438.15	463.55	488.95	514.35	1/4
286.15	311.55	336.95	362.35	387.75	413.15	438.55	463.95	489.35	514.75	17/64
286.54	311.94	337.34	362.74	388.14	413.54	438.95	464.34	489.74	515.15	9/32
286.94	312.34	337.74	363.14	388.54	413.94	439.34	464.74	490.14	515.54	19/64
287.34	312.74	338.14	363.54	388.94	414.34	439.74	465.14	490.54	515.94	5/16
287.74	313.14	338.54	363.94	389.34	414.74	440.14	465.54	490.94	516.34	21/64
288.13	313.53	338.93	364.33	389.73	415.13	440.53	465.93	491.33	516.73	11/32
288.53	313.93	339.33	364.73	390.13	415.53	440.93	466.33	491.73	517.13	23/64
288.93	314.33	339.73	365.13	390.53	415.93	441.33	466.73	492.13	517.53	3/8
289.32	314.72	340.12	365.52	390.92	416.32	441.72	467.12	492.52	517.92	25/64
289.72	315.12	340.52	365.92	391.32	416.72	442.12	467.52	492.92	518.32	13/32
290.12	315.52	340.92	366.32	391.72	417.12	442.52	467.92	493.32	518.72	27/64
290.51	315.91	341.31	366.71	392.11	417.51	442.91	468.31	493.71	519.11	7/16
290.91	316.31	341.71	367.11	392.51	417.91	443.31	468.71	494.11	519.51	29/64
291.31	316.71	342.11	367.51	392.91	418.31	443.71	469.11	494.51	519.91	15/32
291.70	317.10	342.50	367.90	393.30	418.70	444.10	469.50	494.90	520.30	31/64
292.10	317.50	342.90	368.30	393.70	419.10	444.50	469.90	495.30	520.70	1/2
292.50	317.90	343.30	368.70	394.10	419.50	444.90	470.30	495.70	521.10	33/64
292.89	318.29	343.70	369.09	394.50	419.89	445.29	470.69	496.10	521.50	17/32
293.29	318.69	344.09	369.49	394.89	420.29	445.69	471.09	496.49	521.89	35/64
293.69	319.09	344.49	369.89	395.29	420.69	446.09	471.49	496.89	522.29	9/16
294.09	319.49	344.89	370.29	395.69	421.09	446.49	471.89	497.29	522.69	37/64
294.48	319.88	345.28	370.68	396.08	421.48	446.88	472.28	497.68	523.08	19/32
294.88	320.28	345.68	371.08	396.48	421.88	447.28	472.68	498.08	523.48	39/64
295.28	320.68	346.08	371.48	396.88	422.28	447.68	473.08	498.48	523.88	5/8
295.67	321.07	346.47	371.87	397.27	422.67	448.07	473.47	498.87	524.27	41/64
296.07	321.47	346.87	372.27	397.67	423.07	448.47	473.87	499.27	524.67	21/32
296.47	321.87	347.27	372.67	398.07	423.47	448.87	474.27	499.67	525.07	43/64
296.86	322.26	347.66	373.06	398.46	423.86	449.26	474.66	500.06	525.46	11/16
297.26	322.66	348.06	373.46	398.86	424.26	449.66	475.06	500.46	525.86	45/64
297.66	323.06	348.46	373.86	399.26	424.66	450.06	475.46	500.86	526.26	23/32
298.05	323.45	348.85	374.25	399.65	425.06	450.45	475.85	501.25	526.65	47/64
298.45	323.85	349.25	374.65	400.05	425.45	450.85	476.25	501.65	527.05	3/4
298.85	324.25	349.65	375.05	400.45	425.85	451.25	476.65	502.05	527.45	49/64
299.24	324.64	350.04	375.44	400.84	426.25	451.64	477.04	502.45	527.85	25/32
299.64	325.04	350.44	375.84	401.24	426.64	452.04	477.44	502.84	528.24	51/64
300.04	325.44	350.84	376.24	401.64	427.04	452.44	477.84	503.24	528.64	13/16
300.44	325.84	351.24	376.64	402.04	427.44	452.84	478.24	503.64	529.04	53/64
300.83	326.23	351.63	377.03	402.43	427.83	453.23	478.63	504.03	529.43	27/32
301.23	326.63	352.03	377.43	402.83	428.23	453.63	479.03	504.43	529.83	55/64
301.63	327.03	352.43	377.83	403.23	428.63	454.03	479.43	504.83	530.23	7/8
302.02	327.42	352.82	378.22	403.62	429.02	454.42	479.82	505.22	530.62	57/64
302.42	327.82	353.22	378.62	404.02	429.42	454.82	480.22	505.62	531.02	29/32
302.82	328.22	353.62	379.02	404.42	429.82	455.22	480.62	506.02	531.42	59/64
303.21	328.61	354.01	379.41	404.81	430.21	455.61	481.01	506.41	531.81	15/16
303.61	329.01	354.41	379.81	405.21	430.61	456.01	481.41	506.81	532.21	61/64
304.01	329.41	354.81	380.21	405.61	431.01	456.41	481.81	507.21	532.61	31/32
304.40	329.80	355.20	380.60	406.00	431.40	456.80	482.20	507.60	533.00	63/64

FIG. 207 (continued)

Fractions	Inches									
	21	22	23	24	25	26	27	28	29	30
0	533.40	558.80	584.20	609.60	635.00	660.40	685.80	711.20	736.60	762.00
1/64	533.80	559.20	584.60	610.00	635.40	660.80	686.20	711.60	737.00	762.40
1/32	534.20	559.60	585.00	610.40	635.80	661.20	686.60	712.00	737.40	762.80
3/64	534.59	559.99	585.39	610.79	636.19	661.59	686.99	712.39	737.79	763.19
1/16	534.99	560.39	585.79	611.19	636.59	661.99	687.39	712.79	738.19	763.59
5/64	535.39	560.79	586.19	611.59	636.99	662.39	687.79	713.19	738.59	763.99
3/32	535.78	561.18	586.58	611.98	637.38	662.78	688.18	713.58	738.98	764.38
7/64	536.18	561.58	586.98	612.38	637.78	663.18	688.58	713.98	739.38	764.78
1/8	536.58	561.98	587.38	612.78	638.18	663.58	688.98	714.38	739.78	765.18
9/64	536.97	562.37	587.77	613.17	638.57	663.97	689.37	714.77	740.17	765.57
5/32	537.37	562.77	588.17	613.57	638.97	664.37	689.77	715.17	740.57	765.97
11/64	537.77	563.17	588.57	613.97	639.37	664.77	690.17	715.57	740.97	766.37
3/16	538.16	563.56	588.96	614.36	639.76	665.16	690.56	715.96	741.36	766.76
13/64	538.56	563.96	589.36	614.76	640.16	665.56	690.96	716.36	741.76	767.16
7/32	538.96	564.36	589.76	615.16	640.56	665.96	691.36	716.76	742.16	767.56
15/64	539.35	564.75	590.15	615.55	640.95	666.35	691.75	717.15	742.55	767.96
1/4	539.75	565.15	590.55	615.95	641.35	666.75	692.15	717.55	742.95	768.35
17/64	540.15	565.55	590.95	616.35	641.75	667.15	692.55	717.95	743.35	768.75
9/32	540.55	565.95	591.35	616.75	642.15	667.55	692.95	718.35	743.75	769.15
19/64	540.94	566.34	591.74	617.14	642.54	667.94	693.34	718.74	744.14	769.54
5/16	541.34	566.74	592.14	617.54	642.94	668.34	693.74	719.14	744.54	769.94
21/64	541.74	567.14	592.54	617.94	643.34	668.74	694.14	719.54	744.94	770.34
11/32	542.13	567.53	592.93	618.33	643.73	669.13	694.53	719.93	745.33	770.73
23/64	542.53	567.93	593.33	618.73	644.13	669.53	694.93	720.33	745.73	771.13
3/8	542.93	568.33	593.73	619.13	644.53	669.93	695.33	720.73	746.13	771.53
25/64	543.32	568.72	594.12	619.52	644.92	670.32	695.72	721.12	746.52	771.92
13/32	543.72	569.12	594.52	619.92	645.32	670.72	696.12	721.52	746.92	772.32
27/64	544.12	569.52	594.92	620.32	645.72	671.12	696.52	721.92	747.32	772.72
7/16	544.51	569.91	595.31	620.71	646.11	671.51	696.91	722.31	747.71	773.11
29/64	544.91	570.31	595.71	621.11	646.51	671.91	697.31	722.71	748.11	773.51
15/32	545.31	570.71	596.11	621.51	646.91	672.31	697.71	723.11	748.51	773.91
31/64	545.70	571.10	596.50	621.90	647.30	672.70	698.10	723.50	748.91	774.31
1/2	546.10	571.50	596.90	622.30	647.70	673.10	698.50	723.90	749.30	774.70
33/64	546.50	571.90	597.30	622.70	648.10	673.50	698.90	724.30	749.70	775.10
17/32	546.90	572.30	597.70	623.10	648.50	673.90	699.30	724.70	750.10	775.50
35/64	547.29	572.69	598.09	623.49	648.89	674.29	699.69	725.09	750.49	775.89
9/16	547.69	573.09	598.49	623.89	649.29	674.69	700.09	725.49	750.89	776.29
37/64	548.09	573.49	598.89	624.29	649.69	675.09	700.49	725.89	751.29	776.69
19/32	548.48	573.88	599.28	624.68	650.08	675.48	700.88	726.28	751.68	777.08
39/64	548.88	574.28	599.68	625.08	650.48	675.88	701.28	726.68	752.08	777.48
5/8	549.28	574.68	600.08	625.48	650.88	676.28	701.68	727.08	752.48	777.88
41/64	549.67	575.07	600.47	625.87	651.27	676.67	702.07	727.47	752.87	778.27
21/32	550.07	575.47	600.87	626.27	651.67	677.07	702.47	727.87	753.27	778.67
43/64	550.47	575.87	601.27	626.67	652.07	677.47	702.87	728.27	753.67	779.07
11/16	550.86	576.26	601.66	627.06	652.46	677.86	703.26	728.66	754.06	779.46
45/64	551.26	576.66	602.06	627.46	652.86	678.26	703.66	729.06	754.46	779.86
23/32	551.66	577.06	602.46	627.86	653.26	678.66	704.06	729.46	754.86	780.26
47/64	552.05	577.45	602.85	628.25	653.65	679.05	704.45	729.85	755.26	780.66
3/4	552.45	577.85	603.25	628.65	654.05	679.45	704.85	730.25	755.65	781.05
49/64	552.85	578.25	603.65	629.05	654.45	679.85	705.25	730.65	756.05	781.45
25/32	553.25	578.65	604.05	629.45	654.85	680.25	705.65	731.05	756.45	781.85
51/64	553.64	579.04	604.44	629.84	655.24	680.64	706.04	731.44	756.84	782.24
13/16	554.04	579.44	604.84	630.24	655.64	681.04	706.44	731.84	757.24	782.64
53/64	554.44	579.84	605.24	630.64	656.04	681.44	706.84	732.24	757.64	783.04
27/32	554.83	580.23	605.63	631.03	656.43	681.83	707.23	732.63	758.03	783.43
55/64	555.23	580.63	606.03	631.43	656.83	682.23	707.63	733.03	758.43	783.83
7/8	555.63	581.03	606.43	631.83	657.23	682.63	708.03	733.43	758.83	784.23
57/64	556.02	581.42	606.82	632.22	657.62	683.02	708.42	733.82	759.22	784.62
29/32	556.42	581.82	607.22	632.62	658.02	683.42	708.82	734.22	759.62	785.02
59/64	556.82	582.22	607.62	633.02	658.42	683.82	709.22	734.62	760.02	785.42
15/16	557.21	582.61	608.01	633.41	658.81	684.21	709.61	735.01	760.41	785.81
61/64	557.61	583.01	608.41	633.81	659.21	684.61	710.01	735.41	760.81	786.21
31/32	558.01	583.41	608.81	634.21	659.61	685.01	710.41	735.81	761.21	786.61
63/64	558.40	583.80	609.20	634.60	660.00	685.40	710.80	736.20	761.61	787.01

FIG. 207 (continued)

Inches										
31	32	33	34	35	36	37	38	39	40	Fractions
787.40	812.80	838.20	863.60	889.00	914.40	939.80	965.20	990.60	1016.00	0
787.80	813.20	838.60	864.00	889.40	914.80	940.20	965.60	991.00	1016.40	1/64
788.20	813.60	839.00	864.40	889.80	915.20	940.60	966.00	991.40	1016.80	1/32
788.59	813.99	839.39	864.79	890.19	915.59	940.99	966.39	991.79	1017.19	3/64
788.99	814.39	839.79	865.19	890.59	915.99	941.39	966.79	992.19	1017.59	1/16
789.39	814.79	840.19	865.59	890.99	916.39	941.79	967.19	992.59	1017.99	5/64
789.78	815.18	840.58	865.98	891.38	916.78	942.18	967.58	992.98	1018.38	3/32
790.18	815.58	840.98	866.38	891.78	917.18	942.58	967.98	993.38	1018.78	7/64
790.58	815.98	841.38	866.78	892.18	917.58	942.98	968.38	993.78	1019.18	1/8
790.97	816.37	841.77	867.17	892.57	917.97	943.37	968.77	994.17	1019.57	9/64
791.37	816.77	842.17	867.57	892.97	918.37	943.77	969.17	994.57	1019.97	5/32
791.77	817.17	842.57	867.97	893.37	918.77	944.17	969.57	994.97	1020.37	11/64
792.16	817.56	842.96	868.36	893.76	919.16	944.56	969.96	995.37	1020.77	3/16
792.56	817.96	843.36	868.76	894.16	919.56	944.96	970.36	995.76	1021.16	13/64
792.96	818.36	843.76	869.16	894.56	919.96	945.36	970.76	996.16	1021.56	7/32
793.36	818.76	844.16	869.56	894.96	920.36	945.76	971.16	996.56	1021.96	15/64
793.75	819.15	844.55	869.95	895.35	920.75	946.15	971.55	996.95	1022.35	1/4
794.15	819.55	844.95	870.35	895.75	921.15	946.55	971.95	997.35	1022.75	17/64
794.55	819.95	845.35	870.75	896.15	921.55	946.95	972.35	997.75	1023.15	9/32
794.94	820.34	845.74	871.14	896.54	921.94	947.34	972.74	998.14	1023.54	19/64
795.34	820.74	846.14	871.54	896.94	922.34	947.74	973.14	998.54	1023.94	5/16
795.74	821.14	846.54	871.94	897.34	922.74	948.14	973.54	998.94	1024.34	21/64
796.13	821.53	846.93	872.33	897.73	923.13	948.53	973.93	999.33	1024.73	11/32
796.53	821.93	847.33	872.73	898.13	923.53	948.93	974.33	999.73	1025.13	23/64
796.93	822.33	847.73	873.13	898.53	923.93	949.33	974.73	1000.13	1025.53	3/8
797.32	822.72	848.12	873.52	898.92	924.32	949.72	975.12	1000.52	1025.92	25/64
797.72	823.12	848.52	873.92	899.32	924.72	950.12	975.52	1000.92	1026.32	13/32
798.12	823.52	848.92	874.32	899.72	925.12	950.52	975.92	1001.32	1026.72	27/64
798.51	823.91	849.31	874.71	900.11	925.51	950.91	976.31	1001.72	1027.12	7/16
798.91	824.31	849.71	875.11	900.51	925.91	951.31	976.71	1002.11	1027.51	29/64
799.31	824.71	850.11	875.51	900.91	926.31	951.71	977.11	1002.51	1027.91	15/32
799.71	825.11	850.51	875.91	901.31	926.71	952.11	977.51	1002.91	1028.31	31/64
800.10	825.50	850.90	876.30	901.70	927.10	952.50	977.90	1003.30	1028.70	1/2
800.50	825.90	851.30	876.70	902.10	927.50	952.90	978.30	1003.70	1029.10	33/64
800.90	826.30	851.70	877.10	902.50	927.90	953.30	978.70	1004.10	1029.50	17/32
801.29	826.69	852.09	877.49	902.89	928.29	953.69	979.09	1004.49	1029.89	35/64
801.69	827.09	852.49	877.89	903.29	928.69	954.09	979.49	1004.89	1030.29	9/16
802.09	827.49	852.89	878.29	903.69	929.09	954.49	979.89	1005.29	1030.69	37/64
802.48	827.88	853.28	878.68	904.08	929.48	954.88	980.28	1005.68	1031.08	19/32
802.88	828.28	853.68	879.08	904.48	929.88	955.28	980.68	1006.08	1031.48	39/64
803.28	828.68	854.08	879.48	904.88	930.28	955.68	981.08	1006.48	1031.88	5/8
803.67	829.07	854.47	879.87	905.27	930.67	956.07	981.47	1006.87	1032.27	41/64
804.07	829.47	854.87	880.27	905.67	931.07	956.47	981.87	1007.27	1032.67	21/32
804.47	829.87	855.27	880.67	906.07	931.47	956.87	982.27	1007.67	1033.07	43/64
804.86	830.26	855.66	881.06	906.46	931.86	957.26	982.66	1008.07	1033.47	11/16
805.26	830.66	856.06	881.46	906.86	932.26	957.66	983.06	1008.46	1033.86	45/64
805.66	831.06	856.46	881.86	907.26	932.66	958.06	983.46	1008.86	1034.26	23/32
806.06	831.46	856.86	882.26	907.66	933.06	958.46	983.86	1009.26	1034.66	47/64
806.45	831.85	857.25	882.65	908.05	933.45	958.85	984.25	1009.65	1035.05	3/4
806.85	832.25	857.65	883.05	908.45	933.85	959.25	984.65	1010.05	1035.45	49/64
807.25	832.65	858.05	883.45	908.85	934.25	959.65	985.05	1010.45	1035.85	25/32
807.64	833.04	858.44	883.84	909.24	934.64	960.04	985.44	1010.84	1036.24	51/64
808.04	833.44	858.84	884.24	909.64	935.04	960.44	985.84	1011.24	1036.64	13/16
808.44	833.84	859.24	884.64	910.04	935.44	960.84	986.24	1011.64	1037.04	53/64
808.83	834.23	859.63	885.03	910.43	935.83	961.23	986.63	1012.03	1037.43	27/32
809.23	834.63	860.03	885.43	910.83	936.23	961.63	987.03	1012.43	1037.83	55/64
809.63	835.03	860.43	885.83	911.23	936.63	962.03	987.43	1012.83	1038.23	7/8
810.02	835.42	860.82	886.22	911.62	937.02	962.42	987.82	1013.22	1038.62	57/64
810.42	835.82	861.22	886.62	912.02	937.42	962.82	988.22	1013.62	1039.02	29/32
810.82	836.22	861.62	887.02	912.42	937.82	963.22	988.62	1014.02	1039.42	59/64
811.21	836.61	862.01	887.41	912.81	938.21	963.61	989.01	1014.42	1039.82	15/16
811.61	837.01	862.41	887.81	913.21	938.61	964.01	989.41	1014.81	1040.21	61/64
812.01	837.41	862.81	888.21	913.61	939.01	964.41	989.81	1015.21	1040.61	31/32
812.41	837.81	863.21	888.61	914.01	939.41	964.81	990.21	1015.61	1041.01	63/64

FIG. 208. Split Gauges in Decimal Sizes (*Steel Wire Gauge*)

Gauge	—	¼	½	¾
7/0	.490	.483	.476	.469
6/0	.4615	.454	.446	.438
5/0	.4305	.421	.412	.403
4/0	.3938	.386	.378	.370
3/0	.3625	.355	.347	.339
2/0	.331	.325	.319	.313
1/0	.3065	.301	.295	.289
1	.283	.278	.273	.268
2	.2625	.258	.253	.248
3	.2437	.239	.235	.230
4	.2253	.221	.216	.212
5	.207	.203	.200	.196
6	.192	.188	.185	.181
7	.177	.173	.170	.166
8	.162	.159	.155	.152
9	.1483	.145	.142	.138
10	.135	.131	.128	.124
11	.1205	.117	.113	.109
12	.1055	.102	.099	.095
13	.0915	.089	.086	.083
14	.080	.078	.076	.074
15	.072	.070	.067	.065
16	.0625	.060	.058	.056
17	.054	.052	.051	.0491
18	.0475	.0459	.0443	.0426
19	.0410	.0394	.0379	.0363
20	.0348	.0340	.0332	.0325
21	.0317	.0309	.0301	.0294
22	.0286	.0279	.0272	.0265
23	.0258	.0251	.0244	.0237
24	.0230	.0224	.0217	.0211
25	.0204	.0198	.0193	.0187
26	.0181	.0179	.0177	.0175
27	.0173	.0170	.0168	.0165
28	.0162	.0159	.0156	.0153
29	.0150	.0148	.0145	.0143
30	.0140	.0138	.0136	.0134
31	.0132	.0131	.0130	.0129
32	.0128	.0126	.0123	.0121
33	.0118	.0115	.0111	.0108
34	.0104	.0102	.0100	.0097
35	.0095	.0094	.0093	.0091
36	.0090	.0089	.0087	.0086
37	.0085	.0084	.0083	.0081
38	.0080	.0079	.0078	.0076
39	.0075	.0074	.0073	.0071
40	.0070	.0069	.0068	.0067
41	.0066	.0065	.0064	.0063
42	.0062	.0061	.0060	.0059

FIG. 209. Arbor Diameters for Spring made from Music Wire[a,b]

Wire Diam. (inch)	Spring Outside Diameter (inch)												
	1/16	3/32	1/8	5/32	3/16	7/32	1/4	9/32	5/16	11/32	3/8	7/16	1/2
	Arbor Diameter (inch)												
.008	.039	.060	.078	.093	.107	.119	.129	...	...	....	...	...	....
.010	.037	.060	.080	.099	.115	.129	.142	.154	.164	....	...	...	....
.012	.034	.059	.081	.101	.119	.135	.150	.163	.177	.189	.200	...	....
.014	.031	.057	.081	.102	.121	.140	.156	.172	.187	.200	.213	.234	....
.016	.028	.055	.079	.102	.123	.142	.161	.178	.194	.209	.224	.250	.271
.018	...	.053	.077	.101	.124	.144	.161	.182	.200	.215	.231	.259	.284
.020	...	.049	.075	.096	.123	.144	.165	.184	.203	.220	.237	.268	.296
.022	...	.046	.072	.097	.122	.145	.165	.186	.206	.224	.242	.275	.305
.024	...	.043	.070	.095	.120	.144	.166	.187	.207	.226	.245	.280	.312
.026	...	...	.067	.093	.118	.143	.166	.187	.208	.228	.248	.285	.318
.028	...	...	.064	.091	.115	.141	.165	.187	.208	.229	.250	.288	.323
.030	...	...	.061	.088	.113	.138	.163	.187	.209	.229	.251	.291	.328
.032	...	...	.057	.085	.111	.136	.161	.185	.209	.229	.251	.292	.331
.034	...	...	...	.082	.109	.134	.159	.184	.208	.229	.251	.292	.333
.036	...	...	...	.078	.106	.131	.156	.182	.206	.229	.250	.294	.333
.038	...	...	...	.075	.103	.129	.154	.179	.205	.227	.251	.293	.335
.041	...	...	...	...	.098	.125	.151	.176	.201	.226	.250	.294	.336
.0475	...	...	...	...	.087	.115	.142	.168	.194	.220	.244	.293	.337
.054	...	...	...	...	...	.103	.132	.160	.187	.212	.245	.287	.336
.0625	...	...	...	...	...	...	.108	.146	.169	.201	.228	.280	.330
.072	...	...	...	...	...	...	...	.129	.158	.186	.214	.268	.319
.080	...	...	...	...	...	...	...	...	.144	.173	.201	.256	.308
.0915	...	...	...	...	...	...	...	...	...	....	.181	.238	.293
.1055	...	...	...	...	...	...	...	...	...	....	...	.215	.271
.1205	...	...	...	...	...	...	...	...	...	....	...	...	.215
.125	...	...	...	...	...	...	...	...	...	....	...	...	.239

Wire Diam. (inch)	Spring Outside Diameter (inches)													
	9/16	5/8	11/16	3/4	13/16	7/8	15/16	1	1 1/8	1 1/4	1 3/8	1 1/2	1 3/4	2
	Arbor Diameter (inches)													
.022	.332	.357	.380	...	...	...	...	...	...	...	...	...	...	...
.024	.341	.367	.393	.415	...	...	...	...	...	...	...	...	...	...
.026	.350	.380	.406	.430	...	...	...	...	...	...	...	...	...	...
.028	.356	.387	.416	.442	.467	...	...	...	...	...	...	...	...	...
.030	.362	.395	.426	.453	.481	.506	...	...	...	...	...	...	...	...
.032	.367	.400	.432	.462	.490	.516	.540	...	...	...	...	...	...	...
.034	.370	.404	.437	.469	.498	.526	.552	.557	...	...	...	...	...	...
.036	.372	.407	.442	.474	.500	.536	.562	.589	...	...	...	...	...	...
.038	.375	.412	.448	.481	.512	.543	.572	.600	.650	...	...	...	...	...
.041	.378	.416	.456	.489	.522	.554	.586	.615	.670	.718	...	...	...	...
.0475	.380	.422	.464	.504	.541	.576	.610	.643	.706	.763	.812	...	...	...
.054	.381	.425	.467	.509	.550	.589	.625	.661	.727	.792	.850	.906	...	...
.0625	.379	.426	.468	.512	.556	.597	.639	.678	.753	.822	.889	.951	1.06	1.17
.072	.370	.418	.466	.512	.555	.599	.641	.682	.765	.840	.911	.980	1.11	1.22
.080	.360	.411	.461	.509	.554	.599	.641	.685	.772	.851	.930	1.00	1.13	1.26
.0915	.347	.398	.448	.500	.547	.597	.640	.685	.776	.860	.942	1.02	1.16	1.30
.1055	.327	.381	.433	.485	.535	.586	.630	.683	.775	.865	.952	1.04	1.20	1.35
.1205	.303	.358	.414	.468	.520	.571	.622	.673	.772	.864	.955	1.04	1.22	1.38
.125	.295	.351	.406	.461	.515	.567	.617	.671	.770	.864	.955	1.05	1.23	1.39

[a]Example: To coil a spring with ½-in. OD using 0.016 in. diameter wire, the arbor diameter should be 0.271 in.

[b]Source: *Machinery's Handbook* 20th Edition © 1975 by Industrial Press, Inc. Reprinted with permission.

FIG. 210. English/Metric (SI) Stress Conversion Table[a,b]

1000 to 100 000 psi							
ksi		kg/sq mm	MPa	ksi		kg/sq mm	MPa
—	0	—	—	71.10	50	35.16	344.7
1.42	1	0.70	6.89	72.52	51	35.86	351.6
2.84	2	1.41	13.79	73.94	52	36.57	358.5
4.27	3	2.11	20.68	75.37	53	37.27	365.4
5.69	4	2.81	27.57	76.79	54	37.97	372.3
7.11	5	3.52	34.47	78.21	55	38.68	379.2
8.53	6	4.22	41.37	79.63	56	39.38	386.1
9.95	7	4.92	48.26	81.05	57	40.08	393.0
11.38	8	5.63	55.16	82.48	58	40.79	399.9
12.80	9	6.33	62.05	83.90	59	41.49	406.8
14.22	10	7.03	68.95	85.32	60	42.19	413.7
15.64	11	7.74	75.84	86.74	61	42.90	420.6
17.06	12	8.44	82.74	88.16	62	43.60	427.5
18.49	13	9.14	89.63	89.59	63	44.30	434.4
19.91	14	9.85	96.53	91.01	64	45.01	441.3
21.33	15	10.55	103.4	92.43	65	45.71	448.2
22.75	16	11.25	110.3	93.85	66	46.41	455.1
24.17	17	11.95	117.2	95.27	67	47.12	462.0
25.60	18	12.66	124.1	96.70	68	47.82	468.8
27.02	19	13.36	131.0	98.12	69	48.52	475.7
28.44	20	14.06	137.9	99.54	70	49.23	482.6
29.86	21	14.77	144.8	100.96	71	49.93	489.5
31.28	22	15.47	151.7	102.38	72	50.63	496.4
32.71	23	16.17	158.6	103.81	73	51.34	503.3
34.13	24	16.88	165.5	105.23	74	52.04	510.2

100 000 to 200 000 psi							
ksi		kg/sq mm	MPa	ksi		kg/sq mm	MPa
142.20	100	70.32	689.5	213.30	150	105.49	1 034
143.62	101	71.03	696.4	214.72	151	106.19	1 041
145.04	102	71.73	703.3	216.14	152	106.89	1 048
146.47	103	72.43	710.2	217.57	153	107.59	1 054
147.89	104	73.14	717.1	218.99	154	108.30	1 062
149.31	105	73.84	724.0	220.41	155	109.00	1 069
150.73	106	74.54	730.8	221.83	156	109.70	1 076
152.15	107	75.25	737.7	223.25	157	110.41	1 082
153.58	108	75.95	744.6	224.68	158	111.11	1 089
155.00	109	76.65	751.5	226.10	159	111.81	1 096
156.42	110	77.36	758.4	227.52	160	112.52	1 103
157.84	111	78.06	765.3	228.94	161	113.22	1 110
159.27	112	78.76	772.2	230.36	162	113.92	1 117
160.69	113	79.47	779.1	231.79	163	114.63	1 124
162.11	114	80.17	786.0	233.21	164	115.33	1 131
163.53	115	80.87	792.9	234.63	165	116.03	1 138
164.95	116	81.58	799.8	236.05	166	116.74	1 145
166.38	117	82.28	806.7	237.47	167	117.44	1 151
167.80	118	82.98	813.6	238.90	168	118.14	1 158
169.22	119	83.68	820.5	240.32	169	118.85	1 165
170.64	120	84.39	827.4	241.74	170	119.55	1 172
172.06	121	85.09	834.3	243.16	171	120.25	1 179
173.48	122	85.79	841.2	244.58	172	120.96	1 186
174.91	123	86.50	848.1	246.01	173	121.66	1 193
176.33	124	87.20	855.0	247.43	174	122.36	1 200

35.55	**25**	17.58	172.4	106.65	**75**	52.74	517.1	177.75	**125**	87.90	861.8	248.85	**175**	123.07	1 207
36.97	**26**	18.28	179.3	108.07	**76**	53.45	524.0	179.17	**126**	88.61	868.7	250.27	**176**	123.77	1 213
38.39	**27**	18.99	186.2	109.49	**77**	54.15	530.9	180.60	**127**	89.31	875.6	251.69	**177**	124.47	1 220
39.82	**28**	19.69	193.1	110.92	**78**	54.85	537.8	182.02	**128**	90.01	882.5	253.12	**178**	125.18	1 227
41.24	**29**	20.39	199.9	112.34	**79**	55.56	544.7	183.44	**129**	90.72	889.4	254.54	**179**	125.88	1 234
42.66	**30**	21.10	206.8	113.76	**80**	56.26	551.6	184.86	**130**	91.42	896.3	255.96	**180**	126.58	1 241
44.08	**31**	21.80	213.7	115.18	**81**	56.96	558.5	186.28	**131**	92.12	903.2	257.38	**181**	127.29	1 248
45.50	**32**	22.50	220.6	116.60	**82**	57.67	565.4	187.70	**132**	92.83	910.1	258.80	**182**	127.99	1 255
46.93	**33**	23.21	227.5	118.03	**83**	58.37	572.3	189.13	**133**	93.53	917.0	260.23	**183**	128.69	1 262
48.35	**34**	23.91	234.4	119.45	**84**	59.07	579.2	190.55	**134**	94.23	923.9	261.65	**184**	129.39	1 269
49.77	**35**	24.61	241.3	120.87	**85**	59.77	586.1	191.97	**135**	94.94	930.8	263.07	**185**	130.10	1 276
51.19	**36**	25.32	248.2	122.29	**86**	60.48	593.0	193.39	**136**	95.64	937.7	264.49	**186**	130.80	1 282
52.61	**37**	26.02	255.1	123.71	**87**	61.18	599.8	194.81	**137**	96.34	944.6	265.91	**187**	131.50	1 289
54.04	**38**	26.72	262.0	125.14	**88**	61.88	606.7	196.24	**138**	97.05	951.5	267.34	**188**	132.21	1 296
55.46	**39**	27.43	268.9	126.55	**89**	62.59	613.6	197.66	**139**	97.75	958.4	268.76	**189**	132.91	1 303
56.88	**40**	28.13	275.8	127.98	**90**	63.29	620.5	199.08	**140**	98.45	965.3	270.18	**190**	133.61	1 310
58.30	**41**	28.83	282.7	129.40	**91**	63.99	627.4	200.50	**141**	99.16	972.2	271.60	**191**	134.32	1 317
59.72	**42**	29.54	289.6	130.82	**92**	64.70	634.3	201.92	**142**	99.86	979.1	273.02	**192**	135.02	1 324
61.15	**43**	30.24	296.5	132.25	**93**	65.40	641.2	203.35	**143**	100.56	986.0	274.45	**193**	135.72	1 331
62.57	**44**	30.94	303.4	133.67	**94**	66.10	648.1	204.77	**144**	101.27	992.9	275.87	**194**	136.43	1 338
63.99	**45**	31.65	310.3	135.09	**95**	66.81	655.0	206.19	**145**	101.97	999.7	277.29	**195**	137.13	1 344
65.41	**46**	32.35	317.2	136.51	**96**	67.51	661.9	207.61	**146**	102.67	1 007	278.71	**196**	137.83	1 351
66.83	**47**	33.05	324.1	137.93	**97**	68.21	668.8	209.03	**147**	103.38	1 014	280.13	**197**	138.54	1 358
68.26	**48**	33.76	331.0	139.36	**98**	68.92	675.7	210.46	**148**	104.08	1 020	281.56	**198**	139.24	1 365
69.68	**49**	34.46	337.8	140.78	**99**	69.62	682.6	211.88	**149**	104.78	1 027	282.98	**199**	139.94	1 372
71.10	**50**	35.16	344.7	142.20	**100**	70.32	689.5	213.30	**150**	105.49	1 034	284.40	**200**	140.65	1 379

[a]Look up stress to be converted in boldface column. If in ksi (1000 psi), read kg/mm^2 and MPa in right-hand column. If in kg/mm^2 or MPa, read ksi in left-hand column. Example: 100 000 psi = 689.5 MPa = 70.32 kg/mm^2.

[b]Courtesy International Nickel Co.

FIG. 210. (continued) English/Metric (SI) Stress Conversion Table[a]

200 000 to 300 000 psi

ksi		kg/sq mm	MPa	ksi		kg/sq mm	MPa
284.40	**200**	140.65	1 379	355.50	**250**	175.81	1 724
285.82	**201**	141.35	1 386	356.92	**251**	176.51	1 731
287.24	**202**	142.05	1 393	358.34	**252**	177.22	1 737
288.67	**203**	142.76	1 400	359.77	**253**	177.92	1 744
290.09	**204**	143.46	1 407	361.19	**254**	178.62	1 751
291.51	**205**	144.16	1 413	362.61	**255**	179.32	1 758
292.93	**206**	144.87	1 420	364.03	**256**	180.03	1 765
294.35	**207**	145.57	1 427	365.45	**257**	180.73	1 772
295.78	**208**	146.27	1 434	366.87	**258**	181.43	1 779
297.20	**209**	146.98	1 441	368.30	**259**	182.14	1 786
298.62	**210**	147.68	1 448	369.72	**260**	182.84	1 793
300.04	**211**	148.38	1 455	371.14	**261**	183.54	1 800
301.46	**212**	149.09	1 462	372.56	**262**	184.25	1 806
302.89	**213**	149.79	1 469	373.98	**263**	184.95	1 813
304.31	**214**	150.49	1 475	375.41	**264**	185.65	1 820
305.73	**215**	151.20	1 482	376.83	**265**	186.36	1 827
307.15	**216**	151.90	1 489	377.25	**266**	187.06	1 834
308.57	**217**	152.60	1 496	379.67	**267**	187.76	1 841
310.00	**218**	153.30	1 503	381.09	**268**	188.47	1 848
311.42	**219**	154.01	1 510	382.52	**269**	189.17	1 855
312.84	**220**	154.71	1 517	383.94	**270**	189.87	1 862
314.26	**221**	155.41	1 524	385.36	**271**	190.58	1 868
315.68	**222**	156.12	1 531	386.78	**272**	191.28	1 875
317.11	**223**	156.82	1 538	388.20	**273**	191.98	1 882
318.53	**224**	157.52	1 544	389.63	**274**	192.69	1 889

300 000 to 400 000 psi

ksi		kg/sq mm	MPa	ksi		kg/sq mm	MPa
426.60	**300**	210.97	2 068	497.70	**350**	246.13	2 413
428.02	**301**	211.67	2 075	499.12	**351**	246.83	2 420
429.44	**302**	212.38	2 082	500.54	**352**	247.54	2 427
430.86	**303**	213.08	2 089	501.96	**353**	248.24	2 434
432.29	**304**	213.78	2 096	503.39	**354**	248.94	2 441
433.71	**305**	214.49	2 103	504.81	**355**	249.65	2 448
435.13	**306**	215.19	2 110	506.23	**356**	250.35	2 455
436.55	**307**	215.89	2 117	507.65	**357**	251.05	2 461
437.97	**308**	216.60	2 124	509.07	**358**	251.76	2 468
439.40	**309**	217.30	2 130	510.50	**359**	252.46	2 475
440.82	**310**	218.00	2 137	511.92	**360**	253.16	2 482
442.24	**311**	218.71	2 144	513.34	**361**	253.87	2 489
443.66	**312**	219.41	2 151	514.76	**362**	254.57	2 496
445.08	**313**	220.11	2 158	516.18	**363**	255.27	2 503
446.51	**314**	220.82	2 165	517.61	**364**	255.98	2 510
447.93	**315**	221.52	2 172	519.03	**365**	256.68	2 517
449.35	**316**	222.22	2 179	520.45	**366**	257.38	2 523
450.77	**317**	222.93	2 186	521.87	**367**	258.09	2 530
452.19	**318**	223.63	2 193	523.29	**368**	258.79	2 537
453.62	**319**	224.33	2 199	524.72	**369**	259.49	2 544
455.04	**320**	225.03	2 206	526.14	**370**	260.20	2 551
456.46	**321**	225.74	2 213	527.56	**371**	260.90	2 558
457.88	**322**	226.44	2 220	528.98	**372**	261.60	2 565
459.30	**323**	227.14	2 227	530.40	**373**	262.31	2 572
460.73	**324**	227.85	2 234	531.83	**374**	263.01	2 579

319.95	**225**	158.23	1 551	391.05	**275**	193.39	1 896	462.15	**325**	228.55	2 241	533.25	**375**	263.71	2 585
321.37	**226**	158.93	1 558	392.47	**276**	194.09	1 903	463.57	**326**	229.25	2 248	534.67	**376**	264.42	2 592
322.79	**227**	159.63	1 565	393.89	**277**	194.80	1 910	464.99	**327**	229.96	2 255	536.09	**377**	265.12	2 599
324.22	**228**	160.34	1 572	395.31	**278**	195.50	1 917	466.41	**328**	230.66	2 261	537.51	**378**	265.82	2 606
325.64	**229**	161.04	1 579	396.74	**279**	196.20	1 924	467.84	**329**	231.36	2 268	538.94	**379**	266.53	2 613
327.06	**230**	161.74	1 586	398.16	**280**	196.91	1 931	469.26	**330**	232.07	2 275	540.36	**380**	267.23	2 620
328.48	**231**	162.45	1 593	399.58	**281**	197.61	1 937	470.68	**331**	232.77	2 282	541.78	**381**	267.93	2 627
329.90	**232**	163.15	1 600	401.00	**282**	198.31	1 944	472.10	**332**	233.47	2 289	543.20	**382**	268.64	2 634
331.33	**233**	163.85	1 606	402.42	**283**	199.02	1 951	473.52	**333**	234.18	2 296	544.62	**383**	269.34	2 641
332.75	**234**	164.56	1 613	403.85	**284**	199.72	1 958	474.95	**334**	234.88	2 303	546.05	**384**	270.04	2 648
334.17	**235**	165.26	1 620	405.27	**285**	200.42	1 965	476.37	**335**	235.58	2 310	547.47	**385**	270.75	2 654
335.59	**236**	165.96	1 627	406.69	**286**	201.12	1 972	477.79	**336**	236.29	2 317	548.89	**386**	271.45	2 661
337.01	**237**	166.67	1 634	408.11	**287**	201.83	1 979	479.21	**337**	236.99	2 324	550.31	**387**	272.15	2 668
338.44	**238**	167.37	1 641	409.53	**288**	202.53	1 986	480.63	**338**	237.69	2 330	551.73	**388**	272.86	2 675
339.86	**239**	168.07	1 648	410.96	**289**	203.23	1 993	482.06	**339**	238.40	2 337	553.16	**389**	273.56	2 682
341.28	**240**	168.78	1 655	412.38	**290**	203.94	1 999	483.48	**340**	239.10	2 344	554.58	**390**	274.26	2 689
342.70	**241**	169.48	1 662	413.80	**291**	204.64	2 006	484.90	**341**	239.80	2 351	556.00	**391**	274.97	2 696
344.12	**242**	170.18	1 669	415.22	**292**	205.34	2 013	486.32	**342**	240.51	2 358	557.42	**392**	275.67	2 703
345.55	**243**	170.89	1 675	416.64	**293**	206.05	2 020	487.74	**343**	241.21	2 365	558.84	**393**	276.37	2 710
346.97	**244**	171.59	1 682	418.07	**294**	206.75	2 027	489.17	**344**	241.91	2 372	560.27	**394**	277.08	2 717
348.39	**245**	172.29	1 689	419.49	**295**	207.45	2 034	490.59	**345**	242.62	2 379	561.69	**395**	277.78	2 723
349.81	**246**	173.00	1 696	420.91	**296**	208.16	2 041	492.01	**346**	243.32	2 386	563.11	**396**	278.48	2 730
351.23	**247**	173.70	1 703	422.33	**297**	208.86	2 048	493.43	**347**	244.02	2 392	564.53	**397**	279.18	2 737
352.66	**248**	174.40	1 710	423.75	**298**	209.56	2 055	494.85	**348**	244.73	2 399	565.95	**398**	279.89	2 744
354.08	**249**	175.11	1 717	425.18	**299**	210.27	2 062	496.28	**349**	245.43	2 406	567.38	**399**	280.59	2 751
355.50	**250**	175.81	1 724	426.60	**300**	210.97	2 068	497.70	**350**	246.13	2 413	568.80	**400**	281.29	2 758

[a] 1 MPa (megapascal) = 1 MN/m^2 (meganewton per square meter). MN/m^2, N/mm^2, and MPa are equal numerically.

FIG. 211. Temperature Conversion Table[a,b]

	0 to 100						100 to 1000				
C		F	C		F	C		F	C		F
-17.8	0	32	10.0	50	122.0	38	100	212	260	500	932
-17.2	1	33.8	10.6	51	123.8	43	110	230	266	510	950
-16.7	2	35.6	11.1	52	125.6	49	120	248	271	520	968
-16.1	3	37.4	11.7	53	127.4	54	130	266	277	530	986
-15.6	4	39.2	12.2	54	129.2	60	140	284	282	540	1004
-15.0	5	41.0	12.8	55	131.0	66	150	302	288	550	1022
-14.4	6	42.8	13.3	56	132.8	71	160	320	293	560	1040
-13.9	7	44.6	13.9	57	134.6	77	170	338	299	570	1058
-13.3	8	46.4	14.4	58	136.4	82	180	356	304	580	1076
-12.8	9	48.2	15.0	59	138.2	88	190	374	310	590	1094
-12.2	10	50.0	15.6	60	140.0	93	200	392	316	600	1112
-11.7	11	51.8	16.1	61	141.8	99	210	410	321	610	1130
-11.1	12	53.6	16.7	62	143.6	100	212	413.6	327	620	1148
-10.6	13	55.4	17.2	63	145.4	104	220	428	332	630	1166
-10.0	14	57.2	17.8	64	147.2	110	230	446	338	640	1184
-9.4	15	59.0	18.3	65	149.0	116	240	464	343	650	1202
-8.9	16	60.8	18.9	66	150.8	121	250	482	349	660	1220
-8.3	17	62.6	19.4	67	152.6	127	260	500	354	670	1238
-7.8	18	64.4	20.0	68	154.4	132	270	518	360	680	1256
-7.2	19	66.2	20.6	69	156.2	138	280	536	366	690	1274
-6.7	20	68.0	21.1	70	158.0	143	290	554	371	700	1292
-6.1	21	69.8	21.7	71	159.8	149	300	572	377	710	1310
-5.6	22	71.6	22.2	72	161.6	154	310	590	382	720	1328
-5.0	23	73.4	22.8	73	163.4	160	320	608	388	730	1346
-4.4	24	75.2	23.3	74	165.2	166	330	626	393	740	1364
-3.9	25	77.0	23.9	75	167.0	171	340	644	399	750	1382
-3.3	26	78.8	24.4	76	168.8	177	350	662	404	760	1400
-2.8	27	80.6	25.0	77	170.6	182	360	680	410	770	1418
-2.2	28	82.4	25.6	78	172.4	188	370	698	416	780	1436
-1.7	29	84.2	26.1	79	174.2	193	380	716	421	790	1454
-1.1	30	86.0	26.7	80	176.0	199	390	734	427	800	1472
-0.6	31	87.8	27.2	81	177.8	204	400	752	432	810	1490
0.0	32	89.6	27.8	82	179.6	210	410	770	438	820	1508
0.6	33	91.4	28.3	83	181.4	216	420	788	443	830	1526
1.1	34	93.2	28.9	84	183.2	221	430	806	449	840	1544
1.7	35	95.0	29.4	85	185.0	227	440	824	454	850	1562
2.2	36	96.8	30.0	86	186.8	232	450	842	460	860	1580
2.8	37	98.6	30.6	87	188.6	238	460	860	466	870	1598
3.3	38	100.4	31.1	88	190.4	243	470	878	471	880	1616
3.9	39	102.2	31.7	89	192.2	249	480	896	477	890	1634
4.4	40	104.0	32.2	90	194.0	254	490	914	482	900	1652
5.0	41	105.8	32.8	91	195.8				488	910	1670
5.6	42	107.6	33.3	92	197.6				493	920	1688
6.1	43	109.4	33.9	93	199.4				499	930	1706
6.7	44	111.2	34.4	94	201.2				504	940	1724
7.2	45	113.0	35.0	95	203.0				510	950	1742
7.8	46	114.8	35.6	96	204.8				516	960	1760
8.3	47	116.6	36.1	97	206.6				521	970	1778
8.9	48	118.4	36.7	98	208.4				527	980	1796
9.4	49	120.2	37.2	99	210.2				532	990	1814
			37.8	100	212.0				538	1000	1832

[a]Look up temperature to be converted in middle column. If in degrees Celsius read Fahrenheit equivalent in right-hand column; if in Fahrenheit degrees, read Celsius equivalent in left-hand column.
[b]Courtesy International Nickel Co.

FIG. 211. (continued) Temperature Conversion Table[a]

1000 to 2000					
C	↓	F	C	↓	F
538	*1000*	1832	816	*1500*	2732
543	*1010*	1850	821	*1510*	2750
549	*1020*	1868	827	*1520*	2768
554	*1030*	1886	832	*1530*	2786
560	*1040*	1904	838	*1540*	2804
566	*1050*	1922	843	*1550*	2822
571	*1060*	1940	849	*1560*	2840
577	*1070*	1958	854	*1570*	2858
582	*1080*	1976	860	*1580*	2876
588	*1090*	1994	866	*1590*	2894
593	*1100*	2012	871	*1600*	2912
599	*1110*	2030	877	*1610*	2930
604	*1120*	2048	882	*1620*	2948
610	*1130*	2066	888	*1630*	2966
616	*1140*	2084	893	*1640*	2984
621	*1150*	2102	899	*1650*	3002
627	*1160*	2120	904	*1660*	3020
632	*1170*	2138	910	*1670*	3038
638	*1180*	2156	916	*1680*	3056
643	*1190*	2174	921	*1690*	3074
649	*1200*	2192	927	*1700*	3092
654	*1210*	2210	932	*1710*	3110
660	*1220*	2228	938	*1720*	3128
666	*1230*	2246	943	*1730*	3146
671	*1240*	2264	949	*1740*	3164
677	*1250*	2282	954	*1750*	3182
682	*1260*	2300	960	*1760*	3200
688	*1270*	2318	966	*1770*	3218
693	*1280*	2336	971	*1780*	3236
699	*1290*	2354	977	*1790*	3254
704	*1300*	2372	982	*1800*	3272
710	*1310*	2390	988	*1810*	3290
716	*1320*	2408	993	*1820*	3308
721	*1330*	2426	999	*1830*	3326
727	*1340*	2444	1004	*1840*	3344
732	*1350*	2462	1010	*1850*	3362
738	*1360*	2480	1016	*1860*	3380
743	*1370*	2498	1021	*1870*	3398
749	*1380*	2516	1027	*1880*	3416
754	*1390*	2534	1032	*1890*	3434
760	*1400*	2552	1038	*1900*	3452
766	*1410*	2570	1043	*1910*	3470
771	*1420*	2588	1049	*1920*	3488
777	*1430*	2606	1054	*1930*	3506
782	*1440*	2624	1060	*1940*	3524
788	*1450*	2642	1066	*1950*	3542
793	*1460*	2660	1071	*1960*	3560
799	*1470*	2678	1077	*1970*	3578
804	*1480*	2696	1082	*1980*	3596
810	*1490*	2714	1088	*1990*	3614
			1093	*2000*	3632

[a]To convert to the corresponding absolute scales: Kelvin = Celsius + 273; Rankine = Fahrenheit + 459.4.

FIG. 211. (continued) Temperature Conversion Table[a]

2000 to 3000					
C		F	C		F
1093	*2000*	3632	1371	*2500*	4532
1099	*2010*	3650	1377	*2510*	4550
1104	*2020*	3668	1382	*2520*	4568
1110	*2030*	3686	1388	*2530*	4586
1116	*2040*	3704	1393	*2540*	4604
1121	*2050*	3722	1399	*2550*	4622
1127	*2060*	3740	1404	*2560*	4640
1132	*2070*	3758	1410	*2570*	4658
1138	*2080*	3776	1416	*2580*	4676
1143	*2090*	3794	1421	*2590*	4694
1149	*2100*	3812	1427	*2600*	4712
1154	*2110*	3830	1432	*2610*	4730
1160	*2120*	3848	1438	*2620*	4748
1166	*2130*	3866	1443	*2630*	4766
1171	*2140*	3884	1449	*2640*	4784
1177	*2150*	3902	1454	*2650*	4802
1182	*2160*	3920	1460	*2660*	4820
1188	*2170*	3938	1466	*2670*	4838
1193	*2180*	3956	1471	*2680*	4856
1199	*2190*	3974	1477	*2690*	4874
1204	*2200*	3992	1482	*2700*	4892
1210	*2210*	4010	1488	*2710*	4910
1216	*2220*	4028	1493	*2720*	4928
1221	*2230*	4046	1499	*2730*	4946
1227	*2240*	4064	1504	*2740*	4964
1232	*2250*	4082	1510	*2750*	4982
1238	*2260*	4100	1516	*2760*	5000
1243	*2270*	4118	1521	*2770*	5018
1249	*2280*	4136	1527	*2780*	5036
1254	*2290*	4154	1532	*2790*	5054
1260	*2300*	4172	1538	*2800*	5072
1266	*2310*	4190	1543	*2810*	5090
1271	*2320*	4208	1549	*2820*	5108
1277	*2330*	4226	1554	*2830*	5126
1282	*2340*	4244	1560	*2840*	5144
1288	*2350*	4262	1566	*2850*	5162
1293	*2360*	4280	1571	*2860*	5180
1299	*2370*	4298	1577	*2870*	5198
1304	*2380*	4316	1582	*2880*	5216
1310	*2390*	4334	1588	*2890*	5234
1316	*2400*	4352	1593	*2900*	5252
1321	*2410*	4370	1599	*2910*	5270
1327	*2420*	4388	1604	*2920*	5288
1332	*2430*	4406	1610	*2930*	5306
1338	*2440*	4424	1616	*2940*	5324
1343	*2450*	4442	1621	*2950*	5342
1349	*2460*	4460	1627	*2960*	5360
1354	*2470*	4478	1632	*2970*	5378
1360	*2480*	4496	1638	*2980*	5396
1366	*2490*	4514	1643	*2990*	5414
			1649	*3000*	5432

[a]To convert to the corresponding absolute scales: Kelvin = Celsius + 273; Rankine = Fahrenheit + 459.4.

FIG. 212. Approximate Relationships: "Rockwell," "Rockwell" Superficial, and Brinell Hardness Numbers[a]

C	A	D	Brinell	15-N	30-N	45-N
Rockwell C Scale Brale Penetrator 150 kg. Load	Rockwell A Scale Brale Penetrator 60 kg. Load	Rockwell D Scale Brale Penetrator 100 kg. Load	Brinell Numbers 10 mm. Steel Ball 3000 kg. Load	Rockwell Superficial 15-N Scale N Brale Penetrator 15 kg. Load	Rockwell Superficial 30-N Scale N Brale Penetrator 30 kg. Load	Rockwell Superficial 45-N Scale N Brale Penetrator 45 kg. Load
65	84.5	75.0	690	92.5	82.0	72.0
64	83.5	74.0	673	92.0	81.0	71.0
63	83.0	73.0	658	91.5	80.0	70.0
62	82.5	72.5	645	91.0	79.0	69.0
61	82.0	72.0	628	90.5	78.5	68.0
60	81.5	71.0	614	90.0	77.5	66.5
59	81.0	70.5	600	89.5	77.0	65.5
58	80.5	69.5	587	89.0	76.0	64.5
57	80.0	69.0	573	88.5	75.0	63.5
56	79.5	68.0	560	88.0	74.0	62.0
55	79.0	67.0	547	87.5	73.5	61.0
54	78.5	66.5	534	87.0	72.5	60.0
53	78.0	65.5	522	86.5	71.5	59.0
52	77.5	65.0	509	86.0	70.5	57.5
51	77.0	64.0	496	85.5	70.0	56.5
50	76.5	63.5	484	85.0	69.0	55.0
49	76.0	62.5	472	84.5	68.0	54.0
48	75.5	62.0	460	84.0	67.0	53.0
47	75.0	61.0	448	83.5	66.0	52.0
46	74.5	60.5	437	83.0	65.5	51.0
45	74.0	59.5	426	82.5	64.5	49.5
44	73.5	59.0	415	82.0	63.5	48.5
43	73.0	58.0	404	81.5	62.5	47.5
42	72.5	57.5	393	81.0	62.0	46.5
41	72.0	56.5	382	80.5	61.0	45.0
40	71.5	56.0	372	80.0	60.0	44.0
39	71.0	55.0	362	79.5	59.0	43.0
38	70.5	54.5	352	79.0	58.0	42.0
37	70.0	53.5	342	78.5	57.5	40.5
36	69.5	53.0	333	78.0	56.5	39.5
35	69.0	52.0	322	77.5	56.0	38.5
34	68.0	51.5	313	77.0	55.0	37.5
33	67.5	50.5	305	76.5	54.0	36.0
32	67.0	50.0	296	76.0	53.0	35.0
31	66.5	49.0	290	75.5	52.0	34.0
30	66.0	48.5	283	75.0	51.5	33.0
29	65.5	47.5	276	74.5	50.5	31.5
28	65.0	47.0	272	74.0	49.5	30.5
27	64.5	46.0	265	73.5	48.5	29.5
26	64.0	45.5	260	73.0	48.0	28.5
25	63.5	44.5	255	72.5	47.0	27.0
24	63.0	44.0	248	72.0	46.0	26.0
23	62.5	43.0	245	71.5	45.0	25.0
22	62.0	42.0	240	71.0	44.5	24.0
21	61.5	41.5	235	70.5	43.5	23.0
20	61.0	41.0	230	70.0	43.0	22.0

[a]Courtesy International Nickel Co.

FIG. 212. (continued) Approximate Relationships: "Rockwell," "Rockwell" Superficial, and Brinell Hardness Numbers[a]

B	E	F	Brinell	15-T	30-T	45-T
Rockwell B Scale 1/16" Ball 100 kg. Load	Rockwell E Scale 1/8" Ball 100 kg. Load	Rockwell F Scale 1/16" Ball 60 kg. Load	Brinell Numbers 500 or 1000 kg. Load	Rockwell Superficial 15-T Scale 1/16" Ball 15 kg. Load	Rockwell Superficial 30-T Scale 1/16" Ball 30 kg. Load	Rockwell Superficial 45-T Scale 1/16" Ball 45 kg. Load
49	88.5	85.5	92	77.5	51.0	25.0
48	88.0	85.0	90	77.5	50.5	24.0
47	87.0	84.5	88	77.0	50.0	23.0
46	86.5	84.0	87	77.0	49.5	22.0
45	86.0	83.5	86	76.5	49.0	21.0
44	85.5	83.0	85	76.0	48.0	20.0
43	85.0	82.5	83	76.0	47.5	19.0
42	84.5	82.0	82	75.5	47.0	18.0
41	84.0	81.5	81	75.0	46.5	17.0
40	83.0	81.0	80	75.0	46.0	16.0
39	82.5	80.0	79	74.5	45.0	15.0
38	82.0	79.5	78	74.5	44.5	14.5
37	81.0	79.0	77	74.0	44.0	13.5
36	80.5	78.5	76	74.0	43.0	12.5
35	80.0	78.0	75	73.5	42.5	11.5
34	79.5	77.5	75	73.0	42.0	10.5
33	79.0	77.0	74	73.0	41.0	9.5
32	78.0	76.0	74	72.5	40.5	9.0
31	77.5	75.5	73	72.0	40.0	8.0
30	77.0	75.0	72	72.0	39.0	7.0
29	76.5	74.5	71	71.5	38.5	6.0
28	76.0	74.0	71	71.5	38.0	5.0
27	75.0	73.5	70	71.0	37.0	4.0
26	74.5	73.0	69	71.0	36.5	3.0
25	73.5	72.5	68	70.5	36.0	2.0
24	73.0	72.0	67	70.0	35.0	1.0
23	72.0	71.5	66	70.0	34.5	0
22	71.5	71.0	66	69.5	34.0	....
21	71.0	70.0	65	69.0	33.0	....
20	70.0	69.5	65	69.0	32.5	....
19	69.0	69.0	64	68.5	32.0	....
18	68.5	68.5	64	68.5	31.0	....
17	68.0	68.0	63	68.0	30.5	....
16	67.0	67.5	63	68.0	30.0	....
15	66.5	67.0	62	67.5	29.0	....
14	66.0	66.5	62	67.0	28.0	....
13	65.0	66.0	62	67.0	27.5	....
12	64.5	65.0	61	66.5	26.5	....
11	64.0	64.5	61	66.0	25.5	....
10	63.0	64.0	60	66.0	25.0	....
9	62.0	63.5	60	65.5	24.0	....
8	61.5	63.0	59	65.5	23.0	....
7	61.0	62.5	59	65.0	22.0	....
6	60.0	62.0	58	65.0	21.0	....
5	59.5	61.5	58	64.5	20.0	....
4	58.5	60.5	58	64.0	19.5	....
3	58.0	60.0	58	64.0	18.5	....
2	57.5	59.5	57	63.5	17.5	....
1	56.5	59.0	57	63.0	17.0	....
0	....	....	....	63.0	16.0	....

[a]Courtesy International Nickel Co.

FIG. 213. Third and Fourth Powers of Wire Diameters[a,b]

d	d^3	d^4	d	d^3	d^4	d	d^3	d^4	d	d^3	d^4
.001	.000000001	.000000000001	.051	.00013265	.0000067652	.101	.00103030	.00010406	.151	.00344295	.00051989
.002	.000000008	.000000000016	.052	.00014061	.0000073116	.102	.00106121	.00010824	.152	.00351181	.00053379
.003	.000000027	.000000000081	.053	.00014888	.0000078905	.103	.00109273	.00011255	.153	.0035816	.00054798
.004	.000000064	.000000000256	.054	.00015746	.0000085031	.104	.00112486	.00011699	.154	.0036523	.00056245
.005	.000000125	.000000000625	.055	.00016638	.0000091506	.105	.00115762	.00012155	.155	.0037239	.00057720
.006	.000000216	.000000001296	.056	.00017562	.0000098345	.106	.00119102	.00012625	.156	.0037964	.00059224
.007	.000000343	.000000002401	.057	.00018519	.0000105560	.107	.00122504	.00013108	.157	.0038699	.00060757
.008	.000000512	.000000004096	.058	.00019511	.0000113165	.108	.00125971	.00013605	.158	.0039443	.00062320
.009	.000000729	.000000006561	.059	.00020538	.0000121174	.109	.00129503	.00014116	.159	.0040197	.00063913
.010	.000001000	.000000010000	.060	.00021600	.0000129600	.110	.00133100	.00014641	.160	.0040960	.00065536
.011	.000001331	.000000014641	.061	.00022698	.0000138458	.111	.0013676	.00015181	.161	.0041733	.00067190
.012	.000001728	.000000020736	.062	.00023833	.0000147763	.112	.0014049	.00015735	.162	.0042515	.00068875
.013	.000002197	.000000028561	.063	.00025005	.0000157530	.113	.0014429	.00016305	.163	.0043307	.00070591
.014	.000002744	.000000038416	.064	.00026214	.0000167772	.114	.0014815	.00016890	.164	.0044109	.00072339
.015	.000003375	.000000050625	.065	.00027463	.0000178506	.115	.0015209	.00017490	.165	.0044921	.00074120
.016	.000004096	.000000065536	.066	.00028750	.0000189747	.116	.0015609	.00018106	.166	.0045743	.00075933
.017	.000004913	.000000083521	.067	.00030076	.0000201511	.117	.0016016	.00018739	.167	.0046575	.00077780
.018	.000005832	.000000104976	.068	.00031443	.0000213814	.118	.0016430	.00019388	.168	.0047416	.00079659
.019	.000006859	.000000130321	.069	.00032851	.000022667	.119	.0016852	.00020053	.169	.0048268	.00081573
.020	.000008000	.000000160000	.070	.00034300	.0000240100	.120	.0017280	.00020736	.170	.0049130	.00083521
.021	.000009261	.000000194481	.071	.00035791	.0000254117	.121	.0017716	.00021436	.171	.0050002	.00085504
.022	.000010648	.000000234256	.072	.00037325	.0000268739	.122	.0018158	.00022153	.172	.0050884	.00087521
.023	.000012167	.00000027984	.073	.00038902	.0000283982	.123	.0018609	.00022889	.173	.0051777	.00039575
.024	.000013824	.00000033178	.074	.00040522	.0000299866	.124	.0019066	.00023642	.174	.0052680	.00091664
.025	.000015625	.00000039062	.075	.00042188	.0000316406	.125	.0019531	.00024414	.175	.0053594	.00093789

d	d^3	d^4	d	d^3	d^4	d	d^3	d^4	d	d^3	d^4
.026	.000017576	.00000045698	.076	.00043898	.0000333622	.126	.0020004	.00025205	.176	.0054518	.00095951
.027	.000019683	.00000053144	.077	.00045653	.0000351530	.127	.0020484	.00026014	.177	.0055452	.00098051
.028	.000021952	.00000061466	.078	.00047455	.0000370151	.128	.0020972	.00026844	.178	.0056398	.00100388
.029	.000024389	.00000070728	.079	.00049304	.0000389501	.129	.0021467	.00027692	.179	.0057353	.00102663
.030	.000027000	.00000081000	.080	.00051200	.0000409600	.130	.0021970	.00028561	.180	.0058320	.00104976
.031	.000029791	.00000092352	.081	.00053144	.0000430467	.131	.0022481	.00029450	.181	.0059297	.00107328
.032	.000032768	.00000104858	.082	.00055137	.0000452122	.132	.0023000	.00030360	.182	.0060286	.00109720
.033	.000035937	.00000118592	.083	.00057179	.0000474583	.133	.0023526	.00031290	.183	.0061285	.00112151
.034	.000039304	.00000133634	.084	.00059270	.0000497871	.134	.0024061	.00032242	.184	.0062295	.00114623
.035	.000042875	.00000150062	.085	.00061412	.0000522006	.135	.0024604	.00033215	.185	.0063316	.00117135
.036	.000046656	.00000167962	.086	.00063606	.0000547008	.136	.0025155	.00034210	.186	.0064349	.00119688
.037	.000050653	.00000187416	.087	.00065850	.0000572898	.137	.0025714	.00035228	.187	.0065392	.00122283
.038	.000054872	.00000208514	.088	.00068147	.0000599695	.138	.0026281	.00036267	.188	.0066447	.00124920
.039	.000059319	.00000231344	.089	.00070497	.0000627422	.139	.0026856	.00037330	.189	.0067513	.00127599
.040	.000064000	.00000256000	.090	.00072900	.0000656100	.140	.0027440	.00038416	.190	.0068590	.0013032
.041	.000068921	.00000282576	.091	.00075357	.0000685750	.141	.0028032	.00039525	.191	.0069679	.0013309
.042	.000074088	.00000311170	.092	.00077869	.0000716393	.142	.0028633	.00040659	.192	.0070779	.0013590
.043	.000079507	.00000341880	.093	.00080436	.0000748052	.143	.0029242	.00041816	.193	.0071891	.0013875
.044	.000085184	.00000374810	.094	.00083058	.0000780749	.144	.0029860	.00042998	.194	.0073014	.0014165
.045	.000091125	.00000410062	.095	.00085738	.0000814506	.145	.0030486	.00044205	.195	.0074149	.0014459
.046	.000097336	.00000447746	.096	.00088474	.0000849347	.146	.0031121	.00045437	.196	.0075295	.0014758
.047	.000103823	.00000487968	.097	.00091267	.0000885293	.147	.0031765	.00046695	.197	.0076454	.0015061
.048	.000110592	.00000530842	.098	.00094119	.0000922368	.148	.0032418	.00047979	.198	.0077624	.0015370
.049	.000117649	.00000576480	.099	.00097030	.0000960596	.149	.0033079	.00049288	.199	.0078806	.0015682
.050	.000125000	.00000625000	.100	.00100000	.0001000000	.150	.0033750	.00050625	.200	.0080000	.0016000

[a]Tables of third and fourth powers are useful for design calculations as they eliminate the laborious process of solving for the cube and fourth roots often found in equations. Example: What wire diameter d is needed for a compression spring with a D of 1 in. to exert a force P of 50 lb at a stress of 100 000 psi? Answer: From the formula (Fig. 118), $d = \sqrt[3]{(2.55PD)/S} = \sqrt[3]{(2.55 \times 50 \times 1)/100\,000} = \sqrt[3]{0.001275}$; and from this table of cubes, the nearest figure to 0.001275 is 0.00129503 and d = 0.109 in.
[b]Courtesy Spring Manufacturers Institute.

FIG. 214. Table Showing Weights of Round Steel Spring Wire by Decimals and Fractions of an Inch.

Washburn & Moen Wire Gauge No.	Fraction of Inch	Decimal of Inch	Pounds per 100 feet	Feet per pound	Washburn & Moen Wire Gauge No.	Fraction of Inch	Decimal of Inch	Pounds per 100 feet	Feet per pound
	1-2	.5000	66.68	1.500		3-32	.09375	2.344	42.66
7-0		.490	64.04	1.562	13		.092	2.233	44.78
	15-32	.46875	58.61	1.706	14		.080	1.707	58.58
6-0		.462	56.81	1.760	15		.072	1.383	72.32
	7-16	.4375	51.05	1.959	16	1-16	.0625	1.042	95.98
5-0		.431	49.43	2.023	17		.054	.7778	128.60
	13-32	.40625	44.02	2.272	18		.047	.6018	166.20
4-0		.394	41.36	2.418	19		.041	.4484	223.00
	3-8	.3750	37.51	2.666	20		.035	.3230	309.60
3-0		.3629	35.05	2.853	21		.032	.2680	373.10
	11-32	.34375	31.52	3.173		1-32	.03125	.2605	383.90
2-0		.331	29.22	3.422	22		.0286	.2182	458.40
	5-16	.3125	26.05	3.839	23		.0258	.1775	563.3
0		.307	25.06	3.991	24		.0230	.1411	708.7
1		.283	21.36	4.681	25		.0204	.1110	900.9
	9-32	.28125	21.10	4.740	26		.0181	.08738	1144.
2		.263	18.38	5.441	27		.0173	.07983	1253.
	1-4	.250	16.67	5.999	28		.0162	.07000	1429.
3		.244	15.84	6.313		1-64	.0156	.06500	1547.
4		.225	13.54	7.386	29		.0150	.06001	1666.
	7-32	.21875	12.76	7.835	30		.0140	.05228	1913.
5		.207	11.43	8.750	31		.0132	.04647	2152.
6		.192	9.832	10.17	32		.0128	.04370	2288.
	3-16	.1875	9.377	10.66	33		.0118	.03714	2693.
7		.177	8.356	11.97	34		.0104	.02885	3466.
8		.162	7.000	14.29	35		.0095	.02407	4154.
	5-32	.15625	6.512	15.36	36		.009	.02160	4629.
9		.148	5.866	17.05	37		.0085	.01927	5189.
10		.135	4.861	20.57	38		.008	.01707	5858.
	1-8	.125	4.168	24.00	39		.0075	.01500	6665.
11		.120	3.873	25.82	40		.007	.01307	7652.
12		.105	2.969	33.69	41		.0066	.01162	8607.
					42		.0062	.01025	9753.

[a]Courtesy Spring Manufacturers Institute.

FIG. 215. Gauge Thicknesses[a]

Name of Gauge		Steel Wire G. Washburn & Moen or W. & M. Wire G. U. S. Steel W. G.	Music Wire Gauge, M. W. G.	Brown & Sharpe Gauge, B. & S. G. A. W. G.	Stubs' Iron Wire Gauge, W. W. G. Birmingham or B. W. G.
Principal Use		Steel Wire, except Music Wire	Steel Music Wire	Non-ferrous Sheets and Wire	Flats, Steel Plates and Wire
Gauge Number	**Millimeters Decimally**	Thickness, Inch	Thickness, Inch	Thickness, Inch	Thickness, Inch
7/0's	12.45	.4900			
6/0's	11.72	.4615	.004	.5800	
5/0's	10.93	.4305	.005	.5165	.500
4/0's	10.00	.3938	.006	.4600	.454
3/0's	9.208	.3625	.007	.4096	.425
2/0's	8.407	.3310	.008	.3648	.380
0	7.785	.3065	.009	.3249	.340
1	7.188	.2830	.010	.2893	.300
2	6.668	.2625	.011	.2576	.284
3	6.190	.2437	.012	.2294	.259
4	5.723	.2253	.013	.2043	.238
5	5.258	.2070	.014	.1819	.220
6	4.877	.1920	.016	.1620	.203
7	4.496	.1770	.018	.1443	.180
8	4.115	.1620	.020	.1285	.165
9	3.767	.1483	.022	.1144	.148
10	3.429	.1350	.024	.1019	.134
11	3.061	.1205	.026	.0907	.120
12	2.680	.1055	.029	.0808	.109
13	2.324	.0915	.031	.0720	.095
14	2.032	.0800	.033	.0641	.083
15	1.829	.0720	.035	.0571	.072
16	1.588	.0625	.037	.0508	.065
17	1.372	.0540	.039	.0453	.058
18	1.207	.0475	.041	.0403	.049
19	1.041	.0410	.043	.0359	.042
20	.8839	.0348	.045	.0320	.035
21	.8052	.0317	.047	.0285	.032
22	.7264	.0286	.049	.0253	.028
23	.6553	.0258	.051	.0226	.025
24	.5842	.0230	.055	.0201	.022
25	.5182	.0204	.059	.0179	.020

26	.4597	.0181	.063	.0159	.018
27	.4394	.0173	.067	.0142	.016
28	.4115	.0162	.071	.0126	.014
29	.3810	.0150	.075	.0113	.013
30	.3556	.0140	.080	.0100	.012
31	.3353	.0132	.085	.0089	.010
32	.3251	.0128	.090	.0080	.009
33	.2997	.0118	.095	.0071	.008
34	.2642	.0104	.100	.0063	.007
35	.2413	.0095	.106	.0056	.005
36	.2286	.0090	.112	.0050	.004
37	.2159	.0085	.118	.0045	
38	.2032	.0080	.124	.0040	
39	.1905	.0075	.130	.0035	
40	.1778	.0070	.138	.0031	
41	.1676	.0066	.146	.0028	
42	.1575	.0062	.154	.0025	
43	.1524	.0060	.162	.0022	
44	.1473	.0058	.170	.0020	
45	.1397	.0055		.00176	
46	.1321	.0052		.00157	
47	.1270	.0050			
48	.1219	.0048			
49	.1168	.0046			
50	.1118	.0044			
	.1016	.004			
	.09144	.0036			
	.08128	.0032			
	.07112	.0028			
	.06096	.0024			
	.05080	.0020			
	.04064	.0016			
	.03556	.0014			
	.03048	.0012			
	.02540	.0010			

1. Use Steel Wire or W. & M. gauge for all carbon and carbon alloys except music wire.
2. Use Brown & Sharpe gauge for brass and phosphor bronze wire.
3. Use Birmingham gauge for flat spring steels and C. R. strip.
4. Use Piano & Music Wire gauge for Music Wire.

[a]Courtesy Spring Manufacturers Institute.

FIG. 216. Permissible Elevated Temperatures[a]

Spring material	Permissible high temperature		Maximum recommended design stress	
	(°F)	(°C)	(psi)	(MPa)
Brass spring wire	175	80	35 000	240
Phosphor-bronze	225	105	40 000	275
Music wire	250	120	90 000	620
Beryllium-copper	300	150	50 000	345
Hard-drawn MB steel	325	160	60 000	415
Oil-tempered MB steel	350	175	65 000	450
Chrome-vanadium	425	220	70 000	485
Monel 400	425	220	40 000	275
Monel K-500	450	230	45 000	310
Chrome-silicon	475	245	80 000	550
Duranickel	500	260	50 000	345
Stainless 300 series	550	290	55 000	380
Stainless 400 series	600	315	50 000	345
Inconel 600	650	345	50 000	345
High-speed tool steel	775	415	70 000	485
Inconel X-750	850	455	55 000	380
Chrome-moly-vanadium	900	480	60 000	415

[a]Springs used at high temperatures exert less force and have larger deflections than at room temperature. The modulus of elasticity determines deflection and is reduced with high temperatures, thereby causing larger deflections (see Fig. 202).
Design stresses should be kept lower at elevated temperatures too, because corrosive conditions usually occur, especially with steam.
Compression and extension springs have a 5 percent loss of force (or 5 percent increase in deflection if force is constant) at the temperatures and torsional stresses (corrected for curvature) shown. The stresses may be increased 75 percent for torsion and flat springs.
All materials are in the spring temper, heated properly for high-temperature service.

FIG. 217. Elastic Limit of Spring Materials

Spring material	ASTM	Elastic limit (%)[b]	
		Bending[c]	Torsion[d]
Music wire	A228	65-70	45-50
Hard-drawn MB	A227	60-70	45-55
Oil-tempered MB	A229	80-90	40-50
Chrome-vanadium	A231	88-93	65-75
Chrome-silicon	A401	88-93	65-75
Silicon-manganese	—	78-86	55-65
Stainless 300 series	A313	65-75	45-55
17-7 PH	A313	72-78	53-57
420	—	65-75	45-55
431	—	72-76	50-55
Phosphor-bronze	B159	75-80	45-50
Spring brass	B134	75-80	45-50
Beryllium-copper			
Hard-drawn	B197	68-72	48-52
Precipitation hardened	—	73-77	50-55
Inconel 600	B166	65-70	40-45
Inconel X-750	—	65-70	40-45
Monel 400	B164	65-70	38-42
Monel K-500	—	65-70	38-42
Duranickel 300	—	65-70	38-42
Ni-Span C 902	—	65-70	38-42

[a]Elastic limit is based on a percentage of the minimum tensile strength for round wire in the spring temper condition as obtained by cold drawing, oil tempering, or precipitation hardening, whichever is necessary, and stress relieved after coiling into a spring.
[b]Or proportional limit or proof stress.
[c]As used for torsion springs.
[d]As used for compression springs.
Note: The elastic limit can be raised about 15 percent by heating, pressing, and cold work.

FIG. 218. Metric Conversions for Spring Design

For compression, extension, and flat spring forces:

1 g = 0.0353 oz = 0.0022 lb = 0.0098 N
1 kg = 35.2740 oz = 2.2046 lb = 9.8066 N
1 oz = 28.3495 g = 0.0283 kg = 0.2780 N
1 lb = 453.5924 g = 0.4536 kg = 4.4482 N
1 N = 3.5969 oz = 0.2248 lb = 102. g = 0.102 kg

For torsion spring forces in torque:

1 oz-in. = 28.3495 g in. = 72.0079 g-cm = 0.0625 lb-in. = 7.0616 N-mm = 0.0071 N-m
1 lb-in. = 16. oz-in. = 1152.1247 g-cm = 1.1521 kg-cm = 453.5924 g-in. = 112.9848 N-mm = 0.1130 N-m
1 N-m = 141.6119 oz-in. = 8.8507 lb-in. = 10.1972 kg-cm = 0.1020 kg-m
1 kg-cm = 13.8880 oz-in. = 0.8680 lb-in. = 0.0981 N-m

For stress:

1 Pa = 1 N/m^2 = 0.000145 lb/in.^{2}
1 lb/in.^{2} = 6894.7570 Pa = 0.0703 kg/cm^2 (100 000 psi = 689.47570 MPa)
1 kg/cm^2 = 98 066.5 Pa = 14.2233 psi
1 MPa = 1 000 000 Pa = 1 MN/m^2 = 145.0377 psi

For lengths:

1 mm = 0.03937 in.
1 cm = 0.3937 in.
1 in. = 25.4 mm = 2.54 cm
1 m = 39.37 in.

Note: 1. Use lowercase letters for metric terms.
2. Preferred usage is as follows: force should be in newtons; torque should be in newton-metres (or newton-millimetres); stress should be in pascals (or megapascals).

FIG. 219. Wire Diameters in Millimetres[a,b]

0.90	2.2	5.3	11.8
0.95	2.3	5.6	12.0
1.00	2.4	6.0	12.5
1.05	2.6	6.3	13.0
1.10	2.8	6.5	14.0
1.15	3.0	6.7	15.0
1.20	3.2	7.0	16.0
1.30	3.4	7.5	17.0
1.40	3.6	8.0	18.0
1.50	3.8	8.5	19.0
1.60	4.0	9.0	20.0
1.70	4.2	9.5	21.0
1.80	4.4	10.0	22.0
1.90	4.6	10.6	23.0
2.0	4.8	11.0	24.0
2.1	5.0	11.2	25.0

[a]The American Society for Testing and Materials has suggested that wire diameters be used in millimetres.
[b]Courtesy American Society for Testing and Materials.

Subject Index